The Pattern Recognition Basis
of Artificial Intelligence

T0180445

The Pattern Recognition Basis of Artificial Intelligence

Donald R. Tveter

IEEE
COMPUTER
SOCIETY

Los Alamitos, California

Washington • Brussels • Tokyo

Library of Congress Cataloging-in-Publication Data

Tveter, Donald R.
    The pattern recognition basis of artificial intelligence / Donald R. Tveter.
        p.   cm.
    Includes bibliographical references and index.
    ISBN 0-8186-7796-1
    1. Artificial intelligence. 2. Pattern perception.   I. Title.
Q335.T9 1998
006.3—dc21

                                                                97-32259
                                                                CIP

IEEE Computer Society Press Order Number BP07796
Library of Congress Number 97-32259
ISBN 0-8186-7796-1

Additional copies may be ordered from:

IEEE Computer Society Press	IEEE Service Center	IEEE Computer Society	IEEE Computer Society
Customer Service Center	445 Hoes Lane	13, Avenue de l'Aquilon	Ooshima Building
10662 Los Vaqueros Circle	P.O. Box 1331	B-1200 Brussels	2-19-1 Minami-Aoyama
P.O. Box 3014	Piscataway, NJ 08855-1331	BELGIUM	Minato-ku, Tokyo 107
Los Alamitos, CA 90720-1314	Tel: +1-908-981-1393	Tel: +32-2-770-2198	JAPAN
Tel: +1-714-821-8380	Fax: +1-908-981-9667	Fax: +32-2-770-8505	Tel: +81-3-3408-3118
Fax: +1-714-821-4641	mis.custserv@computer.org	euro.ofc@computer.org	Fax: +81-3-3408-3553
Email: cs.books@computer.org			tokyo.ofc@computer.org

Publisher: Matt Loeb
Developmental Editor: Cheryl Baltes
Advertising/Promotions: Tom Fink
Production Editor: Lisa O'Conner

To the Glory of God and to my parents.

Contents

Preface

I believe you will find this book has a refreshing new perspective on a subject that needs a new perspective. In particular, these are some of the unique viewpoints that are found in this book:

First, the book takes the viewpoint that symbol processing techniques are not working out very well and these methods alone will never be able to give the depth and breadth of capabilities found in human beings. I believe the solution is to introduce new foundational principles to AI and these new principles get a prominent place in this book. At the moment, the new principles that are under consideration are: connectionist/neural networking methods, case-based and memory-based methods, and picture processing. Even though there is a movement to include quantum mechanics, I do not cover this possibility except to mention it in passing. All this is not to say that symbol processing can be thrown out, only that a good solution to intelligence certainly consists of many more techniques than just symbol processing.

Second, that the methods of AI are best seen as different ways of doing pattern recognition. One way to do pattern recognition is simply to store lots of cases and when a new problem comes up you compare it to all your stored cases. At the other end of the spectrum, Classical Symbol Processing AI has taken the viewpoint that cases should be condensed or compressed down to a relatively small set of rules and then the cases should be thrown away. A system should work with only its condensed knowledge. In between these two extremes are neural networks, especially backprop type networks, where the networks develop rulelike capabilities but they can also store cases. As much as possible, the book compares these three basic methods using actual AI programs.

Third, the structure of the book starts at the bottom of human abilities with vision and other simple pattern recognition abilities and moves on to the higher levels of problem solving and game playing and finally to the level of natural language and understanding of the world. At the higher levels you need more than a simple pattern recognition approach, you really need some more complex computer architecture that includes methods for structuring thoughts.

Fourth, there is the matter of what constitutes AI. One of the early major goals for the subject was to produce systems that were as capable of dealing with the real world as human beings. The early researchers knew this would take time and that they had to start with much smaller systems and then scale them up as time went on. None of the systems have shown any promise of scaling up. The smaller systems can be highly useful, but for the most part they do not display all the characteristics of intelligence that people would demand of an intelligent system, although they typically show some characteristics of intelligence. Because of this, to a great extent AI has become "advanced algorithms" or "advanced programming techniques," covering topics that deal with specialized areas of human achievement like game playing or natural language processing and where there may be little or no evidence of intelligent behavior on the part of the program. Unfortunately, in many cases, in order to do a good job within these specialized areas, you need a system that has all the flexibility and knowledge of a whole human being. Therefore, to a certain extent I have chosen topics that consider what methods are necessary in a system that has all the flexibility of a human being. This also means that certain other traditional AI methods that would normally come up in an AI course have been ignored.

The main target groups for this book are juniors, seniors, and first year graduate students in computer science or psychology. Other people without computer science and psychology backgrounds should still be able to learn a lot about AI from the book. In the way of mathematics, a very small amount of linear algebra and calculus is called for. In linear algebra a knowledge of vectors, matrices, and matrix multiplication is sufficient. Calculus is only used for the SAINT integration program and for a derivation of the back-propagation algorithm in the appendix and even this derivation can be skipped if necessary. In the way of programming abilities, the ACM's CS-2 course that includes linked lists, trees, and recursion is adequate.

Some Prolog is introduced as a useful notation for describing the symbol processing algorithms. If you want to do more Prolog than this small amount, I have what used to be an additional chapter on Prolog available free on the Internet. If you prefer to have your students use Lisp instead, there is a Lisp chapter there as well. I do believe doing some symbolic programming is important to give students the flavor of the symbol processing approach. If you want them to do a lot of such programming, then instead of starting with Chapter 1, start with Section 4.2 and then go on to either the Internet Prolog or Lisp chapter and then start at Chapter 1.

The book is organized in a manner in which I think the reader will slowly get an intuitive feeling for the principles of AI. (I like to think that the book reads like a novel. A few reviewers have agreed with this assertion and one has disagreed.) Throughout the book, the applications of the basic principles are demonstrated by examining some classic AI programs in detail. To a large extent, people learn by seeing specific examples and then generalizing from them (a case-based approach).

I recommend that you read all the exercises even if you do not want to do them in a formal way. I feel this is worthwhile since many exercises raise issues that are not discussed in the text. Some of them are worth debating in class. Some exercises can be done without any programming, but many of them are programs. Some of these can reasonably be done in general purpose languages like Pascal, C, or Fortran. Instructors, take note that you cannot simply go to a chapter and assign exercises 1–5 one day, 6–10 the next, and so on. That would be altogether too much work for a student. Instead, you need to look at each exercise and choose only those that are appropriate for your type of student. Most exercises are as easy or hard as they look, except some long or hard ones that may seem short or easy have warnings attached to them.

More material is available online including neural networking software for Unix and DOS/MS-Windows and an outline and commentary on the topics. As time goes on there will be a list of frequently asked questions. If there is more material that deserves attention and fits in with the perspective of this book, please let me know and I will consider adding it online. For online information, see my Pattern Recognition Basis of AI page at:

http://www.mcs.com/~drt/basisofai.html

or send me an email note at:

drt@mcs.com.

Chapter 1

Artificial Intelligence

1.1 Artificial Intelligence and Intelligence

The goal of artificial intelligence is to try to develop computer programs, algorithms, and computer architectures that will behave very much like people and will do those things that in people would require intelligence, understanding, thinking, or reasoning. There are two important aspects to this study. First, there is the very grand goal of finding out how intelligence and human thinking works so that the same or similar methods can be made to work on a computer. This makes the subject on a par with physics where the goal is to understand how the whole universe of matter, energy, space, and time works. A second goal of AI is more modest: it is to produce computer programs that function more like people so that computers can be made more useful and so they can be made to do many things that people do and perhaps even faster and better than people can do them. These will be the problems that this book deals with, the grand aspect and the modest one.

1.1.1 Intelligence

To begin the consideration of *artificial* intelligence, it would be appropriate to start with some definition of *intelligence*. Unfortunately, giving a definition of intelligence that will satisfy everyone is not possible and there are critics who claim that there has been no intelligence evident in artificial intelligence, only some modestly clever programming. Thus, to begin this book, we must briefly delay looking at the normal sort of material that would be found in the first chapter of a textbook and instead look first at the controversy that surrounds the definitions of intelligence and artificial intelligence. Looking at this debate will not settle the issues involved to everyone's satisfaction and readers will be left to form their own opinions about the nature of intelligence and artificial intelligence. To begin looking at the definition of intelligence, we will start with aspects of intelligence where there is no disagreement and then move on to the issues that are hotly debated.

Everyone agrees that one aspect of human intelligence is the ability to respond correctly to a novel situation. Furthermore, in giving intelligence tests where the goal is to solve problems, people who quickly give the correct answer will be judged as more intelligent than people who respond more slowly. Then on a long test, the "smarter" or more "intelligent" people will get more correct answers than less smart, less intelligent people. Within this process there is an important aspect to consider. In order to be able to respond correctly to a novel situation, the situation cannot be too novel. Thus, if the situation at hand is to do some calculus problems, you cannot expect people who have never done any calculus to manage to respond at all. Familiarity with the subject area is necessary to be able to demonstrate intelligence. Knowledge gained by experience is essential. Then you can look (or be?) more intelligent in a certain area simply by having more experience with that area.

The matter of possessing a certain amount of knowledge about a subject area can be quite subtle. For instance, adults ordinarily assume that it is easy to tell one person from the next simply by looking at them. It is assumed that adults have some kind of universal pattern recognition ability. However, it is often the case according to many media reports that when Americans visit some foreign country, especially, say, China, Americans often report that all the Chinese look the same. Of course, Chinese do not think that all Chinese look the same because native Chinese have had an extensive amount of experience recognizing Chinese faces. And to turn the tables, when Chinese students come to America they often report that all Americans look the same.[1] Thus, even a "simple" task like recognizing faces is not some kind of universal ability that adults develop but it is an ability that is developed to work within their own specific environment and which will not work very well outside that environment.

In addition to knowledge, speed, and experience, another key element of intelligence is the ability to learn. Everyone agrees that an intelligent system must be able to learn since obviously any person or program that cannot learn or which "mindlessly" keeps repeating a mistake over and over again will seem stupid. In fact, since as people learn a new task they get faster and faster at it, some people might require programs to get faster and faster as well.

If intelligence consisted of only storing knowledge, doing pattern recognition, solving problems, and the ability to learn, then there would not be any problem in saying that programs can be intelligent. But there are other qualities that some critics believe are necessary for intelligence. Some of them are intuition, creativity, the ability to think, the ability to understand or to have consciousness, and feelings. Needless to say it is hard to pin down many of these vague quantities, but this has not stopped artificial intelligence researchers and critics of AI from debating the points ad infinitum. Now we will mention some of the more prominent arguments.

1.1.2   Thinking

The issue of whether or not a machine could think might be decided quite easily by determining *exactly* how people think and then showing that the machine operates internally the same way or so close to the same way that there is no real difference between a human thinker and a machine thinker. For instance, some AI researchers have proposed that thinking consists of manipulating large numbers of rules, so if that is all that a person does and the machine does the same thing, it too, should be regarded as thinking. Or, for another example, it has been suggested that thinking in people involves quantum mechanical processing. If this is the case, then an ordinary computer could not think but it is always possible that the right kind of quantum mechanical computer could think. Settling the issue this way may be simple, but it will be a long time before we know enough about human

---

[1] This comes from an informal survey by the author of Chinese students.

thinking to settle it this way. In the meantime, some people have proposed a weaker test for thinking: the Turing test.

1.1.3 The Turing Test for Thinking

Turing [238] and his followers believe that if a machine behaves very much like a person who is thinking, then the term thinking should apply to what the machine is doing as well. People who argue the validity of this test believe it is the running of an algorithm on a computer that constitutes thinking and it should not matter whether the computer is biological or electronic. This viewpoint is called the *strong AI* viewpoint. On the other hand, people who believe that electronic computing can only simulate thinking are said to have the *weak AI* viewpoint.

The most common version of the Turing test is the following (for Turing's original version, see Exercise 1.1): Put a person or a sophisticated computer program designed to simulate a person in a closed room. Give another person a teletype connection to the room and let this person interrogate the occupant of the closed room. The interrogator may ask the occupant any sort of question, including such questions as, "Are you human?" "Tell me about your childhood." "Is it warm in the room?" "How much is 1087567898 times 176568321?" In this last question a digital computer has a decided advantage over a human being in terms of speed and accuracy so that the designers of the simulated human being must come up with a way to make it as slow and unreliable as people are at doing arithmetic. In the case of "Are you human?" the machine must be prepared to lie. It is given, of course, that if the occupant of the sealed room is a person, the person is thinking. If after a short period of time the questioner could be fooled into thinking that the occupant was a person when it actually was a machine, it should be fair to say that the machine must also be thinking.

With a sufficiently complex computer and computer program, it would be a virtual certainty that *many* naive questioners will be unable to determine after a *short* period of time whether or not the occupant of the sealed room is a human being or a machine simulating a human being. However, it also seems a virtual certainty that more determined and sophisticated questioners will find ways to tell the difference between a machine and a human being in the sealed room (for instance see Exercise 1.1).

Notice also that the Turing test is relatively weak in that to a large extent it is a test of knowledge: if a computer failed to pass the Turing test *because it did not know something that a human being should know* it is no reason to claim that it is not thinking! Thinking is something that is independent of knowledge.

1.1.4 The Chinese Room Argument

An important argument against the strong AI viewpoint is the Chinese room argument of Searle [196, 197]. In this thought experiment the occupant of the Turing test room has to communicate in Chinese with the interrogator and Searle modifies the Turing test in the following way. Searle goes into the closed room to answer questions given to him despite the fact that he does not know any Chinese. He takes with him into the room a book with a Chinese understanding algorithm in it plus some scratch paper on which to do calculations. Searle takes input on little sheets of paper, consults the book that contains the algorithm for understanding Chinese, and by following its directions he produces some output on another sheet of paper. We assume that the output is good enough to fool almost anyone into thinking that the occupant of the Chinese room understands the input and therefore must be thinking. But Searle, who does not understand any Chinese does not understand the input and output at all, so he could not be thinking or understanding. Thus merely executing an algorithm, even if it gets the right answers, should not constitute understanding or thinking.

Believers in strong AI then reply that while Searle does not understand, it is the whole room, including Searle, the algorithm in the book, and the scratch paper that is understanding. Searle counters this by saying he could just as well memorize the rules, do away with the pencil and paper, and do all the calculations in his head, but he still would not be understanding Chinese. Searle takes the point even farther by noting that a room full of water pipes and valves operated by a human being could, in principle, appear to understand without actually understanding as a real Chinese person would understand if that person was in the Turing test room.

The point of the argument is that merely executing some algorithm should not constitute understanding or thinking, understanding and thinking require something more. Searle supposes that the something more in people comes from having the right kind of hardware, the right kind of biology and chemistry.

1.1.5 Consciousness and Quantum Mechanics

Another criticism of the strong AI viewpoint is that intelligence, thinking, and understanding require consciousness. Of course, no one can give a solid definition of consciousness or a foolproof test for it. To the critics of strong AI, consciousness seems to be something that is orthogonal to computing, orthogonal to ordinary matter, but something that people and perhaps higher animals have. The strong AI position on consciousness is that it is something that will emerge in a system when a sufficiently complex algorithm is run on a sufficiently complex computer.

Recently, Roger Penrose, a mathematical physicist, has written two popular books [148, 149] giving his criticism of the strong AI viewpoint. He argues that intelligence requires consciousness and consciousness involves a nonalgorithmic element, an element that no ordinary computer running an algorithm can duplicate. Furthermore, according to Penrose, the nonalgorithmic element involves quantum mechanical effects. Lockwood [101], Wolf [263], and Nanopoulos [133] also speculate on how the mind might operate quantum mechanically and how consciousness might arise from quantum mechanical effects.

1.1.6 Dualism

The 17th century philosopher and mathematician, Rene Descartes, was a proponent of the idea that there is more to a human being than just plain matter, there is an additional component, a spiritual component, often called "mind-stuff." In his conception, the spiritual and material components of the mind can interact with each other. A few researchers such as Eccles and Popper (see [34]) take this position now. If thinking, consciousness, and intelligence require a spiritual component, then it may be difficult or impossible to get a machine to behave much like a human being.

With all this disagreement on what constitutes intelligence, thinking, and understanding, it will be some time before satisfactory definitions are worked out.

1.2 Association

The principle of association may be the most important principle used in intelligence. Briefly put, given that a set of ideas is present in the mind, this set will cause some new idea to come to mind, an idea that has been associated with the set of ideas in the past. This most important principle has been known for hundreds or even thousands of years, but perhaps the best early detailed description was given by a famous 19th century philosopher/psychologist, William James, in his two-volume series, *The Principles of Psychology* [72] (and the abridged one volume version, *Psychology* [73]). In effect, James and other psychologists and philosophers had a very high-level solution to the AI problem by the 19th century, however, they were unfortunate in that there were no computers available at the time with which they could make their ideas more concrete.

In the excerpt below, James gives the principles of association:

> *When two elementary brain processes have been active at the same time, or in immediate succession, one of them, upon reawakening, tends to propagate its excitement into the other.*
>
> But, as a matter of fact, every elementary process has unavoidably found itself at different times excited in conjunction with *many* other processes. Which of these others it shall now awaken becomes a problem. Shall b or c be aroused by the present a? To answer this, we must make a further postulate, based on the fact of *tension* in nerve tissue, and on the fact of summation of excitements, each incomplete or latent in itself, into an open resultant. The process b, rather than c, will awake, if in addition to the vibrating tract a some other tract d is in a state of subexcitement, and formerly was excited with b alone and not with a.

Instead of thinking of summing up "tension" in nerve tissue, today we would think of summing up voltages or currents.

These principles can be better understood with Figure 1.1. If we activate the idea, a, and its only associations in the past are with the idea b with an activation strength of 0.25 and with c with an activation strength of 0.40, then c must come to mind. Think, as James does, as if "tension" or electrical current is flowing from a into both b and c. This is shown in Figure 1.1(a). On the other hand, if at some point in time, b had been associated with a and d, and if a *and* d come to life, the idea c must come to mind, as the sum of currents flowing into it is the greatest. This is shown in Figure 1.1(b). We will often speak of ideas "lighting up" or being "lit." This is in accord with conventional terminology where people often say that ideas "light up" in their mind. We will also talk about the "brightest" (highest rated) idea as the one that comes to mind.

Some examples of this summation process are worth looking at. One example from James is what occurs when the old poem, "Locksley Hall," is memorized. Two different lines of the poem are as follows:

I, the heir of all *the ages* in the foremost files of time,

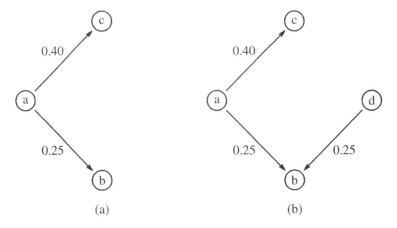

Figure 1.1: Summation of excitements.

and

> For I doubt not through *the ages* one increasing purpose runs.

We focus in on the words, "the ages." If a person had memorized this poem and started reciting the first line, and got to the phrase, "the ages," why should he continue with the words, "in the foremost files of time," rather than "one increasing purpose runs?" The answer is simple. While "the ages" points to, or suggests, *both* "in the foremost files of time" and "one increasing purpose runs," there are those words *before* "the ages" that also, but to a smaller extent, point to "in the foremost files of time." The summation of the "the ages" with "I, the heir of all" produces a larger value for "in the foremost files of time" than for "one increasing purpose runs."

A second example from James is the following:

> The writer of these pages has every year to learn the names of a large number of students who sit in alphabetical order in a lecture-room. He finally learns to call them by name, as they sit in their accustomed places. On meeting one in the street, however, early in the year, the face hardly ever recalls the name, but it may recall the place of its owner in the lecture-room, his neighbors' faces, and consequently his general alphabetical position: and then, usually as the common associate of all these combined data, the student's name surges up in his mind.

The principles of association also form the basis for most TV game shows. The principles have been seen there in their purest form in the shows *Password, Password Plus,* and *Super Password.* In the simplest version, *Password,* there are two teams of two players each. One player on each team is given a secret word, short phrase, or name and the object of the game is for this person to say a word that will induce the player's teammate to say the secret word, phrase, or name. Whichever team gets the right answer scores some points. For example, in one game the secret name was "Jesse James." The first clue given in the game was "western" but the other person's response was "John Wayne." This is fairly reasonable since in many people's minds, John Wayne is very closely associated with Westerns. Some other reasonable responses might be "cowboys," "Indians," or "eastern." Since the response was wrong, the other team gets a chance and this time the clue was "train." Adding together the clues "western" and "train," some reasonable responses might be "Santa Fe," "Union Pacific," or "Central Pacific," all famous western train companies, but again the contestant got the wrong answer. Finally, after the clues, "frank," "brother," and "robber" were given, a contestant got the right answer, "Jesse James," a famous train robber in the Old West who had a brother named Frank. The game is summarized in Figure 1.2.

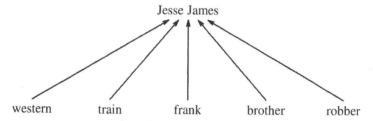

Figure 1.2: A simple game of *Password*.

It is easy to model the process of combining ideas in the following manner. Suppose we assign numeric values to the strength of the associations between ideas. Suppose the associations with "western" are:

John Wayne	0.40
cowboys	0.30
Indians	0.25
Santa Fe	0.25
Jesse James	0.15
Frank James	0.10
Union Pacific	0.05
Central Pacific	0.05

and the associations for "train" are:

Amtrak	0.30
electric	0.25
Santa Fe	0.25
Union Pacific	0.25
Central Pacific	0.25
Jesse James	0.15

If you want to combine the effects of two different clues, like "western" and "train," one simple solution is to simply add up how much is being contributed to each of the other ideas in the lists of ideas. In Figure 1.3 we show how "western" and "train" combine to activate all the ideas with which they have been associated in the past. The numbers to the right in Figure 1.3 show the summations. For instance, "western" contributes 0.25 to "Santa Fe"

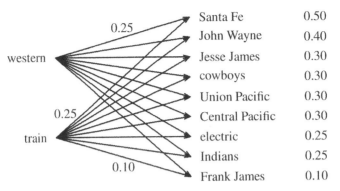

Figure 1.3: How clues in *Password* can combine to produce possible answers. Only a few of the association strengths are shown.

and "train" also contributes 0.25 to "Santa Fe." When it then comes to guessing an answer, the idea with the highest rating is "Santa Fe."

For a final example of these principles we now look at their actual use in a simple AI program. Walker and Amsler [245, 246] created a program called FORCE4 whose purpose is to look at newspaper stories and figure out roughly what the story is about. The program does not acquire any kind of detailed understanding of what the article is about, but only tries to put it in a general category, such as weather, law, politics, manufacturing, and so forth. It makes use of a set of subject codes assigned to specialized word senses in the Longman Dictionary of Contemporary English. For instance, the word "heavy" is often associated with food (coded as FO), meteorology (ML), and theater (TH). "Rainfall" is associated with meteorology (ML). "High" is associated with motor vehicles (AU), drugs and drug experiences (DGXX), food (FO), meteorology (ML), religion (RLXX), and sounds (SN). "Wind" suggests hunting (HFZH), physiology (MDZP), meteorology (ML), music (MU), and nautical (NA).

One story given to FORCE4 was the following:

Heavy rainfall and high winds clobbered the California coast early today, while a storm system in the Southeast dampened the Atlantic Seaboard from Florida to Virginia.

Travelers' advisories warned of snow in California's northern mountains and northwestern Nevada. Rain and snow fell in the Dakotas, northern Minnesota and Upper Michigan.

Skies were cloudy from Tennessee through the Ohio Valley into New England, but generally clear from Texas into the mid-Mississippi Valley.

For each important word, the program counts one point for each of its associated ideas.

When you apply this procedure to the above story, you get the following counts:

10	ML (Meteorology)
4	GOZG (Geographical terms)
4	DGXX (Drugs and drug experiences)
3	NA (Nautical)
2	MI (Military)
2	FO (Food)
2	GO (Geography)
1	TH (Theatre)

The results show that it is a weather-related story. Walker [245] reports that when over 100 news stories were submitted to the program,

> The results were remarkably good: FORCE4 works well over a variety of subjects—law, military, sports, radio and television—and several different formats—text, tables and even recipes.

1.3   Neural Networking

When scientists became aware that nerve cells pass around electrical pulses, most of them assumed that this activity was used for thinking. Relatively little is known about how networks of nerve cells operate, and determining how networks of nerve cells operate is a major part of the field of neural networking. The second major part of neural networking research centers on the study of computer models of simplified nerve cells. In this book we will deal almost exclusively with the computer-based models.

1.3.1   Artificial Neural Networks

Artificial neural networks represent a way of organizing a large number of simple calculations so that they can be executed in parallel. The calculations are performed by relatively simple processors typically called nodes, artificial neurons, or just neurons. Artificial neurons have a number of input connections and a number of output connections. The input connections serve to activate, or excite, or we may say, "light up" a neuron or they might also try to turn off or inhibit a neuron. An excited neuron then passes this excitement on to other neurons through its output connections. Figures 1.2 and 1.3 can be regarded as diagrams of neural networks. In Figure 1.3 there are the input nodes, "western" and "train," and the outputs are "Santa Fe," "John Wayne," and so forth. The connections between inputs and outputs are called *weights* in neural networking terminology and the value of a weight is what we call the "strength of association."

A simple artificial neuron is shown in Figure 1.4. For artificial neurons, each connection has associated with it a real value called a weight and each neuron has an activation value. The typical algorithm for activating an artificial neuron, j, given a set of input neurons, subscripted by i, and the set of weights, w_{ij}, works as follows. First, find the quantity, net_j, the total input to neuron j by the following formula:

$$net_j = \sum_i w_{ij} o_i$$

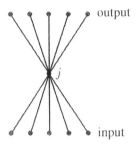

Figure 1.4: A simple artificial neuron, j, with inputs from one set of neurons and outputs to another set.

When the activation value of neuron i, o_i, times the weight w_{ij} is positive, then unit i serves to activate unit j. On the other hand, when the value of unit i times the weight w_{ij} is negative, the unit i serves to inhibit unit j. The activation value of neuron, j, is given by some function, $f(net_j)$. The function, f, may be called an *activation function, transfer function,* or *squashing function*. One simple activation function is simply to let the sum of the inputs, net_j, be the activation value of the neuron. This is how the *Password* network in the last section worked. A second common activation function is to test if net_j is greater than some threshold (minimum) value, and if it is, the neuron turns on (usually with an activation value of +1), otherwise it stays off (usually with an activation value of 0). The neural network in Figure 1.5 computes the exclusive-or function and it uses the activation function, $1/(1 + e^{-net_j})$. While this function only reaches 0 and 1 at $-\infty$ and ∞ respectively, when the outputs are close enough to 0 or 1 they are counted as being the same as 0 or 1. There are many other possible activation functions that can be used.

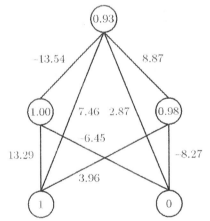

Figure 1.5: This simple network computes the exclusive-or function. The two inputs are made on the bottom layer and the top layer has the answer to within 0.1 of the desired value. In this case the input values are 1 and 0 and the output is 0.93.

Usually the neurons in artificial neural systems have the units arranged in layers as shown in Figures 1.4 and 1.5. These networks have the input layer at the bottom, a *hidden layer* in the middle, and an output layer at the top. The hidden layer gets its name from the fact that if you view the network as a black box with inputs and outputs that you can monitor, you cannot see the hidden inner workings. When the flow of activation or inhibition goes from the input up to higher-level layers only, the network is said to be a *feed-forward* network. Most often the connections between units are between units in adjacent layers, but it is also possible to have connections between nonadjacent layers and connections within each layer. If there are connections that allow the activation or inhibition to spread down to earlier layers, the network is said to be *recurrent*.

Learning in artificial neural systems is accomplished by modifying the values of the weights connecting the neurons and sometimes by adding extra neurons and weights. There have been a number of learning algorithms that have been proposed and tested for neural networks but the most powerful and most generally useful algorithm is the back-propagation algorithm described in Chapter 3.

Currently, neural networks can be used for many pattern recognition applications, such as recognizing letters, rating loan applications, choosing moves to make in a game, and they can even do simple language processing tasks. One system has been used to automatically drive a van along interstate highways. So far at least, networks are not very well suited to doing complex symbol processing tasks like arithmetic, algebra, understanding natural language, or any task that requires more than a single step of pattern recognition. To produce a system capable of doing much of what human beings can do will require at the very least more complex models and a multitude of different subsystems, each tuned to perform slightly different tasks and all working together.

1.3.2 Biological Neural Networks

As mentioned above, relatively little is known about how biological neural networks operate. For quite a long time it has been assumed that the neurons in the human brain act like the artificial neurons in that they pass around electrical signals, but whereas an artificial neuron receives a single real-value input from each of its input neurons, the biological neurons pass around simple pulses, pulses that are either present or not present. The number of pulses per second going along a connection is an indication of the weight of the connection—more pulses mean a higher weight, fewer pulses indicate a lower weight. In this theory each neuron acts like a little switch, when enough pulses are input a neuron outputs a pulse. The estimates are that there are around 100 billion neurons in the brain with about 1000 connections per neuron and each neuron switches about 100 times per second. This gives a processing rate of around 10^{16} bits per second.[2] But biological neurons are more complicated than simple switches since they are influenced by chemicals within the cell. One recent discovery is that at least some cells involved in vision are not just sending out plain pulses but in fact are passing around coded messages.[3] The shape of the pulse codes the message.

Theories by Hameroff et al. [57] and [133] have each cell acting as a small computer rather than as just a simple switch. The computing would be done in microtubules that make up the cell's cytoskeleton. In this case they estimate a single neuron is processing

---

[2] This estimate is taken from the article by Hameroff et al. [57] which in turn was taken from [127].

[3] A simple description can be found in "A New View of Vision" by Christopher Vaughan, *Science News*, Volume 134, July 23, 1988, pages 58–60. There are many other articles on this topic. See [171] for a list of other references.

about 10^{13} bits per second and the whole brain would be processing at least about 10^{23} bits per second assuming there is some redundancy.

From time to time people have made estimates of how many bits of information the human brain can store based on certain assumptions, but since it is most certainly not known how information is stored and processed in the brain, none of these estimates can be taken too seriously.

In short, little is known about what is really going on in the human brain but new research may soon shed a lot of light on what is going on. Whatever is happening, it is much more complicated than the processing done in the current set of artificial neural network models.

1.4 Symbol Processing

For most of the history of artificial intelligence the symbol processing approach has been the most important one. There are several reasons why symbol processing has been the dominant approach to the subject. First, there were a number of highly impressive symbol processing programs done in the early 1960s. Two of these early systems are described in Chapter 8, the SAINT program of Slagle that could do symbolic integration and the Geometry Theorem Proving system of Gelernter and others. In addition, it seemed obvious to process natural language this way since language consists of symbols. The second reason symbol processing has been dominant is that it seemed as if it would be a very long time before artificial neural networks could be designed that could do such impressive things.

The advocates of the symbol processing approach to AI have proposed the Physical Symbol System Hypothesis (PSSH) (see [138], [139], and [40]). It states that symbols, structures of symbols, and rules for combining and manipulating symbols and structures of symbols are the necessary *and sufficient* criteria for creating intelligence. This means that these features and only these features are required for producing intelligent behavior. Advocates of PSSH assume that the human brain is doing nothing more than manipulations of collections of symbols. In current computers the manipulations are done sequentially, but advocates of this position assume that human minds actually do parallel processing of symbols. It is the *Physical* Symbol System Hypothesis because advocates assume that there are physical states in the brain that correspond to the kind of structures that a symbol processing computer program uses. PSSH advocates also assume that although neural hardware implements the symbol processing abilities of the brain, this hardware is too low level to have to worry about. So, just as Pascal programmers do not have to worry about integrated circuits, symbol processing can concern itself with symbols and structures of symbols without worrying about the underlying neural hardware. Of course, symbol processing adherents acknowledge that neural networking *is* important for lower-level tasks like vision and movement.

The techniques used in symbol processing are very similar to those used in programming in conventional languages such as Pascal and Fortran, however symbol processing emphasizes list processing and recursion and symbol processing methods use symbols rather than numbers. Because in the beginning almost all AI was done in symbol processing languages, some people have defined artificial intelligence as symbolic computing. The most important computer language for AI programs has been Lisp (for list processing language) and a newer language is Prolog (for programming in logic). For the most part we will use Prolog as a notation for some symbol processing algorithms later in the book because Prolog has some built-in pattern recognition capabilities that Lisp does not have.

Symbols are defined as unique marks on a piece of paper and in a computer each symbol is represented by a different integer. Two symbols can be equal or not equal, but there are no other relations defined between them. Notice then, that even though symbols are implemented as integers in computer programs, symbols are simpler than the integers that represent the symbols. In addition to being used individually, symbols can also be combined into structures of symbols such as lists or trees. One example of this might be the expression:

$$A \& B.$$

Inside a computer we might find this as a linked list:

$$A \rightarrow \& \rightarrow B \rightarrow nil$$

or as a tree:

Another part of the symbol processing approach is the assumption that there are rules which specify how symbols and structures of symbols are manipulated. Logic and arithmetic provide perfect examples of symbols and how they can be manipulated and combined using rules. Take for example the logic expression, $A \& B \& A$. Now within logic, there is a rewrite rule that says that this expression can be rewritten as $A \& A \& B$. Moreover, there is a rule that says that $A \& A$ can be rewritten as just A. Some rules from arithmetic for manipulating symbols and expressions are: x/x can be replaced by 1, $1 * x$ can be replaced by x, and $x + (-x)$ can be replaced by 0.

The use of rules in symbol processing methods is a key element of the symbol processing approach because everyone who studies human behavior agrees that people exhibit rulelike behavior. For an example of rulelike behavior consider the following case. Suppose a child learns the meaning of a sentence like:

The cat is on the mat.

The child, knowing what a cat is and what a mat is, and what a cat on a mat is, seems to deduce some rules (or form a theory) about how sentences are constructed. Thus, the child can apply the rules and come up with statements like the following:

The dog is on the mat.

The boy is on the mat.

The block is on the mat.

The cat is on the floor.

Some other rules that people form would be "if something is a bird then it can fly" or "if you drop something it will fall." Such facts are typically coded in a rule format something like this:

if bird(X) then flies(X)

if drop(X) then fall(X)

where X stands for the something. For another example concerning language processing, some researchers have studied how children learn to construct the past tenses of verbs and they have come to the conclusion that the errors that children make show that they are producing rules.

From examples like the above and others, traditional AI researchers have concluded that people have some kind of unconscious machinery that deduces rules as well as some kind of symbol processing architecture that applies them.

Notice, though, that rules are little neural networks where the input and output units have symbolic labels as in this rule:

if a and b then c.

The corresponding network is shown in Figure 1.6. It is a two layer network with inputs labeled a and b and the output unit labeled c. Let the two weights be 1 and let the threshold for unit c be 1.5. Now if unit a and b are both on, (= 1), net_c will be 2. Since net_c is greater than 1.5, unit c turns on, otherwise it stays off, (= 0). And so it turns out that a key element of symbol processing can be regarded as a form of neural networking.

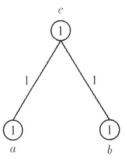

Figure 1.6: The rule, if a and b then c, can be regarded as a neural network where the units a, b, and c can take on the values 0 or 1, the two weights are +1, and the threshold for unit c is 1.5.

Symbol processing techniques have been somewhat successful at doing a number of very narrow but useful tasks involving reasoning and processing natural language.

1.5 Heuristic Search

When people encounter a problem they typically have to do some trial and error work on the problem to find the solution. People look at some of the most likely possible solutions to the problem, not every possible solution. However, a simple computer program is dumb in that it does not have any way of evaluating the possible solutions to determine which are the likely ones. This type of program must do an *exhaustive search* of all the possibilities. Very early on it was recognized that for computer programs to solve problems as human beings do, the programs must be able to look at only the likely possibilities. Programs that use some method to evaluate the possibilities are said to do *heuristic search*.

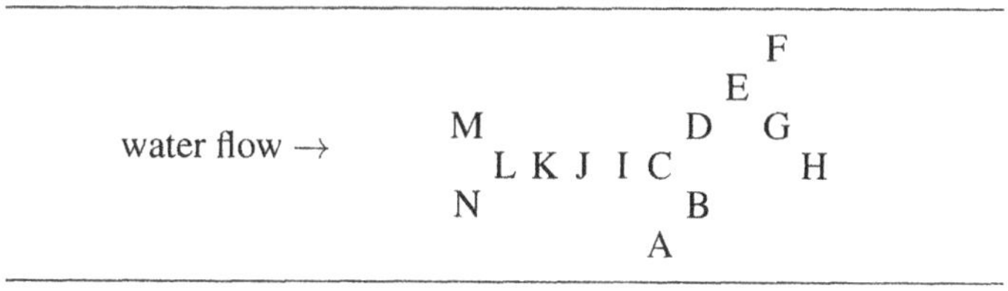

a) a stream where the heuristic succeeds

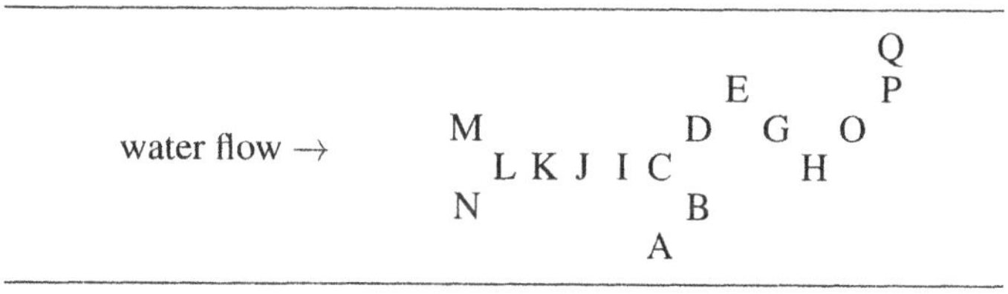

b) a stream where the heuristic fails

Figure 1.7: In trying to cross a stream in a heavy fog by stepping on rocks, one heuristic is to keep trying to move forward and never back up. Rocks are indicated by letters. In a) the heuristic succeeds but in b) never backing up will produce a failure.

As an example, suppose you were hiking through the wilderness and you came upon a small stream that you needed to cross. Suppose there are small flat rocks in the river and you can step from one rock to an adjacent rock. For example, there is the situation shown in Figure 1.7(a) where each rock is indicated by a letter. Suppose you are on the river bank at the bottom and you want to get to the river bank at the top. A human being will "eyeball" the situation and have it solved in a second. The best (and only) path is from A to B to C to D to E to F. How would you have a computer look at the same situation and find that path? To find one computer solution we could make the problem harder. Suppose you come upon this place in the river, but there is fog and the fog is so thick that you can only see one rock ahead of you. You are clearly going to have to start making guesses as to which steps to take. You will realize that *probably* the best thing to do is to keep going forward as much as possible. What sense would it make to go back in the direction that you came from, or up or down the river? You could tell which way was forward by noticing that the river flows from left to right, so when you make a move, you should try to keep upstream on your left and downstream on your right. Therefore, when you get to rock C, the best thing to do is to go on to D and not to I. Again, when you get to rock E, you will go on to F, rather than back up by going to G. A computer program could use the same strategy for finding a path across the river and it would find a path as easily as a person lost in a fog. Both you and the computer were doing a heuristic search of a tree, looking for a goal node. If you did an ordinary search of the tree rather than a heuristic search of the tree, you would find a path across, but probably not nearly as quickly. In an ordinary exhaustive search of the tree, when you get to C, you could try going to I. Follow that path and you could go to N. When you got there, you would fail, but you would back up to try M. When that failed, you would back up and go to C, and so on. The heuristic search is intended to get you across the stream as quickly as possible, but there is a possible problem with this method. If you decide that you *must always* go forward and *never* back up, then there will be some locations where your search will fail because there will not always be such a path available (see Figure 1.7(b)). This illustrates another property of heuristic search: while a heuristic search is usually the fastest way to find an answer, you are not always guaranteed that an answer will be found. Of course, when people get a problem to solve there is no guarantee that they will be able to solve it either. An exhaustive search will get the answer but it may take much longer and sometimes so much longer that the search effectively fails.

1.6   The Problems with AI

The results of decades of experimentation with symbol processing, with and without heuristic search methods, has shown that with these methods computers can do *some* tasks that in people would be regarded as intelligent, such as prove theorems, manipulate mathematical formulas, and understand small amounts of natural language. According to some researchers' definitions of intelligence, these programs display intelligence. On the other hand, such programs typically do not learn from their activities and since learning is a key factor in intelligence critics do not see any intelligence in such programs. Then too, the level of understanding that these programs have is severely limited. For example, if you give a program a statement like "John ate up the street," the program might easily conclude that John was eating asphalt. Or, given that "John was in the 100-meter butterfly," a program might think that John was inside a large insect rather than in a swimming event. People say that such programs that do not have the common sense knowledge that people have are *brittle*. In response to this criticism, most symbol processing researchers say they believe their basic methods are valid and the problems can be eliminated by just producing much larger systems. At the moment a very ambitious project known as CYC (see [99]) is attempting to produce a program with a very large number of facts and rules about the world that hopefully will not be brittle. The early estimate was that the program would need about a million rules. As of 1993,[4] the program had two million and work is continuing at the present time.

1.7   The New Proposals

So while AI has problems, some AI researchers remain optimistic about symbol processing methods but other AI researchers are not and they have started looking into a variety of

---

[4] *Computerworld*, May 10, 1993, pages 104–105.

new proposals. Most of these new proposals come to mind quite easily by just denying the elements of the Physical Symbol System Hypothesis. These ideas are that thinking and intelligence require the use of real numbers, not just symbols; that people use images or pictures, not just structures of symbols; and that they use specific cases or memories, not just rules. In addition, there is another proposal that human thinking involves quantum mechanics and quantum mechanics adds extra capabilities that ordinary computing, not even analog computing, can account for.

1.7.1 Real Numbers

Symbol processing works only with symbols and the only relation defined between symbols is equality, two symbols are equal or not equal. This can work fairly well when the answer to every question is a nice true or false, but in many situations in the real world judgments are fuzzy. The accused person on trial must be proved guilty beyond a reasonable doubt. Music produced by one composer sounds better than the music from another composer. One crime will be judged as more heinous than another. New cars normally look better than older cars. Some government projects are judged as more worthwhile than others. Some chess moves are better than others. It is only natural that researchers think that such judgments are made using some kind of analog computation rather than the simple true/false logic found in symbol processing.

There are a couple of ways to include this analog concept in AI theory, but the most important of these is found in neural networking when the activation values of the units and the weights take on real values. Even though neural networking was present at the beginning of AI research, it was quickly abandoned in favor of symbol processing methods and so it has hardly been investigated until recently. When neural networking methods are applied to AI-type problems it is called *connectionist AI*. One variation on connectionist AI is *parallel distributed processing*, or PDP for short. It uses a specific type of coding within networks.

These methods are fairly good at doing a single step of pattern recognition, but they present connectionist AI researchers with quite a problem as to how to store complicated facts because unlike symbol processing AI where you can use a tree structure to store a fact, in a neural network the facts must be represented as a vector or matrix of real numbers. So for example, if you needed to store away the fact that:

Jack and Jill went home,

it is very straightforward in a digital computer to produce one kind or another of tree structure to represent this such as:

went
and   home
Jack   Jill

So far there is no established good way to represent this tree structure as a vector or matrix of real numbers, although there are some proposals along these lines.

One position on neural networking is that networks do have some features that are required for intelligence, thinking and reasoning, but conventional symbol processing is also necessary. In this case the Physical Symbol System Hypothesis is wrong at the point where it says that symbol processing is *sufficient*. One proposal is that the mind may be basically a connectionist computer architecture, but it simulates a symbol processing architecture to do those tasks that are most suited for symbol processing, while still using connectionist methods for other types of problems.

A more extreme position is that neural networking is necessary and *sufficient* and the only reason that symbol processing methods are somewhat successful is that they just approximate what is happening in the mind. To get better performance, connectionist methods are needed.

1.7.2   Picture Processing

MacLennan [102, 103][5] has argued that the important features of connectionist AI are the use of real numbers, that the large number of neurons in the brain and eye can be treated mathematically the same as fields (as in magnetic, electric, and gravitational) in physics (see [104]) and that image processing or picture processing is going on in the human mind.

For an example of this suppose we are watching the movie "Jack and Jill's Greatest Adventure," with that familiar story:

> Jack and Jill went up the hill to fetch a pail of water. Jack fell down and broke
> his crown and Jill came tumbling after him.

Just watching the movie gives you *images* that are stored away and it is rather difficult to argue that people store these images as some kind of symbolic tree-type representation. Then too, just reading the words will develop images in your mind. Moreover, as Jack starts to fall down you have to predict based on the images that Jack might suffer some damage that will require medical attention, so just working from the images you can do some reasoning. Why use symbols, structures of symbols, and formal rules to do this when picture-based processing will work?

The fact that people do store many memories as pictures and do at least some of their reasoning about the world using pictures ought to be one of the most obvious principles of all, yet it has been neglected, in part due to the predominance of symbol processing and in part due to the fact that processing pictures is hard compared to processing symbols. Unfortunately, at this point in time image processing is still fairly underdeveloped and has not been used in conjunction with representing the real world in programs where the goal of the program is to reason about the real world. Of course, simple image processing has been used by robots and in programs to recognize patterns such as handwritten or typed digits and letters of the alphabet.

1.7.3   Memories

The final key feature of the Physical Symbol System Hypothesis that can be criticized is the idea that people take in large amounts of experience from the real world, condense all these specific instances down to a handful of rules, and then people work from these rules to

---

[5] In addition to discussing how connectionist programs might store knowledge about the real world, MacLennan also reviews the symbol processing position in these papers so they are quite worthwhile and they are online.

solve new problems. An example of this is that when you drop something, call it, X, where X may be a rock, a piece of paper, or a feather. If you have done a lot of experimenting with dropping various things, then you will derive the rule: *if you are holding something and you let go, it will fall straight down*. In a symbol processing representation you are likely to code this as something like:

$$\text{if letgo}(X) \text{ then fall}(X).$$

Yet there is a problem with such a rule because it only applies under certain conditions. If the air is moving and X is a feather or a flat piece of paper, then it will not fall straight down and it may remain in the air for quite some time before reaching the ground. But, if a piece of paper is crumpled up it will fall faster than if it is flat. If the air is moving very rapidly, even a rock will not fall straight down. Then what about the case we have all seen on TV where an astronaut on board a spaceship in orbit around the Earth lets go of something and rather than falling[6] it simply floats in midair? So a humanly coded set of rules is subject to the same problem that comes up in conventional computer programming where you must consider every possible permutation of the input data. Thus a rule-based program where the programmer has neglected to take into account wind velocity could conclude that if someone dropped a feather in a tornado the feather will fall straight to the ground, another example of the brittleness of conventional programs. So far, generating rules from data has not worked especially well either except for very small problem domains, domains much smaller than the real world domain.

One way to eliminate the problems involved with finding and using rules is to just not bother with the rules. If you have done your experiments of dropping various things under various conditions and seen the experiments done in space, then when someone asks you what will happen if you drop something all you have to do is reference your memories of your experiments to get the answer. This idea that people use simple memories to solve many ordinary real world problems is now getting a lot of attention although it is being done in the context of symbol-based methods, not in a picture-based context. These methods are called *case-based* and *memory-based*.

1.7.4 Quantum Mechanics

The possible application of quantum mechanics to thinking comes up in a number of ways. As already mentioned, Penrose in his two well-known popular books [148, 149] argues that consciousness is necessary for intelligence and quantum mechanics is responsible for consciousness, and moreover, that QM contains a nonalgorithmic component that cannot be duplicated by digital computers. Second, Vitiello [240] has proposed a quantum mechanical memory system that has the useful property that no matter how many memories this system has stored, one more can always be added without damaging any of the old memories, so in effect you get an unlimited memory. Finally, there is the idea that quantum mechanics might allow faster than light communication and this would explain the persistent reports of mind reading and predicting the future. For an argument in favor of this see [79]. Nanopoulos [133] has a quantum mechanical theory of brain function that

---

[6] The physics people will explain it by saying the object, the person, and the spaceship are really all falling at the same rate.

fits the psychological theories of William James and Sigmund Freud. Unfortunately, the application of quantum mechanics to thinking is still in a very early stage of development, it is more of a hope than any sort of concrete, testable proposal.

1.8 The Organization of the Book

The book starts with some of the lowest-level vision problems in Chapter 2 and then, generally speaking, the book goes on to cover higher and higher level problems until this progression ends with natural language processing in Chapter 10. The principles found at the beginning in vision systems can be found in slightly different forms all the way up to the highest levels. First, Chapters 2 and 3 illustrate the most important and useful pattern recognition and neural networking methods. Chapters 4 and 5 then give the approximate symbolic equivalents to the material in Chapters 2 and 3. The theme of Chapter 6 is that the methods presented so far are much too simple to produce programs with humanlike behavior. What is really needed is a much more complex architecture *and* a method for storing and retrieving knowledge in that architecture. Chapter 7 is to some extent an extension of the knowledge storage and retrieval problem in that storing, retrieving, and using cases rather than the traditional classical method of using rules is the theme. To a large extent Chapters 8, 9, and 10 are examples and applications of the principles given in Chapters 1 through 7, although, of course, Chapters 8 and 9 also develop the heuristic search theme as well.

One of the key ideas in this book is, of course, that the new methods need to be studied and worked on in order to get programs to achieve human levels of performance in dealing with problems, especially in dealing with the whole range of real world problems. However, all these methods, the symbolic and the neural and the memory-based all represent different ways of doing *pattern recognition*. Pattern recognition can be defined as the ability to classify patterns like the letters of the alphabet, other written symbols, or objects of various sorts; however, pattern recognition can also be used to try and find the more abstract and hidden patterns that exist within economic data or social behavior. Pattern recognition can also be used to describe the process of finding patterns that are close to each other in situations where the goal is not to do formal classification. Pattern recognition is also a formal academic field of study, typically found in electrical engineering or computer science departments, where the goal is once again to recognize patterns, either the visual ones or the more abstract ones.

1.9 Exercises

1.1. In the original Turing test [238], two people, a man and a woman go into the Turing test room while a third person asks them questions through a teletype system. Suppose the man is A and the woman is B but the third person knows them as X and Y. The problem for the third person is to try to determine whether X is male and Y is female or if X is female and Y is male. In the game, Y will try to help the questioner make the correct identifications but X will try to confuse the questioner. Then Turing says:

> We now ask the question, What will happen when a machine takes the part of A in this game? Will the interrogator decide wrongly as often when the game is played like this as he does when the game is played between a man and woman? These questions replace our original, "Can machines think?"

Would this version of the Turing test be any better at identifying thinking than the usual version where the game is simply to determine whether the entity in the Turing test room is a person or a computer?

Consider this too: one person posted a note in the Usenet comp.ai.philosophy newsgroup saying that Turing was noted for being a playful man and that maybe this whole Turing test was just a playful joke.[7]

You might actually want to try the human only version of this test in class. One way that is said to be very effective in determining who is male and who is female is to ask X and Y false questions such as "What is a Lipetz-head screwdriver?" Firschein[8] reports that in his experiments with the test: "Once this false question approach is discovered, few students can successfully fool the class."

1.2. The strong AI position is that certain types of computing are thinking. Suppose this is true. Does this mean that computers will be able to write great music, great poetry, create great art, and so on, or are these things something that only people can do?

1.3. Here are some examples of third and fourth grade arithmetic word problems:

> Matt has 5 cents. Karen has 6 cents. How many cents do they have altogether?

> Kathy is 29 years old. Her sister Karen is 25 years old. How much older is Kathy than Karen?

> If Mary sells 5 pencils at 6 cents each, how many cents will she have altogether?

It is not very hard to get a computer program to do a fair job of reading and solving these problems. These problems are quite simple to program because all children know at this grade level is how to add, subtract, multiply, and divide integers. They do not know about negative numbers or fractions. All they (or a program) have to do is pick off the numbers and then decide which operation to apply. Subtraction must always produce a positive number and division must always produce an integer. Also, the problems are loaded with phrases such as: "have altogether," "how much more," and "at this rate, how many."

Write a program that will look at the words in problems, much as the FORCE4 program did in Section 1.2 and then have it decide on what operation to apply to get the answer. Consult third and fourth grade textbooks for more problems.

Also consider the following alternative strategy. Instead of using individual words alone to choose the operation to apply, try using each *pair* of adjacent words in the problem. For example, "much more" will suggest subtraction and "many altogether" will suggest addition or maybe multiplication. Of course this will produce a longer list of items to store, but see if it produces better results.

---

[7] From Kenneth Colby, UCLA Computer Science Department, Message-ID: <3k4iub$p8n@oahu.cs. ucla.edu>, 14 Mar 1995 09:14:51 -0800.

[8] "Letters to the Editor," Oscar Firschein, *AI Magazine*, Fall 1992.

In fact, it is easy to create a fairly small program that can *learn* to do these problems. Give the program some sample problems and let it break the text down into pairs of words. For the first problem above, you get the pairs, "Matt has," "has 5," "5 cents," "cents," and so forth. Associate addition with each of these pairs. Expose your program to many such problems, and then test it with some of the problems you have trained it on, as well as on some unknown ones, and see how effective it is.

Whether you program this problem or not, you can still evaluate the effectiveness of the techniques that have been suggested as well as suggest more techniques that may work. Consider whether or not you could use these techniques or similar ones to do harder problems like:

> John went to the store and decided to buy 4 pieces of candy at 10 cents each.
> He gave the clerk 50 cents. How much change should he receive?

1.4. Rather than trying to classify arithmetic word problems you may want to classify Usenet news articles on two or more topics. It is probably best to choose articles from two very different newsgroups. After you train your program on the two classes, give the program some additional articles to see how well it classifies them.

1.5. For the network in Figure 1.5, show that when any two binary digits are given to the two input units the correct value (to within 0.1) of the exclusive-or of the two inputs appears on the output unit. Compute by hand and give the hidden unit values as well.

1.6. If we use a neural network where output units have real values for the threshold values and weights, show the networks corresponding to these two rules:

> if a or b then c
> if (a or b) and not c then d

1.7. Is the heuristic search suggested for crossing the river on rocks a realistic model of how people would find a path across the river if there was no fog? If it is not, how do people do it?

1.8. For some extra background on AI, read and summarize one or more of the following articles, all found in the Winter, 1988 issue of *Daedalus*:

> "One AI or Many?" by Seymour Papert,
> "Making a Mind vs. Modeling a Brain" by Stuart and Hubert Dreyfus,
> "Natural and Artificial Intelligence" by Robert Sokolowski,
> "Much Ado About Nothing" by Hilary Putnam,
> "When Philosophers Encounter Artificial Intelligence" by Daniel C. Dennett.

1.9. Stuart and Hubert Dreyfus are two noted critics of artificial intelligence and they give their criticism in the book, *Mind Over Machine* [28]. Read this book and summarize their criticisms and then state whether or not you agree with them and why. (A good due date would be near the end of the course.)

Chapter 2

Pattern Recognition I

In this chapter we will be examining algorithms, especially neural networking algorithms, that can be used for pattern recognition. These algorithms will use the neural and associationist principles discussed in the first chapter. To illustrate the use of the principles we will start by looking at programs that can recognize letters of the alphabet, then look at a method for recognizing words, and finally show how the same principles are involved in higher-level thought as well. One of the results of this study will be that recognizing even simple patterns like letters requires more knowledge about the world than you might at first suspect is necessary. The problems encountered with recognizing letters spill over to higher-level cognitive activities such as understanding natural language and pose difficulties for all AI programs.

2.1   A Simple Pattern Recognition Algorithm

It is easy to apply the pattern recognition ideas described in the first chapter to the recognition of alphabetic characters or other such small patterns. Consider the letter E, shown in Figure 2.1 as a 21×21 matrix consisting of ones and zeros. To aid in identifying the pattern, the area the pattern occupies is also divided into the nine subareas shown in the figure. The solution to recognizing such a pattern is to break it down into its component parts and then form an association between each part and the answer. Let the component parts be the small vertical, horizontal, and diagonal line segments shown in Figure 2.2 and then make a listing of which of these subpatterns are present in each of the nine regions of the unknown pattern. The characteristics of three letters, E, F, and H are listed in a matrix in Figure 2.3. A '1' under a subpattern in a particular region indicates that that subpattern is present in the region and a '0' means it is not there. Each row of the matrix can be regarded as a prototype point that represents the ideal characteristics of each letter.

   When we get an unknown pattern we will also list the subpatterns that are present in it as a column vector, $\vec{x}$. For the letter E in Figure 2.1, $\vec{x}$ would be:

$$\vec{x} = (1,1,0,0,0,1,0,0,0,1,0,0,1,1,0,0,0,1,0,0,0,1,0,0,1,1,0,0,0,1,0,0,0,1,0,0).$$

(where $\vec{x}$ is displayed horizontally for convenience). If we name the matrix of Figure 2.3, A, then if we form the product,

$$\vec{b} = A\vec{x}$$

```
0 0 0 0 0 0 0 0 0 0 0 0 0 0 0 0 0 0 0 0
0 0 0 0 0 0 0 0 0 0 0 0 0 0 0 0 0 0 0 0
0 0 0 0 0 0 0 0 0 0 0 0 0 0 0 0 0 0 0 0
0 0 0 1 1 1 1 1 1 1 1 1 1 1 1 1 1 0 0 0
0 0 0 1 0 0 0 0 0 0 0 0 0 0 0 0 0 0 0 0
0 0 0 1 0 0 0 0 0 0 0 0 0 0 0 0 0 0 0 0
0 0 0 1 0 0 0 0 0 0 0 0 0 0 0 0 0 0 0 0
0 0 0 1 0 0 0 0 0 0 0 0 0 0 0 0 0 0 0 0
0 0 0 1 0 0 0 0 0 0 0 0 0 0 0 0 0 0 0 0
0 0 0 1 0 0 0 0 0 0 0 0 0 0 0 0 0 0 0 0
0 0 0 1 1 1 1 1 1 1 1 1 1 1 1 1 1 0 0 0
0 0 0 1 0 0 0 0 0 0 0 0 0 0 0 0 0 0 0 0
0 0 0 1 0 0 0 0 0 0 0 0 0 0 0 0 0 0 0 0
0 0 0 1 0 0 0 0 0 0 0 0 0 0 0 0 0 0 0 0
0 0 0 1 0 0 0 0 0 0 0 0 0 0 0 0 0 0 0 0
0 0 0 1 0 0 0 0 0 0 0 0 0 0 0 0 0 0 0 0
0 0 0 1 0 0 0 0 0 0 0 0 0 0 0 0 0 0 0 0
0 0 0 1 1 1 1 1 1 1 1 1 1 1 1 1 1 0 0 0
0 0 0 0 0 0 0 0 0 0 0 0 0 0 0 0 0 0 0 0
0 0 0 0 0 0 0 0 0 0 0 0 0 0 0 0 0 0 0 0
0 0 0 0 0 0 0 0 0 0 0 0 0 0 0 0 0 0 0 0
```

1	2	3
4	5	6
7	8	9

Figure 2.1: The letter E as a matrix of zeros and ones. The area is divided into nine subareas, numbered 1 through 9 as shown on the right.

Figure 2.2: The pattern recognition algorithm will start by looking for each of these vertical, horizontal, or diagonal subpatterns and then will list their presence (1) or absence (0) in a vector.

```
        1       2       3       4       5       6       7       8       9
    | − \ / | − \ / | − \ / | − \ / | − \ / | − \ / | − \ / | − \ / | − \ /

E:  1 1 0 0 0 1 0 0 0 1 0 0 1 1 0 0 0 1 0 0 0 1 0 0 1 1 0 0 0 1 0 0 0 1 0 0

F:  1 1 0 0 0 1 0 0 0 1 0 0 1 1 0 0 0 1 0 0 0 1 0 0 1 0 0 0 0 0 0 0 0 0 0 0

H:  1 0 0 0 0 0 0 0 1 0 0 0 1 1 0 0 0 1 0 0 1 1 0 0 1 0 0 0 0 0 0 0 1 0 0 0
```

letter:	E	F	H
pattern is E:	12	9	6
pattern is F:	9	9	6
pattern is H:	6	6	9

Figure 2.3: The above matrix is derived simply by listing the presence (1) or absence (0) of each feature (vertical line segment, horizontal line segment, and two diagonal line segments) in each of the nine regions of a pattern. When you multiply this matrix, A, times the vector $\vec{x}$ that lists the features in the unknown picture, you get a vector that lists the number of votes for the letters, E, F, and H. This is basically the algorithm of Walker and Amsler in the last chapter. This works fairly well, but it gives the same score (9) for an E and an F when an F is presented to the algorithm.

by matrix multiplication, the ith row of the vector, $\vec{b}$, will give the number of "votes" for the ith letter. For the letter E there will be a total of twelve votes, the letter F will have nine votes and H will have six, so the answer must be that the unknown is the letter E. The scores you get from this algorithm when you submit an ideal letter E, an ideal letter F, and an ideal letter H are also shown in the figure. This algorithm is really the same as the one proposed by Walker and Amsler to categorize newspaper stories except instead of using words to contribute votes, it is the line segments in each region that contribute votes.

This algorithm is a simple one and you may already have noticed a problem with it. If the unknown letter happened to be the letter F, doing the matrix multiplication produces a score of 9 for E and 9 for F. A solution to this problem is to change the weights in the matrix and there are simple algorithms that will find a matrix that will work correctly, however, for now there are two simple changes that can be made to correct the problem. First change the values of the weights so that the perfect letter E will score 1.0, the perfect F will be 1.0, and the perfect H will be 1.0. To do this, divide each element of the first row by 12 and each element in the F and H rows by 9. Second, when an unknown pattern has features that the ideal pattern should not have, the score for the letter should be decreased by some amount. For the first row of the matrix change each 0 to $-1/12$ and for the second and third rows change each 0 to $-1/9$. This gives the matrix shown in Figure 2.4 and it corrects the E-F problem.

Note that with this algorithm even if the unknown is distorted a little from its ideal pattern, so that one or two subpatterns were missing or if there were one or two extra characteristics present, the algorithm is still likely to come up with the correct answer.

$$
\begin{array}{ccccccccc}
1 & 2 & 3 & 4 & 5 & 6 & 7 & 8 & 9 \\
|\ -\ \backslash\ / & |\ -\ \backslash\ / & |\ -\ \backslash\ / & |\ -\ \backslash\ / & |\ -\ \backslash\ / & |\ -\ \backslash\ / & |\ -\ \backslash\ / & |\ -\ \backslash\ / & |\ -\ \backslash\ /
\end{array}
$$

$$\tfrac{1}{12}\ \tfrac{1}{12}\ \tfrac{-1}{12}\ \tfrac{-1}{12}\ \tfrac{1}{12}\ \tfrac{1}{12}\ \tfrac{-1}{12}\ \tfrac{-1}{12}\ \tfrac{1}{12}\ \tfrac{1}{12}\ \tfrac{-1}{12}\ \tfrac{-1}{12}\ \tfrac{1}{12}\ \tfrac{1}{12}\ \tfrac{-1}{12}\ \tfrac{-1}{12}\ \tfrac{1}{12}\ \tfrac{1}{12}\ \tfrac{-1}{12}\ \tfrac{-1}{12}\ \tfrac{1}{12}\ \tfrac{1}{12}\ \tfrac{-1}{12}\ \tfrac{-1}{12}\ \tfrac{1}{12}\ \tfrac{1}{12}\ \tfrac{-1}{12}\ \tfrac{-1}{12}\ \tfrac{1}{12}\ \tfrac{1}{12}\ \tfrac{-1}{12}\ \tfrac{-1}{12}\ \tfrac{1}{12}\ \tfrac{1}{12}\ \tfrac{-1}{12}\ \tfrac{-1}{12}$$

$$\tfrac{1}{9}\ \tfrac{1}{9}\ \tfrac{-1}{9}\ \tfrac{-1}{9}\ \tfrac{1}{9}\ \tfrac{1}{9}\ \tfrac{-1}{9}\ \tfrac{-1}{9}\ \tfrac{1}{9}\ \tfrac{1}{9}\ \tfrac{-1}{9}\ \tfrac{-1}{9}\ \tfrac{1}{9}\ \tfrac{1}{9}\ \tfrac{-1}{9}\ \tfrac{-1}{9}\ \tfrac{1}{9}\ \tfrac{1}{9}\ \tfrac{-1}{9}\ \tfrac{-1}{9}\ \tfrac{1}{9}\ \tfrac{1}{9}\ \tfrac{-1}{9}\ \tfrac{-1}{9}\ \tfrac{1}{9}\ \tfrac{1}{9}\ \tfrac{-1}{9}\ \tfrac{-1}{9}\ \tfrac{1}{9}\ \tfrac{1}{9}\ \tfrac{-1}{9}\ \tfrac{-1}{9}\ \tfrac{1}{9}\ \tfrac{1}{9}\ \tfrac{-1}{9}\ \tfrac{-1}{9}$$

$$\tfrac{1}{9}\ \tfrac{-1}{9}\ \tfrac{-1}{9}\ \tfrac{-1}{9}\ \tfrac{-1}{9}\ \tfrac{-1}{9}\ \tfrac{-1}{9}\ \tfrac{-1}{9}\ \tfrac{1}{9}\ \tfrac{-1}{9}\ \tfrac{-1}{9}\ \tfrac{-1}{9}\ \tfrac{1}{9}\ \tfrac{-1}{9}\ \tfrac{-1}{9}\ \tfrac{-1}{9}\ \tfrac{1}{9}\ \tfrac{-1}{9}\ \tfrac{-1}{9}\ \tfrac{-1}{9}\ \tfrac{1}{9}\ \tfrac{-1}{9}\ \tfrac{-1}{9}\ \tfrac{-1}{9}\ \tfrac{1}{9}\ \tfrac{-1}{9}\ \tfrac{-1}{9}\ \tfrac{-1}{9}\ \tfrac{1}{9}\ \tfrac{-1}{9}\ \tfrac{-1}{9}\ \tfrac{-1}{9}\ \tfrac{1}{9}\ \tfrac{-1}{9}\ \tfrac{-1}{9}\ \tfrac{-1}{9}$$

letter:	E	F	H
pattern is E:	1.000	0.667	0.000
pattern is F:	0.750	1.000	0.333
pattern is H:	0.250	0.333	1.000

Figure 2.4: The above matrix works better than the previous one.

Notice too, that the pattern in Figure 2.1 can be moved around some within the 21×21 matrix and the algorithm still gives the same set of 36 values and therefore will get the same answer.

This algorithm can also be presented as a neural algorithm. A diagram of the algorithm as a neural network is shown in Figure 2.5. This network has 36 nodes in the bottom layer, one for each of the 36 pattern features, and 3 units for the possible answers, E, F, and H in the output layer. The weights come from the values found in the matrix A, in Figure 2.4. The output values are computed in the standard neural way, where the value of each output unit, o_k is computed as follows:

$$o_k = \sum_{j=1,36} w_{j,k} i_j$$

and where the i_j are the 36 input values and $w_{j,k}$ is the weight on the connection between input unit j and output unit k. The portion of the algorithm where you first have to search for the presence or absence of a feature in the unknown is still not 'neural,' however, in the next section we will show how this too can be organized as a neural networking algorithm. The change will be that more layers can be added below the input layer in Figure 2.5 to find the values for the 36 input units.

2.2   A Short Description of the Neocognitron

The program in the last section was a simple illustration of how a program can recognize patterns, however, there have been many more sophisticated programs designed to recognize letters and other symbols. One recent important set of experiments has been done by Fukushima. His first program was the Cognitron [41]. This was followed by later programs, all called the Neocognitron [42, 43, 44, 45, 46, 47]. Each version of the Neocognitron varies slightly in its details. In this section we will mention some of the key features of the Neocognitron without studying its weight adjusting algorithm.

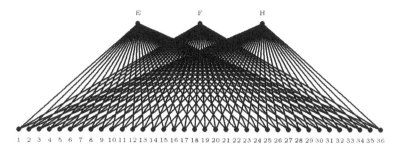

Figure 2.5: A neural interpretation of the algorithm. The bottom layer contains nodes that represent the features found in the unknown letter. Each node will then be 1 or 0 depending on whether the feature was present or not. The top layer contains three nodes for the three letters to be recognized. Whichever one lights up brightest is the answer. The lines connecting nodes in the two layers contain the weights shown in Figure 2.4.

2.2.1 Detecting Short Lines

In the last section we assumed that the short line segments were found by a very conventional algorithm. Finding short line segments in a figure is actually quite easy to do neurally but it requires a large number of neurons and interconnections. First, suppose the characters we have to identify are again E, F, and H, and again they are contained in a 21×21 matrix of zeros and ones as in Figure 2.6. The network we will use will have an input layer, two layers called the S and C layers (following Neocognitron terminology), and an output layer that gives the identity of the pattern. This network is shown in Figure 2.7 and it is much smaller than a typical Neocognitron network. The typical Neocognitron network uses many more layers and we will look at a larger network later, but for now the network in Figure 2.7 will suffice.

In this network we will have it look for the four short line segments shown in Figure 2.8. To do this there will be four sets of S layer neurons, one set dedicated to finding each of the four types of line segments. Each set is represented by a square matrix of neurons as shown in Figure 2.7. Each neuron in each of the four sets connects to the 9 neurons in a 3×3 area of the input matrix just beneath it. Figure 2.9 shows the area of the input matrix that one S level neuron connects to. In this figure the set of S level neurons is looking for the small horizontal line pattern, as a set of three black squares in the middle line of a 3×3 matrix. In the Neocognitron the interconnection weights are trained to turn on when they see these different line segments, but here let us suppose that they are handwired and an S layer cell turns on when it finds the correct pattern below it.

When a picture is given to the network, all the S layer cells look for their respective patterns at every possible location. Now the C layer cells look at the cells on the S layer. Suppose, as in the algorithm in Section 2.1, that we want to flag the existence of each type of line segment in each of the nine areas of the input matrix so that we end up with the vector of 36 values we used in that algorithm. Let each of the C cells connect to a 7×7 area of the S level matrices and have a C cell turn on if it receives any activation from one or more of the 49 S cells it connects to. For instance, in Figure 2.9 the C cell at the bottom and center of the picture turns on because at least one of its S cells is on. Notice then, that the whole pattern on the input matrix could be shifted several rows or columns and

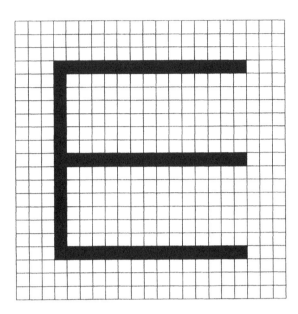

Figure 2.6: The letter E where black squares are a 1 and white squares are a 0.

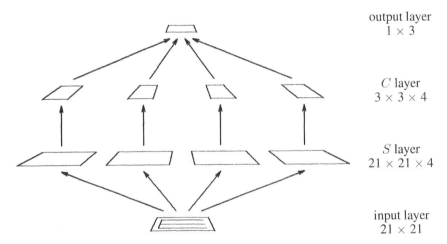

Figure 2.7: This Neocognitron-like network has four layers: the input layer to hold the pattern to be identified, an S layer to scan for the four subpatterns, a C layer to group the subpatterns into nine regions, and finally the output layer. There are too many neurons to display individually and each rectangle represents a matrix consisting of a large number of neurons. Also, there are many more interconnections between neurons than can be drawn here so whole groups of these connections will be represented by arrows. The input layer has 21×21 neurons. The S layer contains four sets of 21×21 neurons each, and the C layer contains four sets of 3×3 neurons each. Each of the four sets in these layers is used for each of the four subpatterns that must be recognized. The C layer will contain the 36 values that would be on the input layer of Figure 2.5, however the values will be in a different order. The output layer will contain three neurons, one each for E, F, and H.

Figure 2.8: These are the four patterns the network will be looking for.

the same C cell will still turn on. In the typical Neocognitron this tolerance for a shifted pattern is taken into account over many layers rather than just two.

2.2.2 A Typical Neocognitron

Figures 2.10 and 2.11 give the outlines of one particular Neocognitron [43] designed to recognize the digits 0 through 9. It has four pairs of S and C layers and an input layer for a total of nine layers. The S_1 layer looks for the twelve different subpatterns shown in Figure 2.12 that represent the patterns you get from horizontal, vertical, and diagonal lines and from the four lines in Figure 2.13. Notice that these latter four lines shown in Figure 2.13 give rise to eight different patterns that the network needs to look for. Since the twelve patterns of Figure 2.12 only represent lines at eight possible angles, the C_1 layer only has to have eight sets of neurons, not twelve. Each of the four pairs of patterns on the right in Figure 2.12 will map into one C_2-layer cell. Each of the eight sets of C_1-layer neurons is 11×11 and these cells turn on when any of the S_1-layer cells they connect to are on. This arrangement is used to take into account patterns that are slightly translated within the input matrix. Figure 2.11 shows a vertical slice of the network and how the units in one layer connect to the previous layer. Most of the C_1 neurons connect to a 5×5 area of the S_1 layer while the C_1 neurons near the edge connect to fewer S_1 neurons.

At the S_2 and C_2 layers the network looks for more complex patterns that can be formed from the short line segments detected by the S_1 layer. Figure 2.14 shows some of the 38 features that this pair of layers looks for. Again, some of the 38 S_2 features that are found can be mapped into just 22 categories just as the 12 subpatterns of layer S_1 can be mapped into just eight categories. Again, to take into account the possible translations of input figures, the 11×11 array in the S_2 layer is mapped into just a 7×7 array in the C_2 layer.

At the S_3 and C_3 layers still more complex features are detected like those in Figure 2.15.

2.2.3 Training the Neocognitron

The Neocognitron can be trained in either of two ways. First, it can be *trained with a teacher* and in this mode the human trainer selects specific patterns that each layer of neurons needs to learn, for instance, 12 at the S_1-C_1 layer, 38 at the S_2-C_2 layer, and so on. In training layer S_1, one of the 12 patterns in Figure 2.12 is placed on the input layer, the trainer selects a cell, called the seed cell, in one of the arrays and trains it to respond to the pattern on the input layer, then all the cells in the array get the same set of weights. The training is done by adjusting the weights in the network but we will not go into that aspect of it. Next, another of the 12 patterns is selected and the weights are adjusted and so on. The same procedure is done for layer S_2 only now the patterns in Figure 2.14 are given to the input layer. The training continues this way for layers S_3 and S_4. In *training without a*

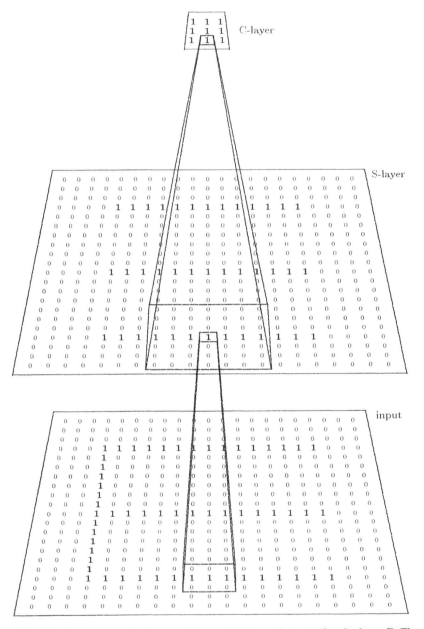

Figure 2.9: This figure shows the input matrix on the bottom that contains the letter E. The next layer up is one of the four S-level sets of neurons, this set is designed to detect the short horizontal line. Each neuron here looks at a 3×3 area of the input matrix. The next layer up is the C layer that contains a 3×3 matrix of neurons that will list which of the nine portions of the input pattern contains a horizontal line. Each of these nine neurons is connected to a 7×7 area in the S-level matrix. A C-layer neuron turns on if it finds at least one occurrence of a 1 within the area it scans. In this case all nine neurons will be 1.

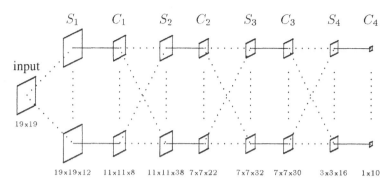

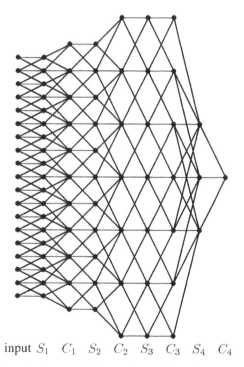

Figure 2.10: The nine layers of one Neocognitron network [43] designed to recognize the digits 0 through 9. The input layer is 19×19. Layer S_1 looks for 12 features, but these 12 map into just 8 in layer C_1. Layer S_2 looks for 38 features, but these map into just 22 in layer C_2. Layer S_3 looks for 32 features that map into 30 in the C_3 layer. Finally, layer S_4 looks for 16 features and the answer appears on layer C_4. S-layer neurons must scan all the C-level arrays of neurons while the C-level sets look at only one S-array of neurons.

Figure 2.11: This slice through the network shows how cells are interconnected. The figure shows that most of the S_1 cells vertically scan three of the input cells except near the edge of the input where only two cells are scanned. Most of the C_1 cells vertically scan five of the S_1 cells, most of the S_2 cells vertically scan three of the C_1 cells, and so on.

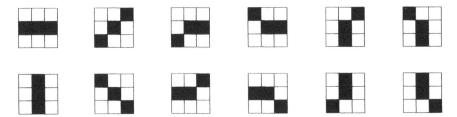

Figure 2.12: These are the twelve patterns that the S_1-layer cells search for and they represent lines with eight different angles. The third pair from the left comes from a line of slope 1/2, the fourth pair comes from a line with slope −1/2. The fifth and sixth pairs come from lines with slopes 2 and −2. So, even though there are twelve patterns, they represent only eight lines, therefore only eight C_1-layer cells are needed.

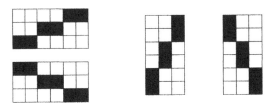

Figure 2.13: These are the four lines with slope 1/2, −1/2, 2, and −2 that give rise to the four pairs of patterns on the right of Figure 2.12.

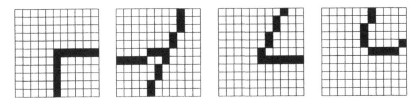

Figure 2.14: These are four of the 38 patterns the S_2 layer of the network will be looking for. Most of the 38 features consist of curved lines and intersections of lines, but some of them (not shown) are simply plain vertical, horizontal, and diagonal lines.

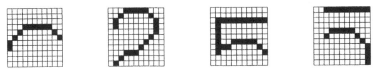

Figure 2.15: These are some of the patterns the S_3 layer looks for.

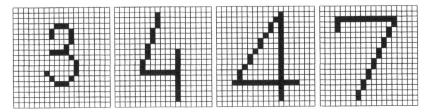

Figure 2.16: These are some of the patterns the S_4 layer looks for.

teacher the Neocognitron can simply be given *only input patterns* like those in Figure 2.16 and *not any of the features shown* in Figures 2.12, 2.14, and 2.15 and it can actually learn to classify them correctly. However, Fukushima reports that training without a teacher takes much much longer than training with a teacher and the program does better at classifying distorted patterns when it is trained by a teacher.

2.2.4 Some Results

Figure 2.17 shows some of the distorted digits that the Neocognitron can correctly classify. An interesting feature in the 1987 version of the Neocognitron is that it can look at a picture containing more than one character and first identify one character and then another. In this scheme one character emerges as dominant and all the neural pathways involved in identifying this character become active and inhibit the pathways needed to recognize other characters. At the flip of a switch these active pathways can be inhibited so that some other pathways will find another character in the input. Figure 2.18 shows a time series of the Neocognitron doing this. While the Neocognitron was developed to do visual pattern recognition, Fukushima expects that it could easily be modified to do speech recognition as well.

2.3 Recognizing Words

Various pattern recognition programs can perform almost as well as people at recognizing individual letters and digits. Among people, badly distorted characters all by themselves can easily be misinterpreted, but people rarely encounter such problems in practice because the characters are usually seen in context and context can give the extra amount of information necessary to identify the pattern. This is an important capability of human data processing systems and it can be added to programs as well, but adding this capability to programs reveals that to do an accurate job of pattern recognition requires that the program know quite a lot about the world. The particular work we are going to look at is "An Interactive Activation Model of Context Effects in Letter Perception," by Rumelhart and McClelland [180, 181].

The work by Rumelhart and McClelland was an effort to model certain results obtained by experiments on people who had to recognize four letter words that are briefly flashed on a screen. In these experiments the words would also have some parts of letters obscured or missing as in Figure 2.19. As for the first three letters in this figure, there is no doubt that they are W, O, and R. Based on these letters, the unknown word may be WORD or WORK or WORM or WORN. Looking at just the features that are visible in the fourth letter we see a vertical line on the left. That suggests the letters E, F, K, R, and others. The diagonal line in the lower right suggests the letter could be a K or an R. The short horizontal line is consistent with the letters K and R and others. Given our knowledge about words, we can safely conclude that the incomplete letter is a K and that the unknown word is the word "WORK." The individual features suggest letters, the letters suggest words, the words suggest possible letters, and the letters even suggest individual features. This phenomenon shows why when people write text (especially computer programs!) it is very common for them to miss many mistakes in the text just because they know what the text is supposed to

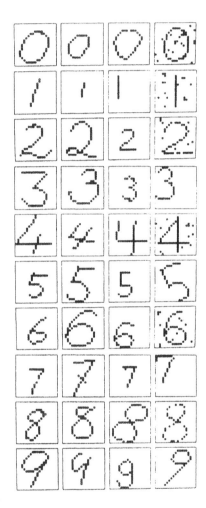

Figure 2.17: These are some of the patterns the Neocognitron has been able to recognize. A 1991 version of the program has been able to recognize 35 characters.

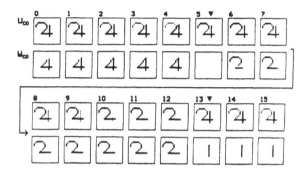

Figure 2.18: A pattern containing 1, 2, and 4. The top row shows the activation state of the input cells. The row below shows another layer of cells that isolates the parts of the input that correspond to particular figures. First, the network "sees" the 4. At time = 5, a switch is flipped that resets the network. The "4" pattern is inhibited and now the network sees the 2. At time = 13, the switch is flipped again and now the network sees the 1.

Figure 2.19: An input to the program, an incomplete version of a word. In reality the program does not "look at" a picture, instead the program gets a vector that flags which line segments are present in each position.

say and they know what the words are supposed to be. To really catch mistakes you need some other person or program to proofread the text.

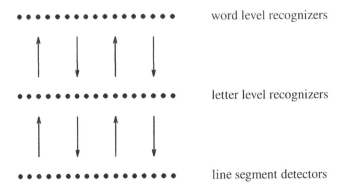

Figure 2.20: Layers of processors for word recognition.

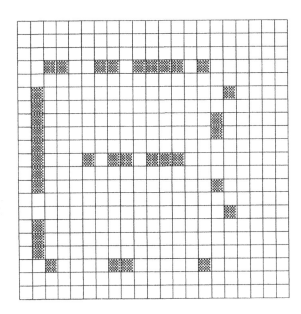

Figure 2.21: Even with quite a lot missing from this letter B it is easy to interpret a small number of dots as forming curved lines or straight lines.

To model the word recognition process we will have three layers of neurons arranged as shown in Figure 2.20. In this arrangement, the detection of certain line segments influences the letters that are recognized and the letters that are recognized influence the words that are recognized. When processing proceeds this way from the primitive data to conclusions it is referred to as *bottom-up* or *data driven* processing. Most pattern recognition has tradition-ally been bottom-up in character. On the other hand, if a person expects a particular word, the lighting up of the word can influence what letters are seen. Furthermore, any bias we may have toward a particular letter will also influence whether or not we see a collection of dots as forming a curved or straight line as in Figure 2.21. Processing that proceeds from expectations is referred to as *expectation driven* or *top-down* processing. When information flows both from the bottom up and from the top down it is called *interactive* processing.

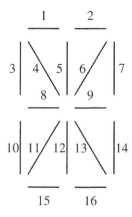

Figure 2.22: The 16 features used to construct the letters.

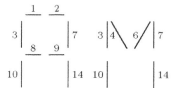

Figure 2.23: The word AM, with its features numbered.

To make this concrete we will construct a network designed to recognize just two words, AM and AN. Each letter will be constructed from a set of 16 short line segments. These segments are shown in Figure 2.22. Figure 2.23 shows how the word AM is constructed using these line segments. Figure 2.24 shows the neural network designed to recognize the two words. The top layer has the nodes, AM and AN, and either AM or AN lights up when their features are present in the middle layer of nodes. The middle layer has nodes A1, M2, and N2 for A in the first position and M and N in the second position. A1 will light up when there are features in the first group of 16 units on the bottom layer that suggest that an A is present. M2 will light up when there are features for an M present in the second group of 16 units and N2 will light up when there are features for an N present in the second group. In Figure 2.24 the connections between nodes that end in a dot are used to indicate an inhibition link (negative weight) and all these connections are bidirectional. The connections without a dot are activation links and they are also bidirectional. For instance, the presence of an M in the second position activates AM and inhibits AN. Notice that the network will have AM and AN inhibit each other. This was done so that when one of these words begins to be recognized the network will inhibit the other. This design makes it possible for just one word to clearly win and this type of network is an example of a *winner-take-all* network. In a winner-take-all network the output nodes compete with each

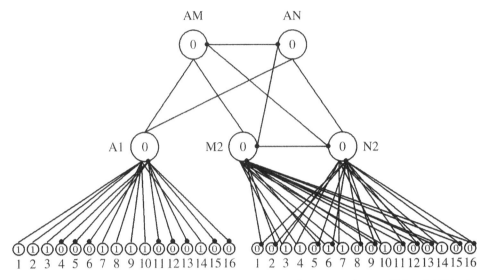

AM   AN

A1   M2   N2

1 2 3 4 5 6 7 8 9 1011 1213 1415 16   1 2 3 4 5 6 7 8 9 1011 1213 1415 16

Figure 2.24: A simple word recognition network. The words to recognize are AM and AN in the top layer. The letters to recognize are A in position 1 (A1), M in position 2 (M2) and N in position 2 (N2) in the middle layer. At the start of the process the middle and upper layer nodes will be 0. The lower layer has the 16 numbered features for A on the left and for M on the right. The connections between nodes that end in a • are inhibition links and the other connections are activation links.

other and the most highly activated node is the winner. To produce top-down activation and inhibition, the word node AM activates the letter nodes A1 and M2 and inhibits N2. The node AN activates A1 and N2 and inhibits M2. The input-feature-level nodes activate the letters that they are part of and inhibit the letters they are not part of. These connections also work in reverse, so as letters turn on they activate or inhibit the input level features. The activation weights will all have a value of +0.02 and the inhibition links will all have weights of –0.045.

In the previous networks we have looked at we took the input and used it to activate higher layers in the network until the activation reached the output layer. Now the activation procedure will be more complicated in several ways. The input features on the lower level will all have initial values of either 0, meaning the feature is not present, or 1, meaning that the feature is present. The other nodes in the network will start out at 0. The value of a node such as M2 will be calculated using a formula that will only increase or decrease its activation *by a small amount*. At the same time that node N2 is calculating its activation value all the other nodes will be doing the same thing. The net result is that all values will go up or down by a small amount (or sometimes they may be unchanged). We then repeat this process over and over again until each node reaches a stable value. The formula we use to light up nodes will keep the values of the nodes between 0 and 1.

Each node in the network is updated using the following algorithm. First, compute the net input to each node, j, at time, t by in the usual way:

$$net_j(t) = \sum_i w_{i,j} a_i(t). \tag{2.1}$$

where $w_{i,j}$ is the value of the connection from node i to node j and $a_i(t)$ is the activation level of node i at time t. To compute the activation value of node j, first, if $net_j(t)$ is greater than 1.0 then set $net_j(t)$ to 1.0 and if it is less than -1.0 set it to -1.0. If the input is positive then the new value of the node at time $t + \Delta t$ will be:

$$a_j(t + \Delta t) = a_j(t) + net_j(t)(1.0 - a_j(t)), \tag{2.2}$$

while if net_j is negative the new value of the node will be:

$$a_j(t + \Delta t) = a_j(t) + net_j(t)a_j(t). \tag{2.3}$$

Notice, how, if the value of a node is close to 1.0, say 0.9, and if the input is large, say 0.9, the new value of the node will not go over 1.0:

$$a_j(t + \Delta t) = 0.9 + 0.9(1.0 - 0.9) = 0.99.$$

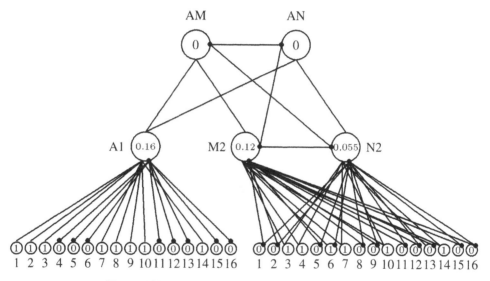

Figure 2.25: The word recognition network after one cycle.

For an example of the process we will give the network the word AM and show how the network lights up. Figure 2.24 shows the values of the nodes for this example at the start of the process and Figure 2.25 shows the values after the first iteration. After 130 cycles the values have stabilized at those shown in Figure 2.26 with A1, M2, and AM clearly the winners. The activation of some nodes as a function of time is shown in Figure 2.27. In another experiment with the network we can turn off the input feature number 3 in the letter M. Figure 2.28 shows that top-down activation from the M node turns it on anyway. In effect, the network is trying to "see" the feature.

This example embodies the most important aspects of the Rumelhart and McClelland algorithm and we can now look at the somewhat more complicated algorithm they used.

In the experiments done by Rumelhart and McClelland they used 1,179 four-letter words. At the letter level there are 26 letter nodes for each of the four positions in a word.

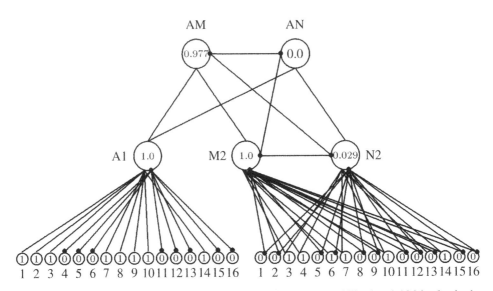

Figure 2.26: The word recognition network after 130 cycles has stabilized and AM is clearly the winner.

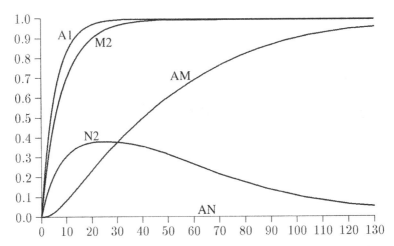

Figure 2.27: The state of the network is computed at 130 discrete time intervals and smooth lines are drawn between the points. A1 rises the fastest because it is supported by eight low-level features and the word AM. M2 rises almost as quickly but it is inhibited slightly by N2 and it is only being supported by six low-level features. N2 initially rises but its low-level support is less than M2 and it is inhibited by M2 and AM so after a while it starts to decline. The word AM slowly lights up during the process but AN never gets enough activation to rise above 0.

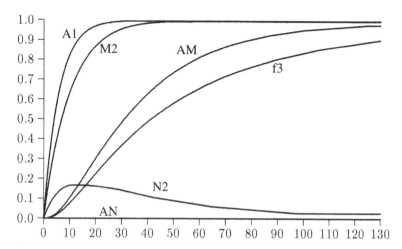

Figure 2.28: In this run, feature number 3 (f3) was missing from the M but the program recognizes the word AM anyway. As M is recognized, top-down activation turns on the missing feature so you could say that it is trying to "see" the missing feature.

The range of values that nodes could take on was from +1.0 down to –0.20. The initial values for the words, called the resting values, were between 0 and –0.05. Common words have resting values greater than the resting values for uncommon words. This bias toward common words is used because, in experiments on people, when an uncommon word is briefly flashed on a screen the human subjects very often see the word as being a more common word that is very similar to the uncommon word. For instance, the word TEEN is relatively common and the word TEEM is relatively rare. When subjects see TEEM flashed on the screen for a very short period of time they may well see it as the word TEEN. With these considerations in mind, the formula for the summation of the inputs for node i, at iteration t, is:

$$net_i(t) = \sum_j \alpha_{ij} e_j(t) - \sum_k \gamma_{ik} i_k(t), \qquad (2.4)$$

where $e_j(t)$ is the activation value of an excitatory neighbor of the node, $i_k(t)$ is the activation value of an inhibitory neighbor of the node, and α_{ij} and γ_{ik} are the weights. When $net_j(t)$ is positive, we calculate a quantity, $\eta_j(t)$, by:

$$\eta_j(t) = net_j(t)(M - a_j(t)), \qquad (2.5)$$

where M is the maximum activation level of the unit. As we said earlier, this value is simply 1.0. When $net_j(t)$ is negative, we calculate $\eta_j(t)$ by:

$$\eta_j(t) = n_j(t)(a_j(t) - m), \qquad (2.6)$$

where m is the minimum activation of the unit. The value for m in their experiments was –0.20. The new value of the activation of node j, at time $t + \Delta t$ is given by:

$$a_j(t + \Delta t) = a_j(t) - \Theta_j(a_j(t) - r_j) + \eta_j(t). \qquad (2.7)$$

The second term on the right of this formula is used to make the activation value of the node decay with time, so that if, for instance, the inputs were all turned off, the nodes would decay to 0 or to their resting values. This becomes important for certain aspects of the experiments that we will not actually be concerned with.

All the parameters for the model were determined by repeating the experiments until the program produced results consistent with the results obtained in the human experiments. The values they found were:

M, the maximum value of a node	1.0
m, the minimum value of a node	−0.20
feature-letter excitation	0.005
feature-letter inhibition	0.15
letter-word excitation	0.07
letter-word inhibition	0.04
word-word inhibition	0.21
letter-letter inhibition	0
word-letter excitation	0.30
word-letter inhibition	0
Θ_j, the decay rate	0.07
r_j, the resting value for letter nodes	0
r_j, the resting value for word nodes	variable

Since the weights between words and letters, features and letters, and so forth are all the same, regardless of the particular word, letter, or feature, the weights for every individual connection do not need to be stored and this saves a lot of memory space. Rumelhart and McClelland's experiments also included another part where the probability that the computer gives a response to a given input is estimated but it is not necessary to describe that portion of their work here.

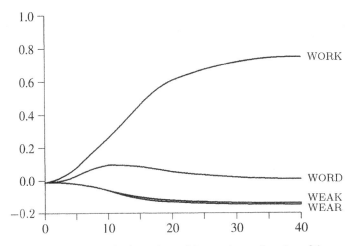

Figure 2.29: The activation values of the words as a function of time.

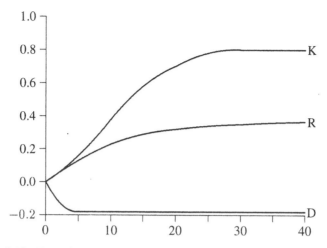

Figure 2.30: The activation values of the letters K, R, and D as a function of time.

Figure 2.29 shows how the words WORK, WORD, WEAK, and WEAR are activated by the algorithm given the incomplete word, WORK, shown in Figure 2.19. WEAK and WEAR, while they have features in common with WORK, are inhibited by the more likely words, WORK and WORD, and these inhibited words are pushed down below their resting levels. WORD first increases a little and then decreases to a near zero value. With respect to the letter activations, as you can see in Figure 2.30, K is the winner. This is because K is receiving support up from the features and down from the word, WORK, while R only receives support from the bottom up.

As we said earlier on, this program was designed to model results obtained by flashing words and portions of words on a screen. Rumelhart and McClelland discuss their results in their two papers on the matter. It is not important for us to discuss those results here. It is enough to say that their method produces very similar results to those obtained from tests done on human subjects.

2.4   Expanding the Pattern Recognition Hierarchy

The top-down and bottom-up effects used in recognizing letters also extend to how we interpret the sounds we hear. Moreover, the interpretation of words we hear and see depends on still higher level patterns, namely sentences and our knowledge of the world. In this section we look at these phenomena and how they make producing an artificial system that is equivalent in capability to a human being very difficult.

2.4.1   Hearing

The last section demonstrated that people can fill in obscured or uncertain visual inputs based on context using their knowledge of words. An analog to this visual phenomenon also occurs in auditory processing where, for example, there is a phenomena known as the "phonemic restoration effect." One example of this phenomenon comes from an experiment where some researchers [250] took a recording of the word "legislature," edited out the "s" sound and replaced it with a click. People who hear this altered recording hear the entire word "legislature" and also report hearing the click as a disembodied sound.

2.4.2 Higher Levels

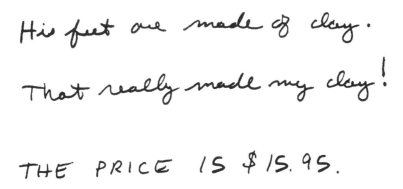

Figure 2.31: The interpretation of identical curves is highly dependent on the context in which they are seen.

The filling in ability of the mind also extends beyond the word level to the higher levels of sentence processing and interpreting events in the world. For instance, suppose we have the sentence:

The ____ broke the window.

where the blank space is either blank or so badly printed as to be completely illegible. First, our knowledge of sentence structure prompts us to think that the blank space should contain a noun. Also, our knowledge of the world prompts us to think that the noun may represent a person or some heavy object such as a rock. Another example of this phenomenon is shown in Figure 2.31 where the identity of a curve is highly dependent on the context in which it is seen. Likewise, the interpretation of sounds is also influenced by the context as shown in this story:

> In Center Harbor, Maine, local legend recalls the day some 10 years ago when Walter Cronkite steered his boat into port. The avid sailor, as it's told was amused to see in the distance a small crowd of people on shore waving their arms to greet him. He could barely make out their excited shouts of "Hello Walter ... Hello Walter."
>
> As his boat sailed closer, the crowd grew larger, still yelling "Hello Walter ... Hello Walter." Pleased at the reception, Cronkite tipped his white captain's hat, waved back, even took a bow.

> But before reaching dockside, Cronkite's boat abruptly jammed aground. The crowd stood silent. The veteran news anchor suddenly realized what they'd been shouting: "Low water . . . low water." [1]

In a more scientific experiment Klatt[2] reports doing the following: He recorded continuous speech and, of course, when he played it back listeners could recognize all the words. However, when he broke up the speech into words and played them back in a random order, listeners could only recognize about 70 percent of the words. The conclusion is that meaning and word order play a significant role in making it possible for people to understand speech.

2.4.3   The Hierarchy

To take into account these effects we can expand on the model of the last section and hypothesize that people have two extra levels of pattern recognizing processors as shown in Figure 2.32. We will designate them as the "sentence level recognizers" and "event level recognizers." Although this hierarchy makes sense, there is as yet no complete system that uses it, so this arrangement must still be considered highly theoretical.

While it was quite easy to model word recognition by having a node for each of 1,179 four-letter words, it is clearly not so simple to adopt that plan for the sentence recognition level. Keeping an example of every kind of sentence is not reasonable. Researchers in natural language processing have generally held that in trying to understand the meaning of sentences the proper way to proceed is to look for patterns of nouns, verbs, adjectives, and so forth. They believed that only after determining these grammatical aspects could you go on to consider the meaning of a sentence. Years of research, however, has found problems with this approach. Consider, for instance, the simple sentence:

Time flies like an arrow.

People readily discern the meaning: time passes quickly. However, there are other interpretations that are possible to a machine that only knows about nouns, verbs, adjectives, and so forth. Flies could be interpreted as a noun, and then time must be an adjective, giving a possible meaning to a naive machine that creatures known as "time flies" like arrows. Or time could be interpreted as an imperative verb and flies as a noun giving the meaning that "You should time flies the same way as you time arrows." It is human knowledge of the world that enables people to figure out the meanings of sentences and to figure out whether or not a particular word is being used as a noun or a verb or an adjective.

It is far from settled exactly how to do sentence recognition and neither does anyone really know how to represent the real world information, however, there is one project done by Waltz and Pollack [247, 157] that can make sense of ambiguous words in sentences by crudely taking into account facts about the world. Their research shows how easily an interactive activation model can potentially be used to sort out meanings of sentences and

---

[1] Don Oldenberg, *Chicago Sun-Times*, June 10, 1987, and the *Washington Post* February 27, 1987, reprinted with permission from the Washington Post.

[2] This report comes from the article by White [257] where the author gives this report as being from a personal communication with Dennis Klatt.

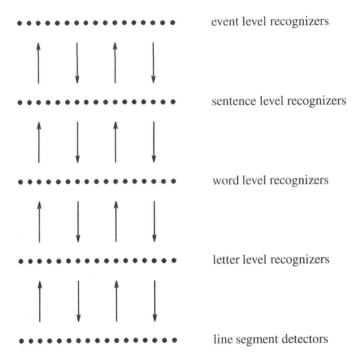

Figure 2.32: Human pattern recognition depends not only on low level features but on higher level patterns as well.

correctly identify nouns, verbs, and so on. One example from Waltz and Pollack uses the sentence:

John shot some bucks.

There are many possible meanings for the words "shot" and "bucks." The possible interpretations for shot are "worn out" (tired, an adjective), "fire" (shoot with a gun, a verb), "bullet" (a unit of ammunition, a noun), or "waste" (as in squander, a verb). "Bucks" could mean either "resist" (opposes, a verb), "throw off" (a bucking horse, a verb), "dollar" (a unit of currency, a noun), or "deer" (male deer, a noun). People who think that John went to Las Vegas will interpret "shot" as "waste" and "bucks" as "dollars" while people who think that John went hunting will interpret "shot" as "fire" and "bucks" as "male deer."

The Waltz and Pollack program works by looking up the possible meanings of all the words in a sentence and constructing an interactive activation network that will be used to settle on a consistent meaning for all the words. The network for "John shot some bucks." is shown in Figure 2.33 and it consists of four parts. First, the top part is a *parse tree*, a way of diagramming a sentence that shows how it can be generated from a formal grammar.[3] The second part of the network is the set of words used in the sentence. The third part is the most complicated. It contains the possible meanings of the words. The meanings of

---

[3] Parse trees and formal grammars will be described in Chapter 10 but understanding these concepts is not really important now.

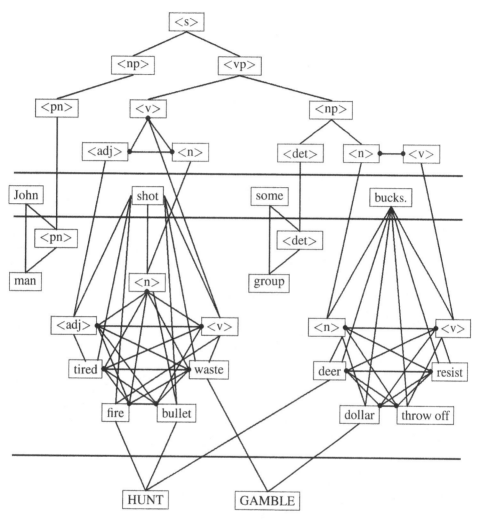

Figure 2.33: The Waltz and Pollack program constructed this network to determine the meaning of the sentence, "John shot some bucks." It consists of four parts: at the top a network designed to resolve the syntax of the sentence, the next layer down lists the words in the sentence and these are activated from left to right, below this are subnetworks to determine the parts of speech of each word (lexical analysis), and finally at the bottom two nodes give the possible contexts in which the sentence is seen. In the syntactic part, the notation, <s> means a sentence; <np>, a noun-phrase; <vp>, a verb-phrase; <v>, a verb; <n>, a noun; <adj>, an adjective; <pn>, a proper noun; and <det>, a determiner.

the words "John" and "some" do not pose a problem, "John" is a proper noun, the name of a man, and "some" is a simple determiner for a group. In the "shot" subnetwork there are nodes to indicate that "shot" is being used as an adjective, a noun, or a verb. Given the context that a verb is the only candidate that could follow the noun "John," it is quite easy to see that when the network is activated the verb meanings of "shot" are the only ones that could come up, so either the "fire" or "waste" nodes will come on. Just as in the "shot" subnetwork where "shot" could only be a verb based on the order of words in the sentence, in the "buck" subnetwork, "bucks" could only be a noun, again based on the order of words. The possible interpretations will be "deer" and "dollars." The bottom part of the network contains only two nodes, "HUNT" and "GAMBLE" and these represent the possible contexts that apply. If "HUNT" is on, the network will ultimately give the hunting interpretation of the sentence, while if "GAMBLE" is on, the gambling meaning will win out.

2.4.4   On the Hierarchy

The simple networking hierarchy proposed here accurately reflects the fact that a human mind takes in all the facts and then tries to make the maximum amount of sense out of them. There is no guarantee that human minds actually achieve this result using simple interactive activation networks. In all likelihood, the algorithms seen here are just simple models of a much more complex architecture.

A second important point that comes from the above considerations is that to produce an artificial intelligence with near human performance capabilities will require giving that artificial intelligence a considerable knowledge of the world. Knowing *a lot* about the world is not something that the early artificial intelligence researchers expected would be necessary to create intelligent machines. This "knowledge" factor clearly makes the problem of creating an artificial intelligence comparable to that of a human being much harder. Even the process of correctly recognizing a letter can get complicated and can require a considerable knowledge of the world. People see what they expect to see and hear what they expect to hear. People perhaps even see, hear, and believe what they *want* to see, hear, and believe.

2.5   Additional Perspective

The pattern recognition methods discussed in this chapter have been chosen because the principles involved are representative of how vision programs work; however, they do not represent all that can be said about this subject so in this final section we will add some perspective by mentioning other issues and systems.

2.5.1   Other Systems

Other algorithms to recognize hand-printed and typed characters are also being developed. A recent system by Le Cun et al. [96] used the back-propagation algorithm (described in the next chapter) and various other techniques and it managed to achieve a 1 percent error rate with a 9 percent rejection rate[4] on hand-printed characters taken from zip codes on actual US mail. A number of other such projects report similar results. In addition, Le Cun et al. constructed a system consisting of a video camera to scan digits which were then processed by a PC equipped with a special digital signal processor board. The system can process 10 to 12 digits per second.

Simple and efficient algorithms also exist to take handwritten characters from a tablet. These algorithms are faster and more accurate than scanning already existing characters because the system can detect the beginning and ending points of individual strokes as the strokes are made. For an example of one, see the Ledeen character recognizer described in Appendix VIII of [221].

In the way of hardware implementations, Mead has produced a "silicon retina" that behaves much like a human retina (see [110] and [111]).

2.5.2   Realism

Some researchers like Fukushima believe that their models of visual processing are fairly realistic, however, it is not really known *exactly* how the eye and nervous system process the data that a retina receives. One interesting result of studying human visual hardware is that it tries to get by with sending a minimum of information to the brain. For instance, suppose we had a retina consisting of a 100×100 array of cells. This comes to 10,000 cells and it represents a lot of information to be passed along the optic nerve. However, the cells of the retina are designed to only send information when they detect a change in the color or intensity of light hitting them. Figure 2.34 shows a pattern, half black and half white. If we place this picture on the retina, then, after about a tenth of a second, most of the cells will stop sending information except for the cells along the border between the two colors. The cells along the border will keep sending information because the eye actually oscillates with a frequency of about 10 cycles per second so the border between the two colors is actually moving back and forth on the cells of the retina. These cells with changing inputs keep reporting while the other cells stop reporting. Now if the retina was 100×100 cells, instead of 10,000 cells reporting, only several hundred will have to report. In the brain, the brain fills in the enclosed areas with the correct color so you continue to see the whole picture despite the fact that most cells have stopped reporting. (For more on this see [82].) This experimental result and many others cannot be explained with the simple artificial neural networking ideas that have been seen in this chapter and much more complex models need to be developed. Some more complex and realistic models of human visual processing can be found in the first four chapters of [54] and they can explain many visual illusions.

Remember, too, as mentioned in Chapter 1, it has been recently discovered that at least some of the messages being passed between neurons involved in vision processing are not simple activation values, they are in fact coded messages.

2.5.3   Bigger Problems

---

[4] When the difference between the ratings for the two highest patterns is not large enough, the network does not attempt to classify the unknown pattern and these very close calls are rejected rather than being classified. By rejecting close calls you lower the error rate.

Figure 2.34: When this pattern is placed on a retina all the cells stop reporting to the brain after about a tenth of a second except for cells that are near the border between black and white.

In this chapter we have been working with simple letters and numerals because it is quite easy to do so; however, recognizing three-dimensional objects and finding objects in a scene has also been an important research area. The principles involved in recognizing more complex objects are the same as those involved in recognizing letters, that is, small edges are detected, these edges are then used to detect more complex features, and the more complex features point toward the identity of the object. Some methods only work from the small features upward while other methods also work from the top down. Perhaps the most well-known method for recognizing 3D objects has been developed by Marr [105], however Grossberg (see [54], Chapter 2) argues that Marr's methods are unrealistic and he has developed other methods to do the processing that he argues are more realistic.

2.6   Exercises

2.1. In Section 2.1 the pattern recognition method was illustrated using only the patterns E, F, and H. Expand this base of patterns to include at least five more letters and determine how well your program works by submitting a few test cases to the program. If you have a computer or a terminal with some graphics capability where you can draw letters on the screen, expand on the previous exercise and write a program so that people can draw letters on the screen and then have the program identify them.

2.2. Figure 2.5 did not have enough room to show the weights on the connections. Using the matrix in Figure 2.4, jot down the the values of the weights that go from input nodes 1, 2, and 3 to the E, F, and H nodes.

2.3. Given the letter E shown in Figure 2.6, produce the four S- and four C-layer matrices the algorithm will produce. One S-layer matrix and one C-layer matrix are shown in Figure 2.9.

2.4. In Section 2.1 the algorithm located the presence of subpatterns or operators in a very conventional way by searching the entire matrix to find them. If you did Exercise 2.1 modify the program so that it uses sets of neurons the way the Neocognitron does. Do this with the four subpatterns and nine regions used in the algorithm in Section 2.1.

2.5. Working by hand, determine what the state of the AM/AN network will be after the second and third iterations, given the values shown in Figure 2.25.

2.6. If the network in Figure 2.24 is given the features of an M, minus feature 6, will the program recognize the word as AM?

2.7. Program either the simplified network algorithm (used with the AM and AN example) or the slightly more complicated version actually used by Rumelhart and McClelland. In either case, use data for the WORK/WORD example where part of the last letter is missing. Use data consisting of the short line segments used in the text. Since this algorithm is useful for other exercises in this chapter you may want to produce a general purpose implementation of the algorithm rather than one that is specifically tailored to the WORK/WORD example.[5]

2.8. If you are interested in the effects that occur in the Rumelhart and McClelland word recognition network that follow the results on human subjects, read those portions of the Rumelhart and McClelland papers that discuss this and then give a brief listing and summary of the effects.

2.9. It seems that the word recognition network of Rumelhart and McClelland may be quite a nice way to recognize misspelled words. Consider how well it could do this if the misspelled words included missing and extra characters. Consider what would be necessary in the way of a sequential algorithm to recognize misspelled words that include missing and extra characters. Compare the two methods.

2.10. Below is a small map consisting of regions a, b, c, d, and e:

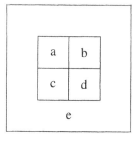

The map can be colored using three colors, say red, green, and blue, such that a region of one color never has a common border with a region of the same color, except possibly, if the regions meet at a point. If, for instance, a is green then b, c, and e cannot be green. The constraints involved in coloring the map can be wired into an interactive activation network. For instance, there can be a node that stands for region a being green, a node for

---

[5] This program is available on the Internet.

region a being red, a node for region c being green, and so on. If the node for a being green lights up, then this should inhibit regions b, c, and e from being green, as well as activating the possibility that b is red, b is blue, c is red, c is blue, e is red, e is blue, and so on. Devise a network and a set of weights that will find the colors of regions b, c, d, and e given that a is green. If you are not given the color of any region can this method still find solutions?

2.11. An interactive activation network can be used to choose tic-tac-toe moves. Create such a network and show how it can play an entire game starting from an empty board. Here is an outline of how this can be done.

First, here is a way to represent the input game board. Suppose the board is the following:

$$\begin{array}{c|c|c} X & & O \\ \hline & O & \\ \hline & & X \end{array}$$

The positions can be represented using 18 nodes divided up as follows:

```
      1 0 0                          0 0 1
      0 0 0                          0 1 0
      0 0 1                          0 0 0
```

Values of the nodes 1	Values of the nodes 10
through 9 representing the	through 18 representing the
presence (1) or absence (0)	presence (1) or absence (0)
of an X in that square.	of an O in that square.

Naturally, for the output layer of the network you will need 9 nodes and the goal will be to light up the one output node that will be the move to make.

To generate moves, some important principles must be wired into the network:

a) First, if a square is already occupied, the network cannot be allowed to select this square as its move.

b) Second, if you already have two marks in a row, column, or diagonal and there is an empty space available in the row, column, or diagonal, you should take the empty space so as to win.

c) Third, if your opponent has two marks in a row, column, or diagonal and there is an empty space available in the row, column, or diagonal, then you must block the win by your opponent.

First, work out a network that will meet these basic conditions. It will be convenient to have a layer of units between the input and output layers. The units in this layer can detect pairs of your nodes and pairs of your opponent's nodes as well as possibly other features of the game board. One additional useful feature you may want to add to your network is to put a threshold on each unit so that it does not even begin to turn on until the input exceeds a certain threshold. This principle makes it easy for units in the middle layer to stay off unless they receive input from two sources.

Besides the basic requirements in b) and c), you will need to devise some way to make other reasonably good moves early in the game when these important conditions are not present. One useful feature here would be to have some of the middle layer units start out on (= 1.0) instead of off. This is especially useful to have when the game board is empty.

In addition, use a random updating scheme. In the Rumelhart and McClelland model, all the units update at once. A certain amount of randomness can be brought into the network by choosing a unit at random and updating it. Then randomly choose another unit and update it and then another, and so on. Demonstrate that your network can decide to make different moves given the same board configuration.

Chapter 3

Pattern Recognition II

In this chapter we will look at more mathematically rigorous approaches to doing pattern recognition. The most important of these will be the back-propagation algorithm. This remarkable algorithm can be used to do a variety of highly useful pattern recognition tasks. A number of applications of this algorithm will be shown.

3.1 Mathematics, Pattern Recognition, and the Linear Pattern Classifier

The pattern recognition algorithms that have been presented here so far were designed by their creators simply because they thought the algorithms would work. Much of AI is simply done that way. You believe something will work and then you test it to see if it does. To more mathematically inclined people this experimental proof of success, by itself, is not completely acceptable. If at all possible, algorithms should be proven correct. The side effects of such proofs should also give an insight into why the algorithm works, under what (if any) special conditions it works, and hopefully too, some way of estimating how long it will take to work.

Most pattern recognition work actually has been a mathematical problem with the following form. Some characteristics of an unknown pattern are measured and these characteristics are listed in a vector we will call $\vec{u}$. If, for example, you were trying to predict the weather, the measurements might include the barometric pressure, wind direction, wind speed, temperature, cloud cover, and humidity. If, for example, you are trying to predict the stock market, you would probably want to list at least the changes for the last few days, the interest rate, inflation rate, price/earnings ratios, and so on. In any case, an operator, we will call it Op, is then applied to the vector $\vec{u}$ and it gives the identity of the unknown pattern. In a mathematical format it is simply:

$$answer = Op(\vec{u}).$$

Much of pattern recognition consists of finding and studying good operators.

3.1.1 The Linear Pattern Classifier

To be more concrete about the matter we will look at an instance of this approach in a very simple case, the linear pattern classifier. Take, for example, a set of four items of the class

A and another set of four items of the class B. These might, for instance, be four examples of the letter E and four examples of the letter F. There will be only two characteristics measured for each item. For items from class A, let the characteristics be:

$$(-6, 4) \qquad (-6, -1) \qquad (-2, -2) \qquad (4, 2)$$

Let the characteristics measured for the items in class B be:

$$(7, -1) \qquad (4, -2) \qquad (-1, -4) \qquad (-4, -7)$$

These eight points are plotted in the xy-plane in Figure 3.1. They have been chosen so that they are *linearly separable*. A linearly separable set of patterns is one in which a line, a plane, or a hyperplane can be drawn between two different sets such that all the patterns in one set are on one side of the line, plane, or hyperplane, while all the patterns in the second set are on the other side of the line, plane, or hyperplane.

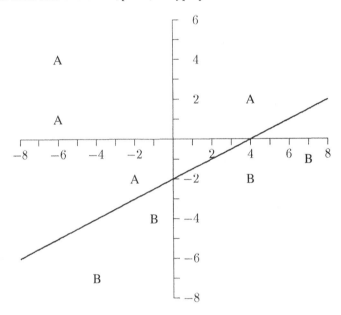

Figure 3.1: A set of linearly separable points.

From just looking at the graph of the points in each class, we can conclude that the line $y = x/2 - 2$ (and many other nearby lines as well) can be used to separate the region of A's from the region of B's. Writing the equation as $x - 2y - 4 = 0$, the coefficients give a *weight vector*, $\vec{w} = (1, -2, -4)$, that can be used to separate the two classes of objects. The algorithm requires that we augment each member of the set of eight patterns by adding a third coordinate with a constant value of +1. The examples of class A are now:

$$(-6, 4, 1) \qquad (-6, -1, 1) \qquad (-2, -2, 1) \qquad (4, 2, 1)$$

and the examples of class B are now:

$$(7, -1, 1) \qquad (4, -2, 1) \qquad (-1, -4, 1) \qquad (-4, -7, 1)$$

When we take $\vec{w}$ and form the dot product of it with a member of class A, we will get a negative value. Dotting it with a member of class B will give a positive value. For example, $(1, -2, -4) \cdot (-6, 4, 1)$ gives -18 and $(1, -2, -4) \cdot (-1, -4, 1)$ gives +3. Figure 3.2 shows how the algorithm can be formulated as a neural network. In the figure, the points $(-6, 4, 1)$ and $(-1, -4, 1)$ are submitted to the network and the output, whether positive or negative, gives the pattern classification.

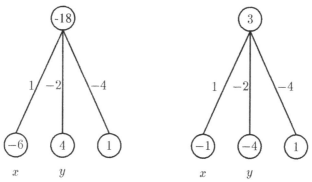

Figure 3.2: The linear pattern classifier can be viewed as a neural algorithm. The output unit looks at its inputs and sums them. The coefficients in the linear pattern classifier are the weights in the network. The coordinates of a point to be classified are input to the units in the input layer. On the left, the network takes in the point, $(-6, 4, 1)$ and puts it in class A. On the right, the same network takes in the point $(-1, -4, 1)$ and puts it in class B.

In the above example it was easy to find a weight vector by looking at the points laid out in two dimensions, however, it is not so easy if the points are in three-, four-, or n-dimensional space. Fortunately, however, a learning algorithm exists that is guaranteed to find a weight vector that can be used to separate linearly separable classes. The training procedure works by taking members from each set and dotting them with an estimate for the weight vector. Any initial value for the weight vector will work. If the weight vector gives the wrong answer, it is changed, but if it gives the correct answer it remains unchanged. The details as to how to change the weights are as follows. If $\vec{w} \cdot \vec{u}$ gives a negative or zero value for a pattern $\vec{u}$, and this is wrong, change the weight vector according to the formula:

$$\vec{w} \leftarrow \vec{w} + c\vec{u}$$

where c is some positive constant. On the other hand, if $\vec{w} \cdot \vec{u}$ gives a non-negative value and the value should be negative, change $\vec{w}$ by:

$$\vec{w} \leftarrow \vec{w} - c\vec{u}.$$

Members of both sets are continually submitted to the algorithm until it comes up with a value for w that works for all the points. It can be shown that for any positive value of c, the training process will converge. For a proof see [142].

As an example of how the weight vector converges to a value that will separate the two classes, we start with the weight vector, $\vec{w} = (0, 0, 0)$ and keep modifying it until it can separate all eight points correctly. The results of the computations are shown in Figure 3.3. The result is a weight vector, $\vec{w} = (4, -5, -7)$ that represents a line quite close to the one we chose by inspection.

PATTERN VECTOR, $\vec{u}$	WEIGHT VECTOR, $\vec{w}$	$\vec{w} \cdot \vec{u}$	CORRECT RESPONSE	NEW WEIGHT VECTOR
-6 4 1	0 0 0	0	—	6 -4 -1
7 -1 1	6 -4 -1	45	+	6 -4 -1
4 -2 1	6 -4 -1	31	+	6 -4 -1
-6 -1 1	6 -4 -1	-33	—	6 -4 -1
-2 -2 1	6 -4 -1	-5	—	6 -4 -1
-1 -4 1	6 -4 -1	9	+	6 -4 -1
4 2 1	6 -4 -1	15	—	2 -6 -2
-4 -7 1	2 -6 -2	32	+	2 -6 -2
-6 4 1	2 -6 -2	-38	—	2 -6 -2
7 -1 1	2 -6 -2	18	+	2 -6 -2
4 -2 1	2 -6 -2	18	+	2 -6 -2
-6 -1 1	2 -6 -2	-8	—	2 -6 -2
-2 -2 1	2 -6 -2	6	—	4 -4 -3
-1 -4 1	4 -4 -3	9	+	4 -4 -3
4 2 1	4 -4 -3	5	—	0 -6 -4
-4 -7 1	0 -6 -4	38	+	0 -6 -4
-6 4 1	0 -6 -4	-28	—	0 -6 -4
7 -1 1	0 -6 -4	2	+	0 -6 -4
4 -2 1	0 -6 -4	8	+	0 -6 -4
-6 -1 1	0 -6 -4	2	—	6 -5 -5
-2 -2 1	6 -5 -5	-7	—	6 -5 -5
-1 -4 1	6 -5 -5	9	+	6 -5 -5
4 2 1	6 -5 -5	9	—	2 -7 -6
-4 -7 1	2 -7 -6	35	+	2 -7 -6
-6 4 1	2 -7 -6	-46	—	2 -7 -6
7 -1 1	2 -7 -6	15	+	2 -7 -6
4 -2 1	2 -7 -6	16	+	2 -7 -6
-6 -1 1	2 -7 -6	-11	—	2 -7 -6
-2 -2 1	2 -7 -6	4	—	4 -5 -7
-1 -4 1	4 -5 -7	9	+	4 -5 -7
4 2 1	4 -5 -7	-1	—	4 -5 -7
-4 -7 1	4 -5 -7	12	+	4 -5 -7
-6 4 1	4 -5 -7	-51	—	4 -5 -7
7 -1 1	4 -5 -7	26	+	4 -5 -7
4 -2 1	4 -5 -7	19	+	4 -5 -7
-6 -1 1	4 -5 -7	-26	—	4 -5 -7
-2 -2 1	4 -5 -7	-5	—	4 -5 -7

Figure 3.3: Responses during training of the simple linear pattern classifier.

3.1.2   ADALINEs and MADELINEs

Some other versions of linear pattern classifiers have also been researched. Widrow researched linear pattern classifiers from the standpoint of using them as adaptive filters in electronics. He called his linear classifier an ADALINE for ADAptive LInear NEuron. Widrow also experimented with putting many ADALINEs together. These were called MADALINEs for Multiple ADALINEs. One very interesting application of an ADALINE was done by Widrow [258] at Stanford in the early 1960s where he used a linear pattern classifier to predict whether or not during the rainy season in San Francisco it would rain today, tonight, or tomorrow. The input data was a set of barometric pressure readings and changes in pressure readings in the Pacific Ocean from Alaska down to almost the equator. The results of the experiment were that the program was able to predict rain for San Francisco just as accurately as Weather Bureau meteorologists. Widrow also designed an "artificial neuron" in which the input weights could be changed by plating or unplating copper on very thin pencil lead.

3.1.3   Perceptrons

Another important researcher into pattern recognition using neuronlike elements was Frank Rosenblatt [177]. His systems of linear neuronlike elements were known as *perceptrons*. (The term perceptron is often applied to the linear pattern classifier as well.) The learning algorithm employed was known as the perceptron convergence procedure. Its learning rule is known as the *delta rule*. (This rule is derived in Appendix A.) The ability of perceptrons and the delta rule and any linear pattern classifier to learn complex functions is limited, however. In 1969, Minsky and Papert produced a book, *Perceptrons* (reprinted and expanded in 1988 [125]), in which they showed some of the limitations of perceptrons. This book is often credited with almost completely stopping research in the field. Since then, however, researchers have produced improved networks with nonlinear activation functions that are capable of learning nonlinearly separable patterns. The back-propagation procedure described in Section 3.4 can learn such patterns.

3.2   Separating Nonlinearly Separable Classes

Most pattern recognition problems cannot be solved by linear neurons because the surfaces separating the patterns are not linear. Since researchers realized the limitations of networks of linear neurons, most pattern recognition research has tried to find ways to separate pattern classes using more complex pattern recognition operators. Thus, there are operators that can take a set of points like the ones shown in Figure 3.4 and separate the points into two different classes using a surface more complicated than a straight line. We will neglect looking at these more complex operators and instead describe a few simple methods for dealing with nonlinearly separable patterns.

3.2.1   The Nearest Neighbor Algorithm

Possibly the simplest way to classify an unknown pattern at the point (x, y) is to compute the Euclidean distance from this point to every other known data point and find its nearest

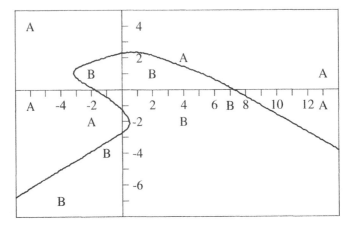

Figure 3.4: A set of points that is not linearly separable. These are the same points as in Figure 3.1 plus the points (13,1) and (13,-1) in class A and the points (-2,1) and (2,1) in class B. The line dividing the two classes is just one way to divide the space.

neighbor. We then assume the unknown has the same identity as its nearest neighbor. For instance, in Figure 3.5 we can assume that the point marked as 'X' is from class A and the point marked as 'Y' is from class B. Figure 3.5 also shows how the space is then divided up using the 12 data points. In actual use, the more data points you have available, the better the classification accuracy will be. If 1,000 points are generated at random from the distribution shown in Figure 3.4 and another 1,000 points are used as test cases, about 98 percent of them will be classified correctly.

Besides classifying an unknown point by finding its nearest neighbor, you can find its k nearest neighbors and let each neighbor contribute one vote toward identifying an unknown. This is known as the *k-nearest neighbor algorithm*. Another variation is to let each of the k nearest neighbors contribute an amount that decreases with its distance from the unknown point.

A nearest neighbor algorithm by Simard, Le Cun, and Denker using a different distance measure (not Euclidean) has managed to do better than all other algorithms on two difficult databases of handwritten digits [208].

3.2.2   Learning Vector Quantization Methods

There has also been a recent series of algorithms called Learning Vector Quantization algorithms: LVQ1, LVQ2, LVQ2.1, and LVQ3 (see [84, 85, 86]) and Decision Surface Mapping (DSM) from Geva and Sitte [51], that use the nearest neighbor algorithm but instead of storing a large number of points and searching through them to find the nearest neighbor, you store only a relatively small number of pattern vectors called *codebook vectors* or *prototype points* that are highly representative of the patterns in each class. In these methods you start with a initial set of prototype points and then move these points around to try to increase the classification performance of the nearest neighbor algorithm. Because the LVQ algorithms are new there have been few applications of them so far, but in a speech recognition

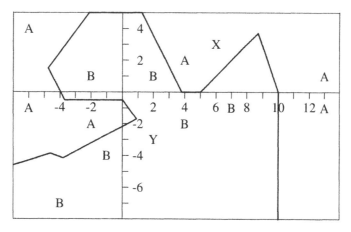

Figure 3.5: Using the nearest neighbor algorithm, point X will be classified as an A and point Y will be classified as a B. This figure also shows how the space is divided using the nearest neighbor classifier with the 12 data points.

problem Kohonen [85] reports better results than with any other algorithm he has tried.

The DSM algorithm is even simpler than the LVQ algorithms and it is reported to train faster and give better results than the LVQ algorithms for some types of problems. It works as follows. Let prototype point i be labeled as p_i and a training set point i, be t_i, then:

Find the nearest prototype point, p_n.

If p_n is the same class as t_i (a right answer), do nothing,

otherwise, move p_n away from t_i using the formula:

$$p_n = p_n - a(t_i - p_n),$$

where a is around 0.3 or less and it slowly decreases with time. Furthermore, find the closest prototype point, p_c, that gives the correct answer and move it closer to t_i using the formula:

$$p_c = p_c + a(t_i - p_c),$$

where again, a decreases slowly with time. Repeat these steps for several passes through the training set.

As an example,[1] we will take 10 prototype points at random from the distribution given in Figure 3.3. These points are shown in Figure 3.6(a). We will then generate 1,000 points for training and 1,000 more points for testing. These points will be the same ones we mentioned using the simple nearest neighbor algorithm. The initial performance on the training and test sets is around 70 percent. We then make seven passes through the training set using the values of 0.3, 0.2, 0.1, 0.05, 0.025, 0.01, and 0.005 for a. The final result is shown in Figure 3.6(b) where about 98 percent of the training and test set points are classified correctly. Notice that with the simple nearest neighbor algorithm, approximately 100 times more prototype points were used to reach the same level of classification accuracy.

---

[1] Results will vary according to the random points generated for training and testing.

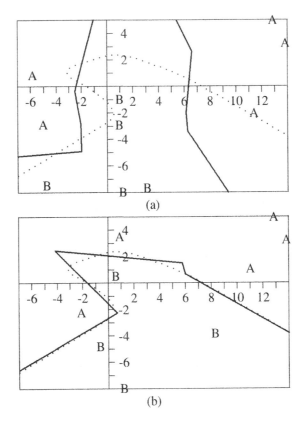

Figure 3.6: To demonstrate how DSM works, 10 prototype points were generated at random and these are shown in part (a). They give a 70% correct classification on the test set of 1000 points generated at random. After applying DSM the points have been moved to the locations shown in part (b) and they now give a 98.5% correct classification on the same test set. One of the B prototype points was moved below the area shown. The dotted lines in each part show the correct boundary.

3.3   Hopfield Networks

We will now look at a theoretically important type of network called the Hopfield network after Hopfield [68] who originated it. It is important because it is possible to prove that under the particular assumed conditions of the network, the network *will* converge to *an* answer, even if it is perhaps not the best possible answer. Knowing that a network will converge to an answer is quite important because some kinds of networks *never* converge. They may just constantly keep changing and so they never do settle down to any conclusion. Proving that a Hopfield network will converge is quite easy to do because there is a mathematical quantity associated with the network called the *computational energy* that will always decrease or, at the very worst, remain constant. When the computational energy of a Hopfield network stops decreasing, the network will be at *an* energy minimum, although not necessarily at the lowest energy minimum. The lowest minimum represents the best answer. To deal with this problem there is another network updating algorithm, the *Boltzman machine* relaxation algorithm that we will cover. Besides the theoretical importance of Hopfield networks, they are also important because they are designed to store items, or memories, in the weights connecting the units and so they are a candidate to model human memory as well.

3.3.1   The Hopfield Network

In a Hopfield network, the processing units take on only the two values 0 or 1. (Another version of the network has them take on the values –1 or 1.) Each unit has a threshold value associated with it such that if the input to a unit exceeds its threshold, the unit will turn on (become 1) or stay on, and if the input to a unit is less than the threshold, it will turn off (become 0) or stay off. The weights on the connections between the units take on the continuous set of values from $-\infty$ to $+\infty$. Also, if the connection weight from unit i to unit j is designated as w_{ij}, then the connection weight from unit j to unit i, w_{ji}, will equal w_{ij}. Such weights are said to be *symmetric*. Finally, the updating of the values for each unit is done *asynchronously*. This means that units in the network sort of take it upon themselves to update when they get the urge to do so (at random). If you "take pictures" of the state of the system at small enough time intervals, you will find only one unit is doing an update at a time. In the interactive activation network the updates for all the units take place at the same time, hence, in that type of network it is said that the updates take place *synchronously*.

The energy function for the n units in a Hopfield network is:

$$E = -\sum_{i<j} w_{ij} s_i s_j + \sum_i \theta_i s_i.$$

where w_{ij} is the weight between unit i and unit j, s_i is the state of unit i, s_j is the state of unit j, and θ_i is the threshold of unit i. To examine the meaning of the energy definition in detail, consider the simple network shown in Figure 3.7. We will first neglect the second term of the energy function by setting all the θ_is equal to 0. Here, the connection between a and b is +3/4, meaning that, if unit a is on, unit b should likely be on as well or if b is on then a should also be on. If we assume that a is currently on and b is currently off, the

Figure 3.7: A simple Hopfield network with four nodes, a, b, c, and d with their values and weights as shown.

contribution of this pair to the energy of the system will be $-3/4 \times 1 \times 0$, but if b is turned on, the contribution will be $-3/4 \times 1 \times 1$. So now, if we do turn on unit b, the energy of the system will decrease by 3/4. Now we look at the link between c and d. The $-7/8$ says that if d is on, it is quite likely that c should be off or if c is on, it is quite likely that d should be off. In terms of the computational energy, if both c and d are on, the contribution of the c-d link is $+7/8$. The energy could be decreased by turning either c or d or both of them off. In effect, the formula contributes penalties to the total computational energy when units that are off should in fact be on, as well as when units that are on should in fact be off. The goal of the algorithm is to remove penalties and therefore decrease the energy of the network.

To see why there is the $\sum_i \theta_i s_i$ term in the energy definition, notice that the difference in the computational energy between the kth unit being on and the kth unit being off is:

$$\Delta E_k = \sum_i w_{ki} s_i - \theta_k$$

If this term is positive, it means that the input to the unit k from the other units ($\sum_i w_{ki} s_i$) is greater than the threshold of the unit k. As it turns out, this is exactly the rule used to decide whether or not the unit k turns on. So, the system can go about its computations just by having each unit look at the inputs from other units and all the time the system will be moving toward a minimum energy state.

As a concrete example of the process we consider the network of Figure 3.7 with the initial values shown. Let the thresholds for all nodes be 1/2. The initial energy of the network will then be 19/8. The node b will be getting 3/4 of an energy unit from node a and 3/4 of an energy unit from node c. Additionally, the θ term will contribute $-1/2$. This total is 1 and is greater than 0, so node b will be turned on. This changes the energy to 11/8. We now look at whether or not to change node c. The input it will be receiving from b is $+3/4$ and the input from d will be $-7/8$ and the threshold term will be $-1/2$, giving a total of $-5/8$, and because this is less than the threshold it will cause c to turn off. The network energy will now be 6/8. Figure 3.8 shows how the energy level of the network changes as the updates are made. Note that the energy levels are quantized. In a very large network the energy levels will approach a continuum. The updating process that moves a network from a higher state of energy to a lower state is also known as a *relaxation algorithm*.

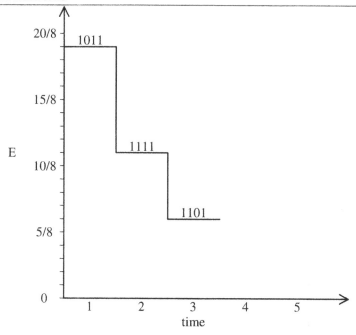

Figure 3.8: Starting with the initial state of 1011, updates to the network lower the energy level. This plot shows the progress after b and c are updated. How far will the energy drop if the updates are continued?

3.3.2 Storing Patterns

```
0  1  1  1  0        1  1  1  1  1         1   2   3   4   5
1  0  0  0  1        0  1  0  1  0         6   7   8   9  10
1  1  1  1  0        0  1  0  1  0        11  12  13  14  15
1  0  0  0  0        0  1  0  1  0        16  17  18  19  20
0  1  1  1  0        0  1  0  1  0        21  22  23  24  25
```

Figure 3.9: Two patterns that will be stored into a Hopfield network, a lowercase e and π. The units are numbered as shown on the right.

When a Hopfield network stores a pattern the pattern will be stored at a minimum of the energy function. Given part of a pattern, the network will update its units and move downhill to the minimum that represents the whole pattern. To illustrate how this works and to show how it is possible for the network to become stuck in a local minimum, let us take two small patterns in a 5 × 5 matrix, a lowercase e and the Greek letter π as shown in Figure 3.9. There will be 25 units and each unit will be connected to all the other units, meaning there will be a total of 25 × 25, or 625 connections, each with a weight. One way to determine the weights in a network is the following.[2] If a unit in the π pattern, say unit 1, is a 1 we

---

[2] Another way is given in the next section.

```
 0  0  0  0  2 -2  2  0  2 -2 -2  0 -2  0  0 -2  2  0  2  0  0  0 -2  0  0
 0  0  2  2  0  0  0 -2  0  0  0  2  0  2 -2  0  0 -2  0 -2 -2  2  0  2 -2
 0  2  0  2  0  0  0 -2  0  0  0  2  0  2 -2  0  0 -2  0 -2 -2  2  0  2 -2
 0  2  2  0  0  0  0 -2  0  0  0  2  0  2 -2  0  0 -2  0 -2 -2  2  0  2 -2
 2  0  0  0  0 -2  2  0  2 -2 -2  0 -2  0  0 -2  2  0  2  0  0  0 -2  0  0
-2  0  0  0 -2  0 -2  0 -2  2  2  0  2  0  0  2 -2  0 -2  0  0  0  2  0  0
 2  0  0  0  2 -2  0  0  2 -2 -2  0 -2  0  0 -2  2  0  2  0  0  0 -2  0  0
 0 -2 -2 -2  0  0  0  0  0  0  0 -2  0 -2  2  0  0  2  0  2  2 -2  0 -2  2
 2  0  0  0  2 -2  2  0  0 -2 -2  0 -2  0  0 -2  2  0  2  0  0  0 -2  0  0
-2  0  0  0 -2  2 -2  0 -2  0  2  0  2  0  0  2 -2  0 -2  0  0  0  2  0  0
-2  0  0  0 -2  2 -2  0 -2  2  0  0  2  0  0  2 -2  0 -2  0  0  0  2  0  0
 0  2  2  2  0  0  0 -2  0  0  0  0  0  2 -2  0  0 -2  0 -2 -2  2  0  2 -2
-2  0  0  0 -2  2 -2  0 -2  2  2  0  0  0  0  2 -2  0 -2  0  0  0  2  0  0
 0  2  2  2  0  0  0 -2  0  0  0  2  0  0 -2  0  0 -2  0 -2 -2  2  0  2 -2
 0 -2 -2 -2  0  0  0  2  0  0  0 -2  0 -2  0  0  0  2  0  2  2 -2  0 -2  2
-2  0  0  0 -2  2 -2  0 -2  2  2  0  2  0  0  0 -2  0 -2  0  0  0  2  0  0
 2  0  0  0  2 -2  2  0  2 -2 -2  0 -2  0  0 -2  0  0  2  0  0  0 -2  0  0
 0 -2 -2 -2  0  0  0  2  0  0  0 -2  0 -2  2  0  0  0  0  2  2 -2  0 -2  2
 2  0  0  0  2 -2  2  0  2 -2 -2  0 -2  0  0 -2  2  0  0  0  0  0 -2  0  0
 0 -2 -2 -2  0  0  0  2  0  0  0 -2  0 -2  2  0  0  2  0  0  2 -2  0 -2  2
 0 -2 -2 -2  0  0  0  2  0  0  0 -2  0 -2  2  0  0  2  0  2  0 -2  0 -2  2
 0  2  2  2  0  0  0 -2  0  0  0  2  0  2 -2  0  0 -2  0 -2 -2  0  0  2 -2
-2  0  0  0 -2  2 -2  0 -2  2  2  0  2  0  0  2 -2  0 -2  0  0  0  0  0  0
 0  2  2  2  0  0  0 -2  0  0  0  2  0  2 -2  0  0 -2  0 -2 -2  2  0  0 -2
 0 -2 -2 -2  0  0  0  2  0  0  0 -2  0 -2  2  0  0  2  0  2  2 -2  0 -2  0
```

Figure 3.10: The matrix produced for the π/e example.

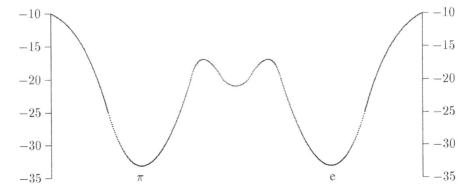

Figure 3.11: Part of the energy landscape for the π/e problem with π and e both at -33 and another shallower minimum at -21 in between. In reality the landscape is 26-dimensional and the energy levels are discrete, not continuous.

take a look at the other 24 units. Unit 2 is on so there should be an activation link with a value of +1 from unit 1 to unit 2 as well as from unit 2 to unit 1. (Notice how this recalls James' rule: "When two ideas have been active at once, or in immediate succession, one of them on re-occurring tends to propagate its excitement into the other.") There also will be +1 links from unit 1 to units 3, 4, 5, 7, 9, 12, 14, 17, 19, 22, and 24. Now when unit 1 is on, unit 6 should be off. This can be arranged by having an inhibition link (value −1) between unit 1 and unit 6. Notice that under this plan, every other 1 in the π pattern will also work to inhibit unit 6 as well as work to inhibit every other unit that is zero. (James did not say this!) Finally, if a pair of units are both zero, the weight between them is set to 1. These weights are conveniently defined by the formula:

$$w_{ij} = (2s_i - 1)(2s_j - 1).$$

Let the weights for the e be the matrix W_1, and the weights for π be the matrix W_2. The weights for the network that stores both the e and the π are obtained by simply adding the two matrices, W_1 and W_2 together:

$$W = W_1 + W_2$$

giving the matrix shown in Figure 3.10. We still need a threshold value for each unit and let us (rather arbitrarily) choose a value of +3 for every unit. Given all these weights, we end up with what we might call an "energy landscape" that contains two large minimas, one for π and one for e, and unfortunately many other low spots that the network can be trapped in. Representing the energy landscape for this problem is rather difficult because it is a 26-dimensional space; however by simplifying it, part of it looks like the one shown in Figure 3.11. The minimas for π and e are at −33 and in between them is another minimum with a value of −21.

Figure 3.12 shows the reconstruction process for the initial state shown in the upper-left-hand corner of the figure. The initial state has the characteristics of an incomplete e pattern and we clamp the units that are 1 at 1 while allowing the other units to update themselves. The relaxation process completes the e when the minimum of −33 is reached. In Figure 3.13, starting with an incomplete π pattern, the relaxation process becomes stuck at an energy value of −18. In fact, if the 1 units were not frozen at 1, the network would turn off unit 1 so it could fall down in the minimum, −21, shown in Figure 3.11. To get the reconstruction process moving again we can set unit 5 equal to 1, in effect giving the network another clue that the pattern is a π, this is shown in Figure 3.14. This raises the energy level to −17, but by updating the rest of the units the π pattern is completed. Because Hopfield networks are oriented toward storing patterns or memories in the network weights, it is sometimes said that when a network falls into a local minimum it is experiencing a kind of *deja vu*, in which it remembers something that it has never really seen.

3.3.3 The Boltzman Machine

In order to try and avoid local minima, researchers have devised a plan, where at the start of the updating procedure a lot of jumps uphill are made at random, and then as time goes on, the chance of taking a step uphill gradually decreases. This is the strategy of an abstract kind of machine, known as a *Boltzman machine* [66]. In it, the high probability of taking

0 1 1 1 0	0 1 1 1 0	0 1 1 1 0	0 1 1 1 0	0 1 1 1 0
1 0 0 0 0	1 0 0 0 0	1 0 0 0 0	1 0 0 0 0	1 0 0 0 0
1 0 0 0 0	1 0 0 0 0	1 0 0 0 0	1 0 0 0 0	1 0 0 0 0
1 0 0 0 0	1 0 0 0 0	1 0 0 0 0	1 0 0 0 0	1 0 0 0 0
0 0 0 0 0	0 0 0 0 0	0 0 0 0 0	0 0 0 0 0	0 0 0 0 0
initial pattern	test 1	test 5	test 7	test 8
$E = 6$	$\Delta E = -9$ $E = 6$	$\Delta E = -9$ $E = 6$	$\Delta E = -9$ $E = 6$	$\Delta E = -9$ $E = 6$

0 1 1 1 0	0 1 1 1 0	0 1 1 1 0	0 1 1 1 0	0 1 1 1 0
1 0 0 0 0	1 0 0 0 1	1 0 0 0 1	1 0 0 0 1	1 0 0 0 1
1 0 0 0 0	1 0 0 0 0	1 1 0 0 0	1 1 1 0 0	1 1 1 1 0
1 0 0 0 0	1 0 0 0 0	1 0 0 0 0	1 0 0 0 0	1 0 0 0 0
0 0 0 0 0	0 0 0 0 0	0 0 0 0 0	0 0 0 0 0	0 0 0 0 0
test 9	test 10	test 12	test 13	test 14
$\Delta E = -9$ $E = 6$	$\Delta E = 3$ $E = 3$	$\Delta E = 3$ $E = 0$	$\Delta E = 5$ $E = -5$	$\Delta E = 5$ $E = -10$

0 1 1 1 0	0 1 1 1 0	0 1 1 1 0	0 1 1 1 0	0 1 1 1 0
1 0 0 0 1	1 0 0 0 1	1 0 0 0 1	1 0 0 0 1	1 0 0 0 1
1 1 1 1 0	1 1 1 1 0	1 1 1 1 0	1 1 1 1 0	1 1 1 1 0
1 0 0 0 0	1 0 0 0 0	1 0 0 0 0	1 0 0 0 0	1 0 0 0 0
0 0 0 0 0	0 0 0 0 0	0 0 0 0 0	0 0 0 0 0	0 0 0 0 0
test 15	test 17	test 18	test 19	test 20
$\Delta E = -13$ $E = -10$	$\Delta E = -13$ $E = -10$	$\Delta E = -13$ $E = -10$	$\Delta E = -13$ $E = -10$	$\Delta E = -13$ $E = -10$

0 1 1 1 0	0 1 1 1 0	0 1 1 1 0	0 1 1 1 0	0 1 1 1 0
1 0 0 0 1	1 0 0 0 1	1 0 0 0 1	1 0 0 0 1	1 0 0 0 1
1 1 1 1 0	1 1 1 1 0	1 1 1 1 0	1 1 1 1 0	1 1 1 1 0
1 0 0 0 0	1 0 0 0 0	1 0 0 0 0	1 0 0 0 0	1 0 0 0 0
0 0 0 0 0	0 1 0 0 0	0 1 1 0 0	0 1 1 1 0	0 1 1 1 0
test 21	test 22	test 23	test 24	test 25
$\Delta E = -13$ $E = -10$	$\Delta E = 7$ $E = -17$	$\Delta E = 7$ $E = -24$	$\Delta E = 9$ $E = -33$	$\Delta E = -17$ $E = -33$

Figure 3.12: Starting with a partial letter "e" shown in the frame in the upper-left-hand corner, the Hopfield network reconstructs the missing parts and the energy minimum is –33.

| 1 1 0 1 0 |
| 0 0 0 0 0 |
| 0 1 0 1 0 |
| 0 0 0 0 0 |
| 0 1 0 1 0 |

initial
pattern
$E = -9$

| 1 1 1 1 0 |
| 0 0 0 0 0 |
| 0 1 0 1 0 |
| 0 0 0 0 0 |
| 0 1 0 1 0 |

test 3
$\Delta E = 9$
$E = -18$

| 1 1 1 1 0 |
| 0 0 0 0 0 |
| 0 1 0 1 0 |
| 0 0 0 0 0 |
| 0 1 0 1 0 |

test 5
$\Delta E = -1$
$E = -18$

| 1 1 1 1 0 |
| 0 0 0 0 0 |
| 0 1 0 1 0 |
| 0 0 0 0 0 |
| 0 1 0 1 0 |

test 6
$\Delta E = -5$
$E = -18$

| 1 1 1 1 0 |
| 0 0 0 0 0 |
| 0 1 0 1 0 |
| 0 0 0 0 0 |
| 0 1 0 1 0 |

test 7
$\Delta E = -1$
$E = -18$

| 1 1 1 1 0 |
| 0 0 0 0 0 |
| 0 1 0 1 0 |
| 0 0 0 0 0 |
| 0 1 0 1 0 |

test 8
$\Delta E = -17$
$E = -18$

| 1 1 1 1 0 |
| 0 0 0 0 0 |
| 0 1 0 1 0 |
| 0 0 0 0 0 |
| 0 1 0 1 0 |

test 9
$\Delta E = -1$
$E = -18$

| 1 1 1 1 0 |
| 0 0 0 0 0 |
| 0 1 0 1 0 |
| 0 0 0 0 0 |
| 0 1 0 1 0 |

test 10
$\Delta E = -5$
$E = -18$

| 1 1 1 1 0 |
| 0 0 0 0 0 |
| 0 1 0 1 0 |
| 0 0 0 0 0 |
| 0 1 0 1 0 |

test 11
$\Delta E = -5$
$E = -18$

| 1 1 1 1 0 |
| 0 0 0 0 0 |
| 0 1 0 1 0 |
| 0 0 0 0 0 |
| 0 1 0 1 0 |

test 13
$\Delta E = -5$
$E = -18$

| 1 1 1 1 0 |
| 0 0 0 0 0 |
| 0 1 0 1 0 |
| 0 0 0 0 0 |
| 0 1 0 1 0 |

test 15
$\Delta E = -17$
$E = -18$

| 1 1 1 1 0 |
| 0 0 0 0 0 |
| 0 1 0 1 0 |
| 0 0 0 0 0 |
| 0 1 0 1 0 |

test 16
$\Delta E = -5$
$E = -18$

| 1 1 1 1 0 |
| 0 0 0 0 0 |
| 0 1 0 1 0 |
| 0 0 0 0 0 |
| 0 1 0 1 0 |

test 17
$\Delta E = -1$
$E = -18$

| 1 1 1 1 0 |
| 0 0 0 0 0 |
| 0 1 0 1 0 |
| 0 0 0 0 0 |
| 0 1 0 1 0 |

test 18
$\Delta E = -17$
$E = -18$

| 1 1 1 1 0 |
| 0 0 0 0 0 |
| 0 1 0 1 0 |
| 0 0 0 0 0 |
| 0 1 0 1 0 |

test 19
$\Delta E = -1$
$E = -18$

| 1 1 1 1 0 |
| 0 0 0 0 0 |
| 0 1 0 1 0 |
| 0 0 0 0 0 |
| 0 1 0 1 0 |

test 20
$\Delta E = -17$
$E = -18$

| 1 1 1 1 0 |
| 0 0 0 0 0 |
| 0 1 0 1 0 |
| 0 0 0 0 0 |
| 0 1 0 1 0 |

test 21
$\Delta E = -17$
$E = -18$

| 1 1 1 1 0 |
| 0 0 0 0 0 |
| 0 1 0 1 0 |
| 0 0 0 0 0 |
| 0 1 0 1 0 |

test 23
$\Delta E = -5$
$E = -18$

| 1 1 1 1 0 |
| 0 0 0 0 0 |
| 0 1 0 1 0 |
| 0 0 0 0 0 |
| 0 1 0 1 0 |

test 25
$\Delta E = -17$
$E = -18$

Figure 3.13: For the incomplete pattern shown in the upper-left-hand corner the network gets stuck in a local minimum with an energy of –18.

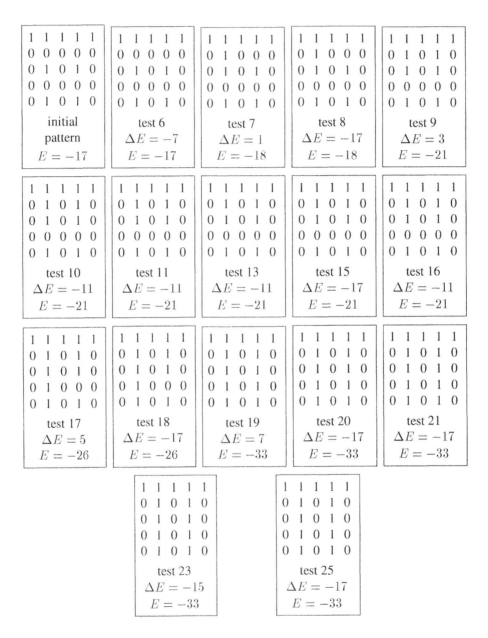

Figure 3.14: With an extra kick the network completes the π pattern. The kick consisted of setting unit 5 = 1.

a step uphill reflects a high "temperature" while as the system "cools down," the chance of a random step uphill decreases until the system "freezes" at a minimum in the energy landscape. This process of cooling the system is often referred to as "simulated annealing" because it is analogous to what happens in metals and other materials when they are melted and then cooled slowly. The atoms or molecules in the material have electric charges that repel and attract each other. When these atoms or molecules lose energy slowly enough they have a chance to align themselves in such a way as to minimize the energy of the solid. In this case, large crystals of the material will be formed, whereas if the material is cooled very rapidly, only very small microscopic crystals will form.

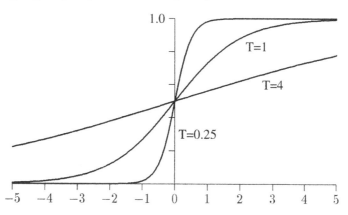

Figure 3.15: The probability of turning a unit on as a function of the unit's input for temperatures of 0.25, 1.0, and 4.0.

A simple modification of the Hopfield updating procedure leads to the Boltzman machine recall plan. The modification is that the probability, p_k, of a unit k becoming 1 (or staying at 1) at temperature T is given by:

$$p_k = \frac{1}{1 + e^{-\Delta E_k/T}}.$$

A graph of this function for several temperatures is shown in Figure 3.15. Notice that at high temperatures and negative values of ΔE_k, the probability of a unit turning on and increasing the computational energy is much greater than for low temperatures. As the temperature goes down, units that do not fit within the pattern are less likely to turn on. Even then, however, sometimes a unit with $\Delta E_k < 0$ will turn on. At very low temperatures you approach the Hopfield relaxation procedure where a unit only turns on if its input is greater than its threshold. The Boltzman procedure in effect lets the network sample a great many portions of the energy landscape. If in so doing the network happens to find a deep minimum, it is unlikely that taking a step upward will bring it out, so thereafter it will continue to settle down, whereas if it is in a shallow minimum, one step up could easily get the network out. The theory behind the Boltzman machine also shows that as the temperature approaches zero *slowly*, the probability that the network will be in the global minimum approaches 1.

The effectiveness of the Boltzman machine recall algorithm is illustrated in Figure 3.16 using the partial π pattern shown in the first frame of Figure 3.13. The network's initial

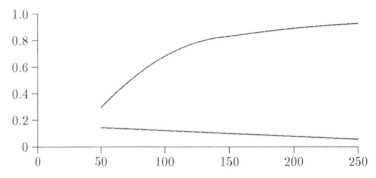

Figure 3.16: Given the partial π pattern shown in the first frame of Figure 3.13, the upper curve shows the probability of being in the global minimum of -33 and the lower curve shows the probability of being in the local minimum of -18 as a function of time. The time is the number of units that were updated at random while decreasing the temperature linearly from 4 to 0.25. When the network is cooled rapidly it often ends up in states other than -33 or -18 so the probabilities do not add to 1.

temperature was 4.0 and it was cooled linearly to 0.25. The amount of time taken to cool the network is simply the number of units that were updated at random as the system was cooled. The results show that the probability of being in the global minimum of -33 is only 0.3 when 50 updates were made but it rises to 0.93 when 250 updates were made.

Another method for overcoming local minima is to redefine the system so that the activation values of the units take on continuous values and the updates are done as in the interactive activation model [69]. This procedure does not guarantee that the system will settle into a minimum, however.

3.3.4   Pattern Recognition

We have been considering how Hopfield networks can do pattern completion in this section, but the methods can easily do pattern recognition as well. In the π/e example we could add two extra units to each network that code for the identity of the pattern. Unit 26 could indicate a π and unit 27 could indicate an e. Now given a pattern, even an incomplete one or one with a small amount of noise present, the network will complete it and identify it as well by turning on unit 26 for a π or unit 27 for an e.

3.3.5   Harmony

Some researchers have also defined a quantity similar to computational energy, called "goodness of fit," which their systems then attempt to maximize. Another similar concept is called "harmony," which a network also attempts to maximize [217]. When the object is to maximize a quantity, the searching process is called *hill climbing*.

3.3.6   Comparison with Human Thinking

The Hopfield/Boltzman models represent a much different way of computing than is found in standard von Neumann machines and they have some aspects that they share with people.

$$
\begin{array}{ccccc}
0 & \blacksquare & \blacksquare & \blacksquare & 0 \\
0 & 0 & 0 & 0 & 0 \\
0 & \blacksquare & 0 & \blacksquare & 0 \\
0 & 0 & 0 & 0 & 0 \\
0 & \blacksquare & 0 & \blacksquare & 0
\end{array}
$$

Figure 3.17: An ambiguous pattern that could be either π or e.

First, the Hopfield/Boltzman machines are, like people, somewhat unpredictable. In the π/e network, depending on the exact order in which the units update themselves, the network could reach different conclusions. Given the pattern in Figure 3.17 that contains only the features common to both π and e, and updating at random, it is clear that sometimes the network will settle down to the e minimum, but with a different sequence of updates it might settle down to the π minimum. This is interesting because if you give this problem or many similar kinds of problems like this to people, some people will find one answer and other people will find another, and even the same person may give one result on one occasion and the other result on another occasion. From the last chapter, recall the sentence, "John shot some bucks." If you do not know whether John was hunting or gambling, your mind could easily slip into either the hunting or the gambling interpretation. Because this algorithm produces the kind of random results you get from people, it makes some researchers feel that this model of artificial thinking may be close to how people's minds actually operate.

Second, another phenomenon that happens in human problem solving is that people often seem to get stuck in local minima. First, they may get stuck on a problem and not be able to find any solution at all, or second, they may find a solution to a problem that satisfies many of the required constraints, but not all of them. When this happens to people the solution is to start over fresh in hopes that a better solution will just pop into their minds. Starting over in Boltzman machine terminology corresponds to reheating the system and hoping that through random updates it will settle into a better solution. It is easy to see how this happens in the Hopfield/Boltzman networks, but it is not at all obvious how this phenomenon can happen in a classical von Neumann architecture running a classical algorithm with classical data structures. The von Neumann architecture is thoroughly predictable and there is no reason why the system should get stuck anywhere in the process or give different responses to the same input at different times. In addition to people not finding correct solutions to problems, when people are rushed to find an answer they will not necessarily find the best solution, but only an adequate solution. Again, in Boltzman machine terminology, being rushed corresponds to cooling the system quickly and then there is an increased chance that the best solution will not be found.

Although the Hopfield/Boltzman algorithms are extremely interesting from the standpoint of being much different than the classical symbol processing algorithms, very little has been done with them so far. One interesting application is by Halici and Sungur [56] where they use the Boltzman machine algorithm to solve the SOMA puzzle problem.

3.4   Back-Propagation

Back-propagation has become the single most useful neural networking algorithm. It is a generalization of the delta rule and its learning rule is sometimes called the generalized delta rule. Back-propagation networks are sometimes referred to as multilayer perceptrons. In this section we give only some informal justification for the formulas, however there is a derivation of the back-propagation algorithm formulas in Appendix A. Until very recently it was thought unlikely that anything like back-propagation occurs in the brain, however there is now some speculation as to how the brain might implement back-propagation (see [224], [65], and [26]). Users of the algorithm call it backprop.

3.4.1   History

Back-propagation has been discovered a number of times. At this time it appears as if the first derivation of the algorithm was by Robbins and Monro in 1951 [176]. Their discovery was reported by White in 1989 [256]. Also in 1989, Hecht-Nielsen [65] noted that it was discovered by Bryson and Ho by 1969 [16]. In 1974, Werbos independently rederived it [252]. Since that time, Werbos has been studying the use of back-propagation in various economic modeling and artificial intelligence problems (for a listing of articles see [253] or [254]). Unfortunately, Werbos' work was not discovered by the AI community until 1987. Back-propagation was again independently rederived by Parker [146, 147] and finally made well known in 1986 by Rumelhart, Hinton, and Williams [180].

3.4.2   The Network

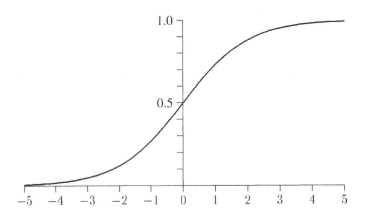

Figure 3.18: A plot of the most commonly used back-propagation activation function, $1/(1 + e^{-x})$.

One of the specific reasons Minsky and Papert gave in their 1969 book, *Perceptrons*, for being pessimistic about neural networking was that networks with any number of layers and where the neurons use a linear activation function (as in the linear pattern classifier) could not learn many important and useful functions. For example, they could not compute the exclusive-or (XOR) of two binary inputs. Back-propagation uses a nonlinear activation function and it can be used to do the XOR problem and other problems where nonlinear curves and surfaces are necessary, such as the problem of separating the nonlinearly separable regions shown in Figure 3.4. The back-propagation algorithm will work with many activation functions but the most commonly used one is:

$$o_j = \frac{1}{1 + e^{-net_j}}$$

where net_j is the sum of the inputs to neuron j and o_j is the activation value of neuron j. This function is shown in Figure 3.18. Functions like this with an S-shaped character are referred to as *sigmoids*. net_j is defined as:

$$net_j = \sum_i w_{ij} o_i + \theta_j$$

where w_{ij} is the weight connecting unit i in a previous layer with unit j and o_i is the activation value of unit i. The term θ_j represents the weight from a *bias unit* that is always on (=1.0) and it functions like the threshold value in other networks.

   This activation function has some important properties. If $net_j = 0$, so that there is no activation at all coming into node j, then $o_j = 0.5$. A value of 0.5 for a node therefore means the unit is "undecided." Also notice that to raise the activation value of a unit to exactly +1, net_j would need to be infinite and to achieve a value of 0, net_j must be negative infinity. Plus and minus infinity are rather hard values to reach so in reality people usually consider the output for a pattern to be correct if the values on its output node(s) that are supposed to be 1 are at least 0.9 and the values of its output node(s) that are supposed to be 0 are below 0.1. Less stringent limits are used at times. When a network is being tested with unknown patterns, any value greater than 0.5 is often considered to be on and any value less than 0.5 is considered to be off. Usually in classification problems the output unit with the largest value is considered to be the answer whether the largest value is greater than 0.5 or not.

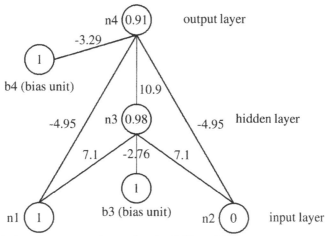

Figure 3.19: A three-layer network to solve the XOR problem with weights produced by back-propagation.

Figure 3.19 shows a network that has been trained to compute the XOR function. The topology of the network was chosen by hand but the weights for the connections have been found by the back-propagation algorithm. The layer of units in between the input and output units is called a *hidden layer* since if the network is viewed as a black box with input units and output units the hidden layer units cannot be seen. Unit b3 is the bias unit for the hidden layer unit, n3, and b4 is the bias unit for the output layer unit, n4. The patterns to be learned and the actual network responses to these patterns are:

n1	n2	n4 (desired)	n4 (actual)
1	0	1	0.91
0	0	0	0.08
0	1	1	0.91
1	1	0	0.10

This particular XOR network has connections from the first to the third layer. Typically, back-propagation networks only have connections between adjacent layers but there is one report that adding extra connections from the input to the output layer improves the algorithm [220]. Networks can also have connections between units in a single layer and the number of hidden layers is unlimited in principle. In practice, three-layer networks work well for most problems. Networks are often described by the number of units in each layer. A network with 19 input units, 12 hidden units, and 7 output units is then described as: 19-12-7. In this description the bias units are not counted. Ordinarily, bias units are not shown in diagrams of networks either.

3.4.3   Computing the Weights

To find the proper weights for a network, all the weights are started out at 0 or small random values. Then, a training pattern is placed on the input level units and this produces a response on the output units. This output pattern computed by the network is compared with what the answer should be and modifications of the weights throughout the network are made in order to make the computed answer come closer to the correct answer. Every weight in the network can be regarded as a variable or a dimension in space. This process of changing weights to try to minimize the error is then called a search through "weight space." The search to try to minimize the error is much the same as trying to minimize energy in a Hopfield network and the process is subject to the same problem as in the Hopfield network: it is possible to come to a local minimum that it is impossible to get out of if you only try to go down. Moreover, a second problem exists in back-propagation: if the weight changes are too large it is possible for the errors to increase as shown in Figure 3.20, whereas in the Hopfield network there was a guarantee that the energy would never increase.

As an example of the weight changing process, we will start with the network shown in Figure 3.21 where all the weights are initially 0. When we place the pattern (1,0) on the input units, the output unit has a value of 0.5 instead of the target value of 1. We now need some formulas that will change the weights so that the next time this pattern is presented,

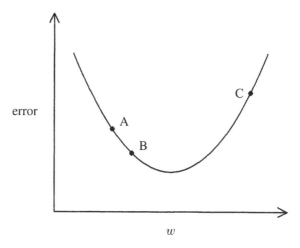

Figure 3.20: The error on the output units plotted as a function of a weight, w in the network. When the network is at point A, the error can be decreased a little by slightly increasing the value of the weight. On the other hand, a large change might put the network at the point C on the other side of the valley where the error is larger. The simplest way to avoid this problem is to make the weight changes fairly small.

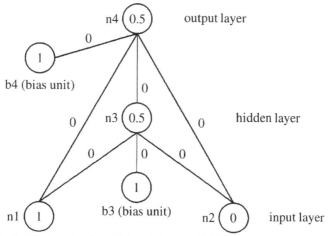

Figure 3.21: To start the training the weights will start at 0. The input unit values are 1 and 0, n3 and n4 are 0.5. The target for n4 is 1.

n4 will be a little closer to 1.0. In order to raise the value of the output unit there are two things that can be done. First, the weights coming from n1, n3, and b4 can all be increased. The second thing that can be done to increase the value of n4 is to raise the activation levels of nodes that feed into n4. In this case, however, the only node that can have its activation level raised is n3. The other nodes are frozen at values of 0 or 1. Raising the activation level of n3 can be done by raising the values of the weights leading into n3. First, we will look at raising the weights leading into n4. The back-propagation formula to modify these weights consists of several definitions:

$$\delta_k = (t_k - o_k)f'(net_k) \tag{3.1}$$

$$f'(net_k) = o_k(1 - o_k) \tag{3.2}$$

$$\Delta w_{jk} = \eta \delta_k o_j \tag{3.3}$$

The quantity δ_k in Equation (3.1) is called the error signal, t_k is the target value for unit k, and o_k is the actual activation value of node k. Note that the difference $(t_k - o_k)$ determines the extent of the error and it is called the simple error. Large differences between what you have as an answer and what you should have will result in larger corrections being made to the weights. The $f'(net_k)$ term given in Equation (3.2) is the derivative of the activation function, f, with respect to net_k. It comes as a consequence of the derivation. Finally, in Equation (3.3), the change to the weight Δw_{jk} that goes from a unit j in a lower layer to unit k in the output layer is the product of δ_k, o_j, the activation value of the unit j, and a small constant η, that controls the learning rate. This constant must be relatively small so that the error decreases. In our XOR example we will choose $\eta = 0.1$ and then the weight changes we compute are as follows:

$$\delta_{n4} = (1 - 0.5) \times 0.5 \times (1 - 0.5) = 0.125,$$

$$\Delta w_{n1n4} = 0.1 \times 0.125 \times 1 = 0.0125,$$

$$\Delta w_{b4n4} = 0.1 \times 0.125 \times 1 = 0.0125,$$

$$\Delta w_{n3n4} = 0.1 \times 0.125 \times 0.5 = 0.00625,$$

$$\Delta w_{n2n4} = 0.1 \times 0.125 \times 0 = 0.$$

Now, without making any weight changes just yet we proceed to the hidden layer units and compute weight changes for weights leading into these units. In this way, we will raise or lower the activation of units in this hidden layer. To do this, we need to give some error values to the hidden layer units. If we now let δ_k be the error found for the kth unit in the output layer (or any layer above), then the δ_j term for the jth hidden unit is:

$$\delta_j = f'(net_j) \sum_k \delta_k w_{jk}.$$

The w_{jk} are the weights that lead from unit j to the output units. The change in the weight w_{ij} between unit i in the layer below the hidden layer and unit j is Δw_{ij}, which is given by:

$$\Delta w_{ij} = \eta \delta_j o_i.$$

Recalling that the error signal for n4 was 0.125, and that the weight between n3 and n4 is still 0, δ_{n3} for the hidden unit n3 will be:

$$\delta_{n3} = 0.5 \times (1 - 0.5) \times 0.125 \times 0 = 0$$

and the weight changes will be:

$$\Delta w_{n1n3} = 0.1 \times 0 \times 1 = 0,$$

$$\Delta w_{b3n3} = 0.1 \times 0 \times 0 = 0,$$

$$\Delta w_{b3n3} = 0.1 \times 0 \times 1 = 0.$$

All the network weights can then be modified by these amounts according to:

$$w_{jk} \leftarrow w_{jk} + \Delta w_{jk}.$$

If the same pattern is now input to the network again, n4 will be 0.507, which is of course a little closer to the correct answer.

The above method is called the "online" or "continuous" update method because changes are made continuously as each pattern is input. Another method for doing weight changes is to collect all the weight changes for all the patterns in the training set and then do them only once for the whole set instead of once for each pattern. The advantage to this method is that less arithmetic needs to be done, but there is a disadvantage to this because then you may need to make more passes through the training patterns. On large problems this second option is usually better. This second option is called the "batch" or "periodic" update method.

Another important consideration when doing weight changes with the periodic update method is that for some pattern sets like the XOR problem all the weight changes will cancel out if all the weights start out at 0 and the result is that no learning takes place. This can be solved by starting out the weights with small random values in the range from about −1 to +1 rather than starting them with 0. The best range of starting weights to use varies from problem to problem and may be larger or smaller than this range. In practice, back-propagation networks are almost always started out with small random values for the weights no matter how the weight changes are managed because this will speed convergence.

3.4.4 Speeding Up Back-Propagation

With $\eta = 0.1$ and starting each weight at 0 and using 32-bit floating point weights, it requires 25,496 iterations through the pattern set to train the network to within 0.1 of its targets. With $\eta = 0.5$ it requires 3,172 iterations and with $\eta = 1$, it requires 1,381 iterations. All these values are much too large to be acceptable for a problem as small as this one. By employing a certain trick it is almost always possible to make a network converge much faster. The trick consists of using a different Δw, we will call it Δw_{better}, that is equal to the Δw calculated for this weight plus α times the value of the Δw that was obtained the last time the weight was updated. In equation form this is:

$$\Delta w_{better} = \Delta w + \alpha \Delta w_{previouschange}.$$

This second extra term is called a "momentum term" because it keeps the process moving in a consistent direction. Typically, α is about 0.9. Using the best combination of η and α the XOR problem can be solved in under 30 iterations.[3] At this time there is no theory that will give the best values for η and α for a given problem, but after you acquire some experience with different types of problems you can usually choose some reasonable values fairly quickly. For a problem with n patterns, a reasonable guess for η is around $1/n$ or $2/n$.

A very large number of methods, both simple and complex, have been devised to speed up back-propagation. Perhaps two of the best are Rprop [172, 173] and Quickprop [37], both of which vary the learning rates automatically for each weight as training goes on.

3.4.5 Dealing with Local Minima

From time to time a network may get stuck in a local minimum and be unable to learn the desired answers. There are not yet any thoroughly researched methods on how best to kick a back-propagation network out of a local minimum. However, one simple method is to assume that this problem has occurred because weights that are positive are too positive and weights that are negative are too negative. Therefore, one simple solution is to take weights that are positive and decrease them by a random amount while weights that are negative can be increased by a random amount. This method can often take a network out of a local minimum.[4] Another method to escape a local minimum is to simply add one or more hidden units to the network. Ash [4] reports that a method he used for automatically adding one node at a time always found a solution for the problems on which he tried the method.

3.4.6 Using Back-Propagation to Train Hopfield/Boltzman Networks

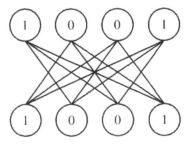

Figure 3.22: Using back-propagation to produce a network with symmetric weights.

It is a simple matter to create a version of back-propagation to produce the kinds of networks with symmetric links that are used in the Hopfield and Boltzman networks. In Figure 3.22 we have a two-layer network with the same number of units in both layers. Given a pattern on the input layer, we will want to produce the same pattern on the output layer.

---

[3] Various other changes can get this down to around 6. See [239].

[4] Unpublished research by the author.

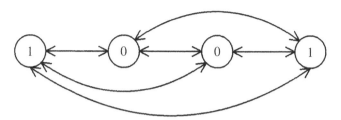

Figure 3.23: The network in Figure 3.22 can be transformed into a Hopfield/ Boltzman type network.

There is one trivial solution to this problem that needs to be eliminated. When input unit 1 is on we can turn output unit 1 on just by having a large weight from input unit 1 to output unit 1; all the weights from the other input units to output unit 1 can be zero. This solution is not acceptable, so to eliminate this possibility, we can either freeze the links from 1 to 1, from 2 to 2, and so forth at zero or eliminate the weights altogether. Now, to enforce the requirement that the weights be symmetric, the weight from input unit 1 to output 2 *will be the same weight* as the weight from input unit 2 to output unit 1, and so on. This procedure also then cuts the number of weights needed in half. Then, whenever we modify the weight from output unit 1 to input unit 2, the weight from output unit 2 to input unit 1 is contained in the same memory location so it is modified at the same time. While we started the problem thinking of it as a two-layer network, this resulting set of interconnections can also be viewed as a Hopfield-style network as shown in Figure 3.23. In training the network, you can use either the traditional back-propagation activation function or a linear one.

3.5 Pattern Recognition and Curve Fitting

It is important to keep in mind that when back-propagation is applied to a pattern recognition problem the algorithm will try to construct a surface that will separate the input data into the correct classes. We will now look at a few examples to illustrate the surfaces that are formed and also note that back-propagation can approximate real-valued functions as well.

3.5.1 Pattern Recognition as Curve Fitting

The XOR problem is a discrete problem because the input and output values are all integers. As such, it does not make sense to try to find XOR of (0.5,0.5) or XOR of any other combination of real-valued inputs. However, if we neglect the fact that such combinations are undefined and go ahead and plot the surface, $z = XOR(x, y)$, for $x = 0$ to 1 and $y = 0$ to 1, we get the surface plotted in Figure 3.24. The network created a valley that runs from (0,0) to (1,1) to solve the problem and the points (0,1) and (1,0) overlook the valley.

In Figure 3.25, we show how a 2-1 network divided up the space for the eight linearly separable points in Figure 3.1. In part (a) of the figure, the line separating the two classes is on a sigmoid-shaped slope at a height of 0.5. Part (b) of the figure shows a cross section of the slope.

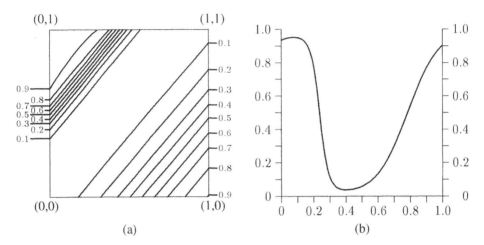

Figure 3.24: On the left in part (a) is a contour plot of the XOR function giving its height as a function of x and y. There is a wide valley between (0,0) and (1,1). On the right in part (b) is a cross section of the surface that runs from (0,1) to (1,0).

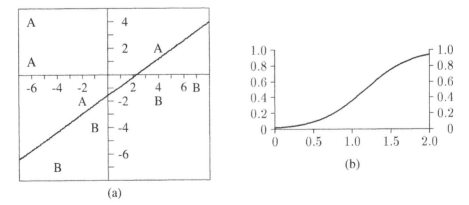

Figure 3.25: A 2-1 back-propagation network produces a sigmoid-shaped surface to separate the eight linearly separable points. The $z = 0.5$ contour line is shown in (a) and part (b) shows a cross section of the surface from $(0, 0)$ to $(2, -2)$.

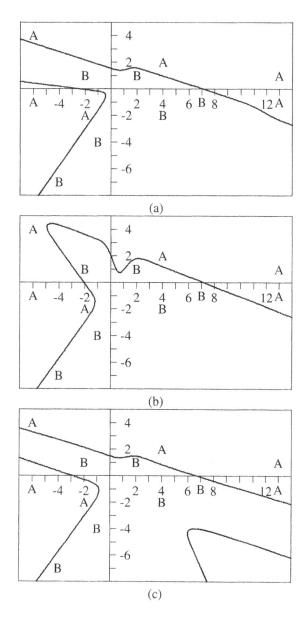

(a)

(b)

(c)

Figure 3.26: These are three different ways three different 2-4-1 back-propagation networks decided to divide up the space using the 12 nonlinearly separable points. If you run a large number of trials, most networks will produce the division shown in part (a), however other divisions also occur as in (b) and (c). Note that in (c), the network made a large area on the lower-right part of class A (less than 0.5), despite the fact that there are no examples of class A nearby.

In Figure 3.26, we show three different ways that different 2-4-1 networks decided to divide the pattern space given the 12 nonlinearly separable points found in Figure 3.4. In all cases the networks tend to see two lines of B points that meet near the center of the space, however, there are major differences between all three solutions. If a very large number of data points was available to train the network, the network is likely to find a mapping that would fit the data very closely.

3.5.2 Approximating Real-Valued Functions

In most instances in this book we will apply back-propagation to problems where the inputs and outputs are 0 or 1, however, it is also possible to use back-propagation for functions where the input and output values can be any real number. For functions with output values beyond the range of the standard sigmoid's range of 0 to 1, the most common solution is to make the output layer activation function linear.

Funahashi [48] has proved that three-layer feed-forward networks can approximate any continuous mapping to any degree of accuracy and Hornik et al. [70] have proven that they can approximate an arbitrary function and its derivatives to any degree of accuracy. Leshno et al. [100] extend the results to show that continuous functions can be approximated if and only if the hidden layer activation function is not a polynomial.

3.5.3 Overfitting

Figure 3.27 shows an example of how backprop can *overfit* a function. The function to be fit is the line, $y = x + 1$ and in the plots this is the straight line shown. The training data consists of seven points that are slightly above or below the ideal straight line and these points are marked with an asterisk. In real world applications it is normal for data points to fall above and below the best line and such data is said to be *noisy*. A test set was made of 301 points along the ideal line.[5] The network chosen was a 1-3-1 network with a linear output unit. After 2,700 iterations of training the fit is quite close to the ideal line, but as training continues the network begins to overfit the data by bending its solution very close to all the points. Notice that while the error continues to go down on the training set, the test set error goes up. Test set error is a much better measure of the performance of a network than the training set error; for serious applications you need a training set and a test set. To minimize overfitting behavior you should have many more training patterns than weights. One way to do this is to use as few hidden layer units as possible. There are other ways to deal with this problem as well.

3.6 Associative Memory and Generalization

An important characteristic of human behavior is that as people gather together a number of cases of some phenomenon or objects, they generalize about them. In the traditional symbol processing approach to artificial intelligence the theory has been that people actually develop explicit rules about the phenomenon or objects. Furthermore, it was generally

---

[5] Using points along the ideal line instead of more noisy points is unrealistic of course, but easy to do.

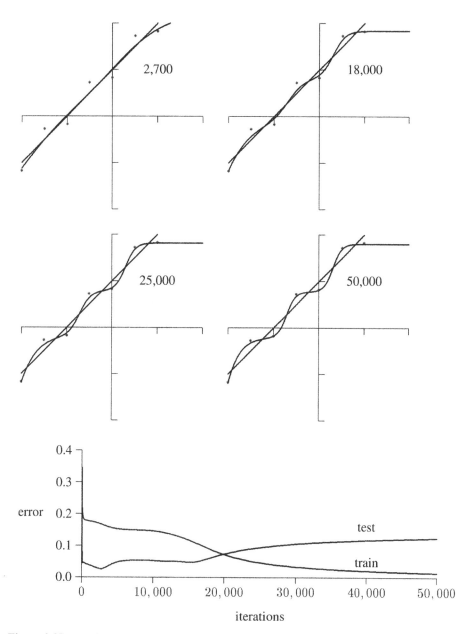

Figure 3.27: The fit for the seven points with a 1-3-1 network at 2,700, 18,000, 25,000, and 50,000 iterations. The graph at the bottom shows the error for the training set and the test set. The minimum on the test set error was at 2,700.

assumed that people then used these rules in something close to a von Neumann style architecture. We noted this in Chapter 1 and the idea will be developed further in the next and later chapters. In this section we will show how neatly neural networks can generalize to give rulelike behavior and do some elementary reasoning without using explicit rules and a von Neumann style architecture. In the neural framework it is the way items are stored in memory that gives rise to generalization and some simple reasoning phenomena. Neural networking researchers believe these methods are more realistic than the traditional symbolic approaches.

3.6.1 Associative Memory

One way to view the XOR back-propagation network introduced in Section 3.4 is that, given some inputs it computes the value of the function, however, another way to view it is as a memory, a device that gives the contents of a memory location when it is given an address (the input pattern). From this perspective it is quite similar to a memory reference operation in a conventional von Neumann style computer, however, when a neural network memory stores and retrieves memories you get some interesting additional effects. Neural networks are a way to implement an *associative memory*. (There are other ways as well.) An associative memory is one that, given a stimulus pattern, produces a response pattern or patterns that have been associated with the stimulus. An associative memory differs from the conventional von Neumann computer memory because an ordered set of stimulus patterns (addresses) is not used and because multiple answers are possible. It is also different in that it can give answers to addresses that are only similar to addresses that it has seen.

```
1010  101  100000000000  01000
1010  010  010000000000  11000
0101  101  001000000000  00001
0100  010  000100000000  00100
1000  101  000010000000  00100
0101  011  000001000000  00001
1010  011  000000100000  11000
0101  010  000000010000  00011
1000  000  000000001000  00100
0100  000  000000000100  00100
1010  100  000000000010  01000
0101  100  000000000001  00001
```

Figure 3.28: The data for an associative memory. The first four bits represent: from Chicago, from New York, Cubs fan, Mets fan. The next three positions represent whether or not the person is a Democrat, a Republican, and likes lemonade. The next 12 bits give each person a unique name. These first 19 bits will be the input pattern (rather like an address). The last 5 bits will be the output of the network and represent whether or not the person is a Sox fan, a Bears fan, likes tennis, is a Yankees fan, and is a Jets fan.

To illustrate some properties of an associative memory we consider an example where the input (address) will be 19 bits. The last 12 bits represent names for 12 different people and the first 7 bits represent properties of these people. Bit 1 is 1 when the person is from

Chicago, bit 2 is 1 when the person is from New York, bit 3 is 1 when the person is a Chicago Cubs baseball fan, bit 4 is 1 when the person is a New York Mets baseball fan. We let bit 5 represent whether or not the person is a Democrat, bit 6 represents whether or not the person is a Republican, and bit 7 represents whether or not the person likes lemonade. The output patterns consist of 5 bits. Bit 1 represents that the person in question is a Chicago White Sox baseball fan, bit 2 represents that the person is a Chicago Bears football fan, bit 3 represents that the person likes tennis, bit 4 represents that the person is a New York Yankees baseball fan, and bit 5 represents that the person is a New York Jets football fan. These inputs and outputs for 12 people are shown in Figure 3.28. The properties of the people involved show the patterns you would expect: people from Chicago are never fans of New York teams and New Yorkers are never fans of Chicago teams. Half the Cubs fans are White Sox fans. Three out of four Mets fans are not Yankees fans. All Cubs fans are Bears fans. All Mets fans are Jets fans. In addition, Chicagoans and New Yorkers who are not baseball fans like tennis. Bits 5, 6, and 7 have little correlation with the last 5 bits, except Republican Cubs fans are also Sox fans. It is easy to train a two-layer network to remember these patterns.

It is not interesting that a network can learn to recall these facts. The interesting part is what happens when you use input patterns, or addresses, that the network has never seen but which are very similar to patterns it has seen. For instance, give the network a nameless person (all name bits = 0) from Chicago who likes the Cubs:

$$1010\ 000\ 000000000000.$$

We get:

$$0.44\ 0.88\ 0.23\ 0.03\ 0.02.$$

The network effectively compares its input with the similar patterns it has experience with and returns a pattern that is close to what those similar input patterns would have produced. We can take these output values to be rough estimates as to whether or not a Cubs fan from Chicago will like the Sox, Bears, tennis, Yankees, and Jets. The results indicate that there is some chance that the person likes the Sox, a high probability that the person likes the Bears, and little or no chance that the person likes the Yankees, Jets, or tennis. This is a rather pleasant surprise considering all the network did was to store "values" at "addresses." If we use Democrat Cubs fans we get:

$$0.11\ 0.88\ 0.14\ 0.01\ 0.02.$$

If we use Republican Cubs fans we get:

$$0.77\ 0.92\ 0.13\ 0.05\ 0.02.$$

For Cubs fans who like lemonade:

$$0.46\ 0.87\ 0.17\ 0.01\ 0.02.$$

Evidently, from the examples it has seen, the network has "come to the conclusion" that Republican Cubs fans also like the Sox while Democrat Cubs fans do not.

We take a look at the network's response to a nameless Mets fan from New York:

$$0.01 \ 0.02 \ 0.18 \ 0.26 \ 0.85,$$

meaning that the person is certainly a Jets fan, just possibly a Yankees fan, but no fan of tennis, the Sox, or the Bears. Also, a nameless New Yorker who does not like the Mets, makes the network think of tennis:

$$0.03 \ 0.05 \ 0.80 \ 0.11 \ 0.26.$$

What the back-propagation algorithm did during training was to notice certain regularities between the input and output sequences. When it now sees "from Chicago" and "likes the Cubs," it activates "likes the Bears" and inhibits "likes tennis," and so forth. The network found that being a Republican, Democrat, or lemonade liker had no correlation with being a Bears fan. The network is behaving *as if* it had *explicitly* come to the conclusion that the following rule applies:

> IF   the person is a Chicagoan and
>   the person is a Cubs fan
> THEN   that person is a Bears fan,

while all it was doing was trying to look up a value at an address. We may say that networks like these have the important ability to generalize from their experience and come to know the important characteristics of a class of objects. Notice also, that the network came up with, we might say, fuzzy sorts of rules, like, that the typical Chicago Cubs fan is sometimes a Sox fan and that a New York Mets fan is not usually a Yankees fan.

One problem with the linear pattern classifier network and the nearest neighbor algorithm is that they assume that there are specific places at which to divide up the space between pattern classes. In real life problems the divisions are not always so easy. For instance, one highly qualified loan officer may decide to grant a loan where another equally qualified one will deny it. In pattern recognition terms, there is a region of space where the decision is uncertain and it is unrealistic to find a hard division between two classes. However, back-propagation networks produce smooth surfaces that will minimize the overall error and then the numbers that come out of the network correspond to an estimate of probability. This shows up in the above network's ability to estimate whether or not Cubs fans are Sox fans or that Mets fans are Yankees fans.

It is important to note, however, that back-propagation networks do not always find the generalizations that you expect them to find. For instance given the 12 nonlinearly separable points used in creating Figure 3.26, three different runs of back-propagation produced three different divisions of the space that no human being is likely to produce. One simple safeguard in using networks is to train a number of them on the same data, submit the same unknown patterns to them for classification, and then average the results. This is the analog to collecting a set of opinions from a number of human experts (say, for instance, doctors) and weighing them. For a mathematical analysis of this see the article by Perrone and Cooper [150].

Of course, another interesting aspect of the problem is that networks that *do not* find the generalizations that people would expect may actually find better generalizations than people or provide people with an entirely different perspective on the situation. Remember, too, that people have often produced incorrect generalizations, such as that the Earth is flat or that the fundamental elements are earth, air, fire, and water. The scientific method of observing data, forming a theory, *and testing as many other cases as possible* is designed to minimize such errors. It cannot be emphasized enough that backprop is not capable of uncovering every sort of regularity that can be found in a set of data. It has very specific limitations. For a discussion of these limitations and some ways they can be overcome see [19].

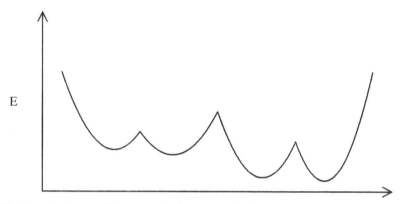

Figure 3.29: A rough picture of the energy landscape near the four Cubs fans. The valley on the left contains the two Democrat Cubs fans and the valley on the right has the two Republican Cubs fans.

In the above sports example we chose to emphasize the memory lookup capability of a network by declaring the input to be an address and the output to be the contents of the address. It is also possible to use back-propagation to produce pattern completion networks for use with the Hopfield and Boltzman recall algorithms. We can put all 24 characteristics of each sports fan on the input layer of a symmetric network and require the same pattern to appear on the output units. After training, we can put part of a pattern on the input units and the network will try to complete the pattern on the output units. A network that simply tries to reproduce its input pattern on the output units is called *auto-associative*. (A network can be auto-associative without being a Hopfield network.) We can also use the Boltzman machine recall algorithm to complete the pattern. For instance, we may give the recall program the pattern:

$$1010 \ 000 \ 000000000000 \ 00000$$

and through random updates of the Boltzman machine relaxation algorithm, the system will settle into a minimum having the characteristics of one of the four Cubs fans. If the cooling is done enough times, the network will recall the characteristics of all four Cubs fans. Thinking in terms of computational energy, the four Cubs fans occupy a valley in the 25-dimensional computational energy landscape. Within that valley is a subvalley for the two Republican Cubs fans and a subvalley for the two Democrat Cubs fans. This is illustrated in Figure 3.29. Each subvalley has two minimums in it. Similar valleys exist for Mets fans and tennis fans.

3.6.2 Local and Distributed Representations

There are two representations for objects that can be used in neural networks. They are local and distributed representations. The term *parallel distributed processing* is derived from the distributed representation. Distributed representations have the important ability to generalize from patterns and we will now look at the differences between local and distributed representations in detail. At first glance, the distributed and local representations appear quite different, yet under certain circumstances the differences are not clear-cut.

A good example of a local representation is the Waltz and Pollack activation network seen in Section 2.4. In that network, different and complex concepts such as hunting, gambling, John, bucks, and deer are each allocated a specific node. By way of contrast, using a distributed representation, we might code the pattern for a deer, not as a single unit, but as a pattern of activity over a whole series of units. For instance the pattern for a deer could be:

$$000\ 000\ 011\ 000\ 111\ 010\ 101\ 0.$$

Each of these 22 units represents a characteristic of animals in general. Some of these features might be color, size, has four legs, has two arms and two legs, has a tail, and so on. If this distributed type of representation is used by people, we would expect there to be many more than 22 such features, including possibly a pattern representing a mental picture of the animal, a pattern representing the sound of the name of the animal, a pattern representing how to speak the name of the animal, and so on. Each feature represents a small part of the animal and in a distributed representation each small feature is called a *microfeature*.

chimpanzees:	111 000 000 111 000 000 000 0
gorillas:	000 111 000 111 000 000 000 0
orangutans:	000 000 111 111 000 000 000 0
dogs:	000 000 000 000 111 111 000 0
cats:	000 000 000 000 111 000 111 0

Figure 3.30: The distributed representations for five animals.

An important aspect of the distributed representation is that it is quite easy to represent many animals by just listing their characteristics in a vector. For instance, we might list the characteristics of chimpanzees, gorillas, orangutans, dogs, and cats as shown in Figure 3.30. If we used this representation in a program it is quite easy to add new animals simply by producing new codes listing their characteristics. This is in marked contrast to what would be necessary if the memory was using a local representation. In that case, for each new animal added, a new node would need to be created and links would need to be created connecting the animal with its characteristics.

If the differences between the two representations were as clear-cut as it sounds from above, all would be well, however, there are situations where the differences are not so obvious. Suppose we have a network as shown in Figure 3.31 with nodes for chimpanzee, gorilla, and orangutan plus nodes for their 22 characteristics as shown in Figure 3.30. The nodes numbered 1 to 22 represent the 22 microfeatures of animals. There are activation and inhibition links between the animal units and the microfeature units and we assume the

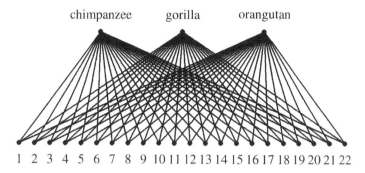

Figure 3.31: A network that seems to use local representations, but if you light up the chimpanzee unit and apply the interactive activation algorithm to the network, a set of feature nodes (1-22) will also come on so 'chimpanzee' ends up looking like a distributed representation after all.

network uses the interactive activation method. If we activate the chimpanzee node then units 1, 2, 3, 10, 11, and 12 will come on giving, in effect, a distributed representation in the network. Thus, distinguishing between local and distributed representations can be a problem at times.

3.6.3   Reasoning within a Network

We can now look at the ability of distributed representations to do some elementary reasoning. We will use the data in Figure 3.30 about chimpanzees, gorillas, orangutans, dogs, and cats and we designate the 22nd bit as indicating whether or not the animal likes onions. The data at the moment shows that none of them do. However, suppose that someone tells us that chimpanzees do in fact like onions. As human beings, we would also suspect that gorillas and orangutans also like onions because these animals are all quite similar. Dogs and cats, however, are very different from apes, so knowing that chimpanzees like onions will not raise our expectations about dogs and cats liking onions by very much, if it raises them any at all. If we later learned that gorillas and orangutans do not like onions we would file that away while still leaving liking onions as a unique characteristic of chimpanzees.

To demonstrate that a network can also do the reasoning process described above we will first train a network using back-propagation to remember the characteristics of the five animals in Figure 3.30. Column 1 of Figure 3.32 shows the activation values for the "likes onions" unit and they are all very close to 0. Now, to have the network learn that chimpanzees do like onions, we retrain this *existing* network of weights with only one pattern rather than with the complete set of five patterns:

$$111 \ 000 \ 000 \ 111 \ 000 \ 000 \ 000 \ 1$$

This pattern is the same as we had before for chimpanzees except in this case we have chimpanzees liking onions. The network will use those units that are 1s in the input layer to turn on the "likes onions" unit in the output layer. Column 2 of Figure 3.32 clearly shows that chimpanzees like onions and that orangutans and gorillas register an increased probability that these animals like onions. There is very little indication that dogs and cats like onions.

	"likes onions" after initial training	"likes onions" after learning that chimps like onions	"likes onions" after learning that gorillas and orangutans do not like onions
chimpanzees	0.02	0.96	0.68
gorillas	0.02	0.37	0.08
orangutans	0.02	0.36	0.07
dogs	0.02	0.05	0.04
cats	0.02	0.05	0.04

Figure 3.32: This data shows how a network can so some elementary reasoning. The slight changes for dogs and cats are due to changes in the weights from the bias units.

We can now tell the network that gorillas and orangutans do not like onions by training it on two patterns, one for gorillas and one for orangutans. Column 3 of Figure 3.32 shows the results. The estimates for gorillas and orangutans liking onions has been lowered to below 0.1, while the rating for chimpanzees remains quite high.

The automatic reasoning ability of the above network was made possible by the fact that each animal was represented as a pattern of activity distributed over a number of units. Knowledge about one animal spills over to the other animals to the extent that their active input units overlap. The more the overlap the more the spillover.

The above demonstration made use of another interesting property of this type of network. All the patterns do not necessarily always have to be added to the memory all at once. They can sometimes be added one at a time and up to a certain point they will not get in each other's way. This is possible because when a new pattern is added there are many weights to be changed and therefore each weight only needs to be changed by a little bit to store the new pattern. In addition, the total effects of all the small changes are likely to cancel each other out. For these reasons, new patterns can often be added without destroying the old ones. There are instances, however, when storing one extra pattern *will* destroy the facts in the original network. With only 22 units for this example, the changes to each weight were still fairly significant. However, by adding one fact at a time, eventually a network will reach a point where old patterns get weaker and weaker and finally the old patterns will be lost forever.

The fact that in parallel distributed networks the generalization of knowledge comes automatically just as a consequence of storing and retrieving knowledge from a network has made some researchers think that this is a better model of generalization than the symbolic approach of constructing actual rules. These networks produce rulelike behavior without the expense of producing actual rules.

3.7   Applications of Back-Propagation

To illustrate the usefulness of back-propagation networks we will now look at some of the early applications as well as mention a few experimental systems. In addition to the applications we mention here, there are very many applications of back-propagation being made to very many types of problems that can be solved by doing a single step of pattern recognition.

3.7.1 Interpreting Sonar Returns

One experimental back-propagation system by Gorman and Sejnowski [52, 53] was designed to classify sonar targets. The two targets used in the experiment were a metal cylinder and a cylindrically shaped rock. Both targets were about five feet long and both were placed on a sandy ocean floor. Sonar returns were collected from the objects at a range of 10 meters and from various angles. Data for each signal was preprocessed in a manner thought to be similar to how human beings hear sound patterns (see [53]) and then it was normalized to fit between 0.0 and 1.0.[6] There were 60 input values. These patterns were used as input to networks with 0, 2, 3, 6, 12, and 24 hidden units. The output layer consisted of two units, one for the rock and one for the cylinder.

	Aspect-Angle Independent Series		Aspect-Angle Dependent Series	
Number of Units in Hidden Layer	Average Performance on Training Sets	Average Performance on Testing Sets	Average Performance on Training Sets	Average Performance on Testing Sets
0	89.4%	77.1%	79.3%	73.1%
2	96.5%	81.9%	96.2%	85.7%
3	98.8%	82.0%	98.1%	87.6%
6	99.7%	83.5%	99.4%	89.3%
12	99.8%	84.7%	99.8%	90.4%
24	99.8%	84.5%	100.0%	89.2%

Figure 3.33: The results of two series of tests done by Sejnowski and Gorman to train networks to distinguish between the sonar returns from a metal cylinder and a cylindrically shaped rock.

Two series of experiments were performed, one set using angle independent data and another using angle dependent data. The results are shown in Figure 3.33. These results were averaged over 130 networks for each number of hidden units. The performance in this series of tests approached 100 percent on the training data and 84.5 percent on testing sets that the program had not seen. Three-layer networks with 12 and 24 units gave the best performance. An analysis by Sejnowski and Gorman found that this relatively poor performance was due to the fact that there were examples of patterns used in the testing set that were not found in the training set. In addition to doing the back-propagation analysis, the authors used another technique to estimate the performance of a nearest neighbor classifier and found it would be 82.7 percent correct on the unknown patterns. In the angle

---

[6] Backprop networks can accept any real values as input, however, the best results come when the magnitude of the inputs is more or less 1.

dependent series of tests the networks performed near 100 percent on the training set and up to 90.4 percent on the test sets. Again, performance reached a peak at 12 hidden units.

In tests on human subjects, the human subjects tended to perform from slightly better than to slightly worse than the networks but there was less testing done with the human subjects than with the networks.

3.7.2   Reading Text

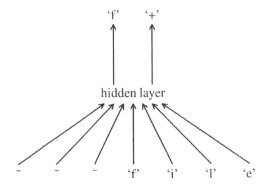

Figure 3.34: The general structure of NETtalk. There are seven letters in the input layer. In this case, the first three letters are blank (˜). The output layer codes for the correct pronunciation of the middle letter of the seven. The left set of output units represents the identity of the phoneme and the right set codes the stress field. In effect, the network has to learn to give the proper response for the input letter 'f' given the fact it is surrounded by three other items on each side.

Another experimental network, this one researched by Sejnowski and Rosenberg, is known as NETtalk [199]. NETtalk consists of a network of processors that learns to read aloud. The program has been interfaced to a speech synthesis system. With the speech synthesis system engaged, the program babbles like an infant to start with, but after 16 hours of practice it learns to read aloud at the level of a 6-year-old child. The general arrangement of NETtalk is shown in Figure 3.34. It is a three-layer network where the input units contain a seven letter portion of a word (called a window) and the output units contain first, a code for the proper pronunciation (the phoneme) for the letter in the middle of the seven input letters and second, a stress field for the middle letter. The stress field takes on the value "1" for primary stress, "2" for secondary stress, "0" for unstressed, "+" for rising, and "-" for falling. Figure 3.34 shows the network looking up the pronunciation for the "f" in "file." The first three characters on the input layer are blank. For patterns selected from the training set, NETtalk produces the correct phoneme 94 percent of the time. For patterns it has not seen before it gets the correct phoneme 78 percent of the time. The number of correctly pronounced words is much less. A system similar to NETtalk was analyzed by Seidenberg and McClelland [198] and they found that it behaved much the same way as people.

Waltz and Stanfill have also attempted this pronunciation task using another method in a system called MBRtalk. MBRtalk is described briefly in Section 7.2. Waltz and Stanfill point out that perfect or near perfect results with the methods employed by NETtalk and MBRtalk are not possible for several reasons. First, while these systems can learn regularities, there are also many words with irregular pronunciations that a system cannot know about unless it has been specifically told about them. Second, English has borrowed many words from other languages. Unless the system knows the language in which the word originated it cannot guess the correct pronunciation. Waltz and Stanfill cite such examples as targET versus filET, piZZa versus fiZZy, and vILLA versus tortILLA.

In addition to MBRtalk and NETtalk, Wolpert [264] has produced a system simpler than MBRtalk that he says "has a generalization error rate of less than 1/3 of that of NETtalk" and this system is described briefly in Section 7.2. Another system for reading text aloud is DECtalk designed by Digital Equipment Corporation. DECtalk was constructed using conventional programming techniques and it is superior to NETtalk, however, DECtalk required 15 years to develop while NETtalk was developed over a summer.

3.7.3   Speech Recognition

Speech recognition is another important area where back-propagation networks may be useful. Waibel et al. [241, 242, 243, 244] have been working on the problem of training networks to recognize spoken sounds. In one early set of experiments a specialized back-prop network was trained to recognize the sounds, b, d, g, p, t, and k and the network managed to get around 98 percent correct on a test set.

3.7.4   Detecting Bombs

One back-propagation-based system in actual use was commissioned by the Federal Aviation Agency and produced by Science Applications International Corporation (see Shea et al. [205, 206]). The system is designed to detect explosives in suitcases by quickly scanning the suitcases in a normal airport environment. In particular, the object is to detect unusual amounts of nitrogen in the suitcases because most ordinary explosives contain large amounts of nitrogen. The system works by first bombarding the luggage with low energy neurons. The neutrons are absorbed by atoms within a suitcase and the atoms then emit gamma rays at various energy levels. Each type of atom will emit a characteristic set of gamma rays which are measured and the results of the measurements are then submitted to a back-propagation network for analysis.

	Percent Explosives Detected	False Alarms
linear discriminant	98.0	11.6
back-propagation	98.0	7.8

Figure 3.35: The explosive detection rates and false alarm rates for a linear discriminant solution and back-propagation.

As in so many real world applications, it was not possible to get perfect detection. Sometimes a bag with explosives will not be detected and sometimes a bag without explosives will be flagged as containing explosives. The goal in the project was to detect at least

90 percent of the suitcases that contain explosives while minimizing the number of false alarms. A practical aspect of such systems is that you can increase the probability of detection but at the cost of increasing the number of false alarms. In operation, every piece of luggage that is suspect must be examined in more detail and this is very time consuming. There were two major approaches that were tried, a linear discriminant analysis system (a linear pattern classifier) and the back-propagation network. Both proved to be equally good at detecting bags with explosives, but the back-propagation system did much better at eliminating false alarms. Some tests were made at John F. Kennedy airport in October 1989 using actual luggage and luggage with simulated explosives and they gave the results shown in Figure 3.35. These results came from a network with one hidden layer. Some experiments were run with a four-layer network where the probability of a false alarm was decreased by one percentage point, but the training required 14,000 to 20,000 cycles, whereas the three-layer network only needed 2,000 to 4,000 cycles.

3.7.5 Economic Analysis

A number of systems have been designed to do economic analysis ranging from stock market analysis, to commodity futures analysis, to rating bonds, to loan underwriting. Kimoto, Asakawa, Yoda, and Takeoka [83] report producing a stock market prediction system that "showed an excellent profit." On the other hand, White [255] attempted to use networks to predict the daily rates of return on IBM stock and he reported that in the experiments he tried, the networks have failed to uncover any predictable fluctuations. Refenes et al. [168] report that neural networks perform better by an order of magnitude than classical statistical techniques for forecasting within the framework of the arbitrage pricing theory model for stock ranking.

In another example of economic analysis, Collard [21] trained a network on commodity futures data using a year's worth of data. The network was then tested on another nine months of data. Collard reports that, given an initial investment of $1,000, the network would have made a $10,301 profit.

Dutta and Shekhar [207] have experimented with networks to rate bonds. They report that their networks function better than the typical mathematical procedures used by bond raters, but the networks do not always produce the same rating that human bond raters produce. They also found that two-layer networks performed as well as networks with hidden layers. Experiments done by Surkan and Singleton [225] found that three-layer networks performed better than discriminant analysis at rating bonds and four-layer networks were better than three-layer networks.

Smith [215] produced a network to do loan underwriting and he reports that "It is delivering an 18 percent increase in profit for its user, compared with the performance of previous decision technology (a point scoring system based on multivariate discriminant analysis)."

McCann [106] used a recurrent backprop network to predict the highs and lows of the gold market.

3.7.6 Learning to Drive

In a project conducted by Pomerleau, Jochem, and Thorpe [160, 161, 162, 163, 164, 74], back-propagation networks have been trained to drive a car down interstate highways, one and two lane roads, and suburban neighborhood streets. The project is called ALVINN for Autonomous Land Vehicle In a Neural Network and it is used to drive a vanlike vehicle called NAVLAB. Pictures from a video camera on the vehicle are digitized to form a 30 × 32 input matrix. This is fed into a hidden layer with 4 units and then to a 30 unit output layer. The output layer units code for how hard the network should steer to the right or left. The system learns to drive by monitoring the driving of a human being for about three minutes. For each different type of road a different network must be trained and the system must monitor which network is giving the best results on how to drive. It switches between networks as necessary to give the best performance. The higher the degree of confidence the system has, the faster the system can drive. ALVINN has managed to drive up to 70 miles per hour and has averaged 60 miles per hour over a 90 mile stretch of highway. Research is now being done on getting ALVINN to handle more complex driving situations such as intersections, off-ramps, and changing lanes. To do this the researchers are working on dynamically moving the camera as well as narrowing its field of view so it can concentrate on the important details of the scene.[7]

3.7.7 DNA Analysis

Lapedes[8] has trained a back-propagation network to predict whether or not short DNA fragments contain codes to produce proteins. Its accuracy rate is 80 percent rather than the 50 percent obtained with conventional pattern recognition techniques. The network was trained by presenting it with 900 examples of fragments that do code for protein production and 900 examples that do not. Lapedes says that the network appears to have learned some fundamental rules about genetics that may possibly have eluded biologists.

3.8 Additional Perspective

In this chapter we have looked at a few of the most common and useful pattern recognition algorithms. For the most part, these algorithms were chosen because they will be useful in later chapters. In particular, back-propagation can be modified to handle an amazing number of tasks. However, over the years a very large number of pattern recognition algorithms have been proposed and used and it is not possible or useful in an AI book to look at any more of them. Even though back-propagation is only a relatively recent addition there are already an amazing number of variations on back-propagation that have been developed and most of these give better results than backprop.

---

[7] Some of these details were taken from a Usenet posting in comp.ai.neural-nets by Dean Pomerleau, Message-ID: <753213973/pomerlea@POMERLEA.BOLTZ.CS.CMU.EDU>, Date: Sat, 13 Nov 1993 18:06:00 GMT. More information on ALVINN and related research can be found in the Robotics Institute Technical Report WWW page at http://www.cs.cmu.edu/afs/cs.cmu.edu/project/alv/member/www/navlab_home_page.html.

[8] This report comes from *Science News*, Volume 132, Number 5, August 1, 1987, page 76.

3.9 Exercises

3.1. Without using a program show that a linear pattern classifier cannot do the XOR problem, that is, it cannot put the XOR outputs of +1 in one class and the XOR outputs of 0 in another class. Let the 0 outputs be –1.

3.2. Without using a program determine whether or not the following patterns can be classified correctly using the linear pattern classifier:

input	output
1 0 0	1
0 0 0	–1
0 1 0	1
1 1 1	–1

3.3. Show that if a three or more layer backprop network uses ONLY the linear activation function that it is equivalent to a two-layer network, so that nothing can be gained in terms of pattern recognition ability by having more than two layers if a network uses linear neurons. This is particularly easy to show if you use matrix algebra.

3.4. Program the linear pattern classifier[9] of Section 3.1 and train it to learn the difference between an E and an F using the data shown below. After training, examine the weight vector to see how the program discriminates between an E and an F.

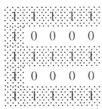   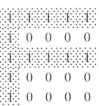

3.5. In Section 2.1 the letter recognition algorithm looked for four different subpatterns in each of nine regions and then applied a formula to determine the unknown. Now, program the same problem but use the simple Euclidean nearest neighbor algorithm described in Section 3.2. Use at least eight different patterns. Test the program with a number of letters that are distorted from their ideal shapes. If you programmed Exercise 2.1 compare those results with the results you get here.

Another variation you may want to try is to collect a large number of patterns of just two very similar hand-printed letters (for instance, E and F, C and G, P and R, F and P, and so forth). Use the nearest neighbor algorithm and put about half the patterns into the database and use the other half as test cases. You can also repeat this with DSM and the back-propagation algorithm described in Section 3.4.

3.6. Starting with the network in Figure 3.7 and the activation values shown there, update the units in this order: b, c, d, a, b, c, d. Also try this series of updates: a, c, d.

---

[9] Available online.

3.7. Program the Boltzman machine and Hopfield network relaxation algorithms.[10] Use the following patterns of the letter J and the number 4:

$$
\begin{array}{cccccccc}
0 & 0 & 0 & 0 & 1 & & & \\
0 & 0 & 0 & 0 & 1 & & & \\
0 & 0 & 0 & 0 & 1 & & & \\
1 & 0 & 0 & 0 & 1 & & & \\
1 & 1 & 1 & 1 & 1 & & & \\
\end{array}
\qquad
\begin{array}{ccccc}
1 & 0 & 0 & 0 & 1 \\
1 & 0 & 0 & 0 & 1 \\
1 & 1 & 1 & 1 & 1 \\
0 & 0 & 0 & 0 & 1 \\
0 & 0 & 0 & 0 & 1 \\
\end{array}
$$

Storing the weights in a 25×25 matrix will be convenient or C programmers can use malloc to create arbitrary size arrays. Run simulations to determine how often the following patterns end up in the global minimum as a function of how long the network stays at each temperature. In addition, vary the threshold value to see how it affects your results.

$$
\begin{array}{ccccc}
0 & 0 & 0 & 0 & 0 \\
0 & 0 & 0 & 0 & 1 \\
0 & 0 & 0 & 1 & 1 \\
0 & 0 & 0 & 0 & 1 \\
0 & 0 & 0 & 0 & 0 \\
\end{array}
\qquad
\begin{array}{ccccc}
0 & 0 & 0 & 0 & 0 \\
0 & 0 & 0 & 0 & 1 \\
0 & 0 & 0 & 0 & 1 \\
1 & 0 & 0 & 0 & 1 \\
0 & 0 & 0 & 0 & 0 \\
\end{array}
$$

3.8. Philosophical Question: Is the concept of finding and getting stuck in a local minimum applicable to other aspects of the world? For instance, do people, societies, and nations get stuck in local minima so that they become incapable of change, even change for the better? Could alcoholism be a local minimum? Are these valid examples and are there other examples? If you think these are valid examples, then how do you get people, societies, and nations out of a local minimum?

3.9. Use backprop to solve the XOR problem, then vary at least one of the weights to see how the error changes. Plot the error for a wide range of values around the minimum value found by the network.

3.10. With the XOR problem, vary η and α and the range of the initial weights and see how few iterations you need to solve the problem. To get reliable estimates, average at least 10 or better still, 25 cases for each parameter setting. If you are more ambitious, do the same for as many different variations of the backprop algorithm as you have available. (Doing a large number of simulations will take some time.)

3.11. Given the following set of weights for the XOR network shown in Figure 3.14, sketch the surface the network found to solve the problem:

```
 7.1252470016   * from n1 to n3
 7.1284189224   * from n2 to n3
-2.7686207294   * n3 bias weight
-4.9701967239   * from n1 to n4
```

---

[10] Also available online.

```
     -4.9693980217   * from n2 to n4
     10.9574527740   * from n3 to n4
     -3.3010675907   * n4 bias weight
```

3.12. Give the assumptions the author made in drawing the line separating the two classes in Figure 3.4 that the back-propagation networks did not make in Figure 3.21.

3.13. Within 6-bit binary integers there are eight palindromes, bit strings that read the same from left to right and right to left. Use back-propagation with two units in the hidden layer to create a network to recognize the 6-bit palindromes. Explain how the network works. You will probably need to train the network on all 64 possible 6-bit patterns. (Also see the next exercise.)

3.14. Neural networks do not always generalize correctly. For example, the data below contains the eight 6-bit palindromes plus some nonpalindromes. Use the patterns and the initial weights shown below and train a back-propagation network with $\eta = 0.5$ and $\alpha = 0.0$. (If you use the integer-based version of the software available with the text it should take exactly 151 iterations to train the patterns to within 0.1 of their targets and the network will not generalize correctly for all 64 patterns. This will happen most of the time for other parameters and initial weights as well.) Test the resulting network on nonpalindromes and see how it classifies them. Try to figure out how the network does classify patterns.

```
m 6 2 1       * use a 6-2-1 network
A dd up       * use the differential step size derivative term and
              * periodic updates
e 0.5         * eta
a 0.0         * alpha
n 15          * the 15 patterns
0 0 0 0 0 0     1
0 1 0 0 1 0     1
0 0 1 0 0 1     0
1 0 1 1 0 1     1
1 0 0 0 1 1     0
0 1 0 1 0 1     0
1 1 1 1 1 1     1
0 0 1 0 0 0     0
0 0 1 1 0 0     1
1 0 0 0 0 1     1
1 1 0 0 0 1     0
1 1 0 0 1 1     1
0 0 0 1 1 1     0
1 1 0 0 0 1     0
0 1 1 1 1 0     1

R             * restore the file of weights
r 152 152     * run 152 iterations (stopping after 151 updates)
```

Here are the weights to start the program with:

```
Or file = pal
```

```
 0.1494140625
-0.3457031250
 0.0507812500
 0.4843750000
-0.1660156250
-0.3867187500
 0.3945312500
-0.0048828125
-0.3603515625
 0.1132812500
-0.2812500000
-0.4521484375
 0.2226562500
 0.2783203125
 0.2304687500
-0.1962890625
 0.0683593750
```

3.15. Train a back-propagation network to learn $sin(x)$.

3.16. Store the data for the 12 sports fans in Section 3.6 in a Hopfield network matrix and save these weights. You can either use the simple Hopfield matrix algorithm given in the text or use the back-propagation software that is available with the text (or any such software). With the back-propagation software, reasonable learning parameters are $\eta = 0.25$ and $\alpha = 0.5$. A threshold value of 3 will work well whichever method you use. Input the weights to a Boltzman machine simulator and give the simulator the pattern that represents a person from Chicago who likes the Cubs and have the program do several dozen or so relaxations. The purpose of this is to have the program complete the Cubs fan pattern by supplying the rest of the data for each Cubs fan. Because the updates are done at random, sometimes the process will "remember" one Cubs fan and sometimes it will remember another. Note how often each person comes up. Explain why some names come up more often than others. If people also remember things using a Boltzman machine algorithm then does your explanation have any implications for human memory?

3.17. Suppose we have a simple two-dimensional pattern classification problem with two classes divided up like so:

If the radius of the circle is 1 and the length of the side of the square is approximately 2.5, then the areas for both classes 1 and 2 are approximately equal. With this configuration it is easy to generate points at random and assign them to one class or another. Train 2-n-2 backprop networks with 50, 100, and 1,000 training points and test the network with 1,000 points. Vary n as well as the algorithm and algorithm parameters to get very close convergence for the training set. While training, test the network at regular intervals to see how well the network is doing. Repeat the training a number of times for each parameter setting to get a good average value for how long the training takes. Graph the results. (This exercise can take a lot of human and computer time.)

3.18. Do Exercise 3.17 with the same training and testing points but use the nearest neighbor algorithm to classify patterns. Compare your results with those from back-propagation. (This will go much faster than Exercise 3.17.)

3.19. Do Exercise 3.17 with the same training and testing points but use the Decision Surface Mapping (DSM) algorithm. Compare your results with those from back-propagation and the simple nearest neighbor approach. (This, too, will go much faster than Exercise 3.17.)

3.20. If you have some real world data of any sort, such as weather data, stock market data, sports data, or scientific data, apply back-propagation and/or DSM and/or the nearest neighbor algorithm to see how well the algorithm can predict answers. A lot of real world data is available via the Internet including the Gorman and Sejnowski sonar data.[11] For other real world data you can look, for instance, in the directory, pub/machine-learning-database on the system ics.uci.edu and the comp.ai.neural-nets FAQ.

---

[11] The sonar data comes with the backprop software found at http://www.mcs.com/~drt/svbp.html.

Chapter 4

Rule-Based Methods

4.1 Introduction

In this chapter we will look at some standard symbol processing techniques, in particular those techniques associated with rule processing. In order to do this we will first look at some elementary Prolog so that it can be used as a notation both in this chapter and in later chapters for symbol processing algorithms. Prolog was chosen for this task rather than the also very popular symbol processing language, Lisp, because Prolog has built-in pattern matching capabilities that Lisp does not have. Prolog gets much of its power by using rules so the language itself is an illustration of rule processing.

In this chapter we will look at examples of a type of AI program known as an *expert system*. An expert system is an AI program that is capable of doing the work of a highly skilled human expert. The programs mentioned in Section 3.7 that rate bonds or analyze sonar echos would be considered expert systems, while a program that can recognize letters of the alphabet would not normally be considered an expert system since any human being can recognize letters of the alphabet. The classification of expert/nonexpert system is made this way even if the internal operation of the letter recognizer and the expert system is the same. A number of famous rule-based expert systems will be described. A cautionary note needs to be kept in mind regarding expert systems. It is that they are rarely as good as genuine human experts. The most famous criticism of the abilities of expert systems comes from Dreyfus and Dreyfus [28] where they list instances of expert systems that are not as good as the best human experts. They maintain that because expert systems are not as good as the best human experts they should not be called *expert* systems at all, but only *competent* systems. Furthermore, they argue that real human experts do not use rules.

4.2 Some Elementary Prolog

Logic has been one of the most appealing approaches used to try to produce artificial intelligence. One of the reasons for this is that people are often viewed as being logical creatures. Another reason is that logic is the means for doing mathematical proofs and many computer scientists are former mathematicians. A third reason is that it is very easy to program computers to do simple examples of logic. The most commonly used form of logic is *predicate calculus*. Predicate calculus statements use a formal functionlike notation to give facts (statements) about a problem. Working with these statements, you can prove the truth or falsity of other statements. Statements that you try to prove correct are often called theorems and the calculating of new results is therefore often called *theorem proving*. It is also called *automated reasoning*. The greatest theoretical success in this area of automated reasoning is a proof technique known as Resolution. The next chapter considers the more general case of Resolution. Prolog is a programming language that uses only a subset of the Resolution technique, it basically consists of processing simple rules of the form:

```
if A and B then C
```

and so it is quite easy to understand a Prolog program without studying Resolution.

4.2.1 Stating Facts

Prolog is a programming language very much unlike conventional languages such as Pascal and C. One of the major differences is that it has few of the traditional control structures found in conventional languages. For another, the usual method of using Prolog is interactive. Statements in Prolog represent facts or rules that are true. Here is a series of facts, stated in Prolog:

```
/* 1 */    likes(matt,mets).
/* 2 */    likes(carol,cubs).
/* 3 */    likes(bob,cubs).
/* 4 */    likes(bob,bears).
/* 5 */    likes(mary,mets).
/* 6 */    likes(mary,yankees).
/* 7 */    likes(nancy,lemonade).
```

The numbers between /* and */ are comments and are not required. These statements are quite English-like and they simply look like sentences where the verb has been removed from the middle of the sentence and placed in the front. All statements and rules must end with a period. We define the first statement to mean "Matt likes the Mets." Usually, most people will translate English to Prolog this way by simply moving the verb to the front of the statement, however, this is not always done, so a Prolog programmer might define the first statement to mean "The Mets like Matt."

4.2.2 Syntax

Prolog programs consist of *terms*. A term is either a *constant*, *variable*, or a *structure*. Constants are *atoms* or *numbers*. Numbers are integers or reals. Atoms consist of any sequence of characters enclosed in single quotes (') such as these atoms:

> 'St. Louis Cardinals'
> '095'
> '+-+'

or an atom can consist of a sequence that begins with a lowercase letter followed by letters, digits, and the underscore character (_) such as these atoms:

> st_louis_cardinals
> stl
> x99

or an atom consists of a sequence of the special characters:

$$+ - * / \backslash \, \hat{} > < = ` \sim : . ? @ \# \$ \&$$

such as:

```
->
:-
?-
```

Variables are another type of term and they start with an uppercase character or an underscore followed by letters, digits, and underscores such as the following:

```
X
Team
_abc
St_Louis

_
```

The variable consisting of just an underscore by itself (_) is called the *anonymous variable* and it is reserved for a special use.

The final type of term is a structure, such as in the facts we have already seen. In the term:

```
likes(matt,mets)
```

`likes` is called the *functor*, and `matt` and `mets` are its components. When a functor is used to structure data it is called a functor (an example will come up later), however, when a functor is used to express facts or rules it is usually called a *predicate*.

4.2.3 Asking Questions

The set of facts, or clauses, 1 through 7 above might be contained in a file and be read by a Prolog command, or the user might type these facts directly into the Prolog system. With this set of facts in memory, we can go on to ask questions such as the following, where the ?- at the beginning of the line represents a prompt for a question that is printed by the system:

```
?- likes(nancy,lemonade).
```

In this request, we have asked the system if Nancy likes lemonade. Prolog attempts to answer this question by simply looking through its database of facts for this particular fact. It starts at the top of the database and tries to match this one question against the set of facts in the database one after another until it either finds the fact, or finds the end of the database. In this case, Prolog responds:

```
yes
```

If the question had been:

```
?- likes(nancy,mets).
```

Prolog would respond with:

```
no
```

This "no" means that, *given the facts that Prolog has available to it, it cannot prove that Nancy likes the Mets. This is different from proving that Nancy does not like the Mets.*

Here is a slightly more complicated question that we can ask the system:

```
?- likes(X,cubs).
```

This question means: "Is there any person, X, who likes the Cubs? If so, report the name of that person." Initially, the variable, X, has no value and is said to be an *uninstantiated* variable. In the pattern matching process an uninstantiated variable can match anything. This request is also a simple pattern matching problem. Prolog starts at the top of the database and looks to see if it can match this pattern with anything. The first fact fails to match. The second fact does match the pattern. The variable, X, now becomes an *instantiated* variable and has the value, "carol" and Prolog prints out:

```
X = carol
```

The interpreter leaves the cursor at the end of this line and waits for the user to respond. If the user types a carriage return, Prolog assumes the user is happy with this one response and gives the prompt for another question. If the user types a ";" at this point, Prolog goes further on into the database looking for another possible solution. We type the ";" and Prolog continues searching from where it left off and X loses the value "carol" and again becomes an uninstantiated variable. The third item matches the pattern so Prolog prints out:

```
X = bob
```

Prolog again waits for a response from the user. A carriage return would end this search for solutions. A ";" will cause Prolog to search further. If we type the latter, it cannot find any more matches so it reports, "no" and prompts the user for another question. As a special case, the anonymous variable matches anything but it is never instantiated to any value. So, to ask if anybody likes the Mets without getting back the name of the person, use:

```
?- likes(_,mets).
```

It is also possible to ask the question, "What does Mary like?" by saying:

```
?- likes(mary,X).
```

and to ask "Who likes what?" by saying:

```
?- likes(X,Y).
```

In this case, likes(X,Y) will end up matching every item in the database.

Prolog is also capable of answering questions such as: "Is there anyone who likes the Cubs and likes the Bears?" This is stated in Prolog as:

```
?- likes(X,cubs),likes(X,bears).
```

The ";" between the two clauses is read as 'and.' Again, in this case, Prolog begins searching the database from the top trying to match the first clause in this question. For the time being, Prolog leaves the second clause alone. Searching through the database, Prolog finds: "likes(carol,cubs)." Having found this, the variable, X, in the statement is set to "carol." The second part of the problem, therefore, becomes: "likes(carol,bears)." To do an orderly search of all the possibilities, Prolog marks the clause, "likes(carol,cubs)," as being as far as it has searched through the database so far in an effort to satisfy the first clause. To keep this straight ourselves, we mark this clause with a '1' as follows:

```
/* 1 */    likes(matt,mets).
/* 2 */    likes(carol,cubs).        1
/* 3 */    likes(bob,cubs).
/* 4 */    likes(bob,bears).
/* 5 */    likes(mary,mets).
/* 6 */    likes(mary,yankees).
/* 7 */    likes(nancy,lemonade).
```

Prolog now works on satisfying the second clause by starting at the top of the database once again. Effectively, Prolog goes off on another call of its pattern matching procedure. Prolog will now search through the database and try to find "likes(carol,bears)." This will fail and so Prolog will return to its first call of its pattern matching procedure and continue trying to match, "likes(X,cubs)," beginning just after the clause we marked with a '1.' The variable, X, is no longer instantiated to the value, "carol." Prolog now finds that "likes(bob,cubs)" matches "likes(X,cubs)" so now X will be instantiated to "bob." Prolog will mark this place in the database, again with a '1' like so:

```
/* 1 */    likes(matt,mets).
/* 2 */    likes(carol,cubs).
/* 3 */    likes(bob,cubs).        1
/* 4 */    likes(bob,bears).
/* 5 */    likes(mary,mets).
/* 6 */    likes(mary,yankees).
/* 7 */    likes(nancy,lemonade).
```

Prolog now goes off on another call of its pattern matching procedure and its problem is to try to find out if "likes(bob,bears)" is true. Prolog quickly finds that it is true and since there are no more clauses in the question to check on, Prolog reports:

```
X = bob
```

and waits for a response from the user. Again, typing a carriage return will end this search process and typing a ";" will continue it. Typing in a ";" is effectively like telling Prolog its answer is a failure and it should continue looking for another answer. We type the ";" and Prolog continues searching on from the fourth clause in the second call of the matching routine. It fails to find another way to match "likes(bob,bears)" and so it returns to the first call of its pattern matching routine. Prolog resumes searching for another Cubs fan starting after the clause marked with the '1.' This fails, and Prolog then reports "no" and asks for another question.

We now want to consider briefly what happens if we ask the longer question:

```
?- likes(X,cubs),likes(X,bears),likes(X,lemonade).
```

That is, is there anybody who likes the Cubs, likes the Bears, and likes lemonade? We have already done part of this problem, so instead of starting at the beginning, we start at the point where the first two clauses have been matched and it is time to go on to the third clause. At this point we will have placed a '1' at the clause "likes(bob,cubs)." Also the second clause in the question has matched the fourth clause in the database, so we mark this fourth clause with a '2':

```
/* 1 */    likes(matt,mets).
/* 2 */    likes(carol,cubs).
```

```
/* 3 */    likes(bob,cubs).              1
/* 4 */    likes(bob,bears).               2
/* 5 */    likes(mary,mets).
/* 6 */    likes(mary,yankees).
/* 7 */    likes(nancy,lemonade).
```

We can now go on to try to look at the third clause, "likes(bob,lemonade)." This search will be happening in the third call of the pattern matching routine. It will, of course, fail and this failure causes Prolog to return to the second call of the pattern matching routine and resume searching the database just after the clause we labeled '2.' This part of the search fails and Prolog returns to the first call of its pattern matching routine, restarting its search at statement 4. Ultimately, this searching fails after more attempts and Prolog reports, "no."

4.2.4 Rules

Prolog can not only handle facts, it can also handle rules. Rules are counted as clauses as well. If we want to write the rule that "All Yankee fans like lemonade," it is:

```
/* 8 */    likes(X,lemonade)  :- likes(X,yankees).
```

This rule can be read as: "If X likes the Yankees, then X likes lemonade" or "X likes lemonade if X likes the Yankees." The conditions on a rule are called the *antecedents* and the conclusion that would then be true is the *consequent*. Using the first seven facts and with this rule placed at the end of the database, we now ask the question:

```
?- likes(Y,lemonade).
```

Prolog starts at the top of the database trying to match this pattern. It first finds that Nancy likes lemonade and prints:

```
Y = nancy
```

and waits for a response from the user. We type a ";" and Prolog goes on to the rest of the database. Prolog will find that the left-hand-side of the rule matches the pattern it is looking for. The two, still uninstantiated variables, X and Y are now said to *share*. When one of them acquires a value, the other one acquires the same value. The problem of figuring out whether or not a person likes lemonade is now the problem of trying to find out if that person is a Yankees fan. We mark our place in the database:

```
/* 1 */    likes(matt,mets).
/* 2 */    likes(carol,cubs).
/* 3 */    likes(bob,cubs).
/* 4 */    likes(bob,bears)
/* 5 */    likes(mary,mets).
/* 6 */    likes(mary,yankees).
/* 7 */    likes(nancy,lemonade).
/* 8 */    likes(X,lemonade)  :- likes(X,yankees).     1
```

and start the pattern matching routine at the top of the list. It will be looking for the pattern "likes(X,yankees)." It will find that Mary likes the Yankees, and so in answer to the question of "likes(Y,lemonade)," it will print out:

```
Y = mary
```

and wait for a response from the user.

Keeping variables straight when processing Prolog rules can be difficult to do so it is probably a good idea to look at how Prolog handles variables internally. For instance, when we asked the question:

```
?- likes(Y,lemonade).
```

internally, it was made into something like this:

```
likes(_1,lemonade)
```

where the Y has been replaced with an interpreter generated variable, _1. When Prolog was searching through its database and it came upon the rule:

```
likes(X,lemonade) :- likes(X,yankees),
```

it created a variable, _2, that stands for X in this rule. The pattern for the rule is now:

```
likes(_2,lemonade) :- likes(_2,yankees).
```

Therefore, even if our original question had been phrased as:

```
?- likes(X,lemonade).
```

it would still have been translated to:

```
likes(_1,lemonade)
```

so Prolog would recognize the X in the question as being a different X than the X in the rule. This means that each variable in a rule is confined to that rule, the same as local variables are confined to the procedures in which they are defined in languages such as Pascal. There are, therefore, no global variables in Prolog. This method of creating new variables for each "call of a rule" also occurs in rules that call themselves recursively.

4.2.5 Recursion

We now turn to some recursive programming in Prolog. The standard example of recursive programming is, of course, the factorial function. Below we show the definition of factorial in Prolog:

```
/* 1 */   factorial(0,1).
/* 2 */   factorial(N,M) :- X is N - 1,
                            factorial(X,Y),
                            M is N * Y.
```

Line 1 says that "factorial of 0 is 1" and line 2 says that "N factorial is M, where M is defined by the clauses on the right hand side of the rule." The notation here for doing arithmetic is a bit unusual. The expression "N – 1" can be written in this infix notation for the convenience of the user, however, internally, Prolog considers it to be:

```
-(N,1),
```

so "N – 1" is just another structure and "–" is called the more general term, functor, rather than a predicate. *Writing N – 1 does not instruct Prolog to do any arithmetic*, rather, it is the 'is' operator that forces arithmetic to be done on the structure "N – 1." The execution of the factorial program proceeds in a straightforward way. If we want Prolog to find 3 factorial, we enter:

```
?- fact(3,A).
```

When Prolog tries to match this against line 1, it fails because it cannot match 3 with 0. It is possible to match 3 with N in the second line of the program. Prolog then sets X to 2 and does the call, fact(2,Y). This recursive call sets up other recursive calls, but eventually, in this current call, Y is set to 2, M becomes 6, and Prolog goes on to report:

```
A = 6
```

From time to time, people have said that the statements in a Prolog program can go into the database in any order. Most of the time that is true, however, recursive definitions like the one for factorial must have their statements go into the database in the correct order or infinite recursion will result.

4.2.6   List Processing

We can now look at some list processing in Prolog. One of the really important features of symbol processing is its reliance on processing lists of symbols. In general purpose languages like Pascal, lists of items are often implemented using arrays and accessed by specifying subscripts. In Lisp and Prolog, arrays have not always been implemented and the standard method of implementing lists is by using linked lists. In Prolog the notation for a list containing one symbol, a, is:

```
[a]
```

and it is shorthand for the structure, .(a,[ ]), where . is a functor called dot and [ ] is the *empty list*. The notation for the list containing the three symbols, a, b, and c is:

```
[a, b, c]
```

which is a shorthand for the structure .(a,.(b,.(c,[ ]))). A diagram of how these lists are stored in memory is shown in Figure 4.1. It is also possible to have lists within lists in Prolog, thereby producing trees. An example of this is:

```
[[a, b], [c, d]]
```

Here, the first item in the list is: [a,b], while the rest of the list is: [[c,d]]. This list is diagrammed in Figure 4.2.

If we find it necessary to work with the first symbol in a list, we would call up a predicate that would try to match the list against the pattern:

```
[First | Rest]
```

The vertical bar, "|" can be viewed as an operator that causes the pattern matcher to try to split a list into a first element on the left and the whole rest of the list on the right. Thus, in matching this pattern against [a, b, c], First would become instantiated to the value, a, while Rest would be instantiated to the remainder of the list, or: [b, c]. When we finish

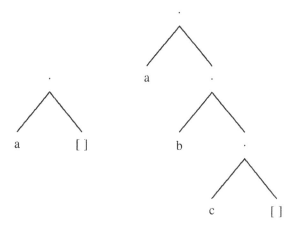

Figure 4.1: The internal tree structure of the lists [a] and [a,b,c]. The notation [a] is shorthand for the structure .(a,[ ]), and [a,b,c] is shorthand for the structure .(a,.(b,.(c,[ ]))), where . is a functor and [ ] is the empty list.

whatever processing needs to be done with a and we want to move on to the next element in the list, the standard means of doing so is to recursively pass this list, [b, c], on to the same predicate. This new call of the predicate again splits the first element off this shorter list. Splitting one element at a time off the list, we eventually encounter the empty list.

As an example of this recursive processing of lists, we will look at a predicate, member, that determines whether or not a particular constant is a member of a particular list. For instance, to ask if the constant, c, is a member of the list, [a,b,c], we will write:

```
?- member(c, [a,b,c]).
```

Member will be defined like so:

```
member(X, [First | Rest])  :- X = First.
member(X, [First | Rest])  :- member(X,Rest).
```

The first line can be translated to English as, "X is a member of the list, [First|Rest], if X and First are equal" or "X is a member of the list [First|Rest], if X is the first symbol in the list." The second line can be translated as: "X is a member of the list, [First|Rest] if X is a member of the list, Rest." In working the problem, member(c,[a,b,c]), Prolog tries to start with the first definition of member. It can match X with c so X is instantiated to c. It then moves on to try to match the pattern in the second argument of member with the pattern:

```
[First | Rest]
```

This is easily done by setting First = a and Rest = [b,c]. Prolog moves on to make the test, X = First and fails since c is not the same as a. The whole statement fails and Prolog moves on to the second definition for member. Again, X = c, First = a, and Rest = [b,c], but now, the test to perform is a recursive call of member. In this call, we will be asking if c is a member of the list, [b,c]:

```
member(c, [b, c]).
```

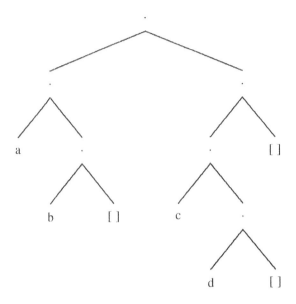

Figure 4.2: The tree structure of the list, [[a,b],[c,d]].

Prolog starts at the top of the list of definitions for member and finds that the first one fails because c is not equal to b. In the second statement, we find X = c, First = b, and Rest = [c]. We do another recursive call of member and now we will have X = c, First = c, and Rest = [ ]. This match succeeds and so ultimately, Prolog reports: yes.

In case we had asked the question:

```
?- member(d, [a, b, c])
```

Prolog will behave as above but it would make another recursive call to member, this time the call will be:

```
member(d, [])
```

This call will fail because when Prolog attempts to pattern match the empty list against [First|Rest], it finds it cannot split the empty list into a "first part" and a "rest part."

For another example of list processing, we consider the problem of looking at a list of integers and producing a list of all the positive numbers in it. For instance, if we name the predicate that creates the list of positive numbers, pos, and we give Prolog the question:

```
?- pos([-1, 44, 97, -300, 10], X)
```

we want Prolog to respond:

```
X = [44, 97, 10]
```

This is easily done with the following definition for pos:

```
/* 1 */  pos([],[]).
/* 2 */  pos([A|B],[A|C]) :- A > 0, pos(B,C).
/* 3 */  pos([A|B],C) :- pos(B,C).
```

Line 1 says that the list of positive numbers in an empty list is the empty list. Line 2 says that if you split the list of integers into a first element, A, and the rest of the list is B, and A is greater than 0, then the list of positive numbers in the list [A|B] is the list with A at the front followed by C ([A|C]), where C is the list of positive numbers found in B. If this second line should fail, the third line applies, and it says that the list of positive numbers in the list [A|B] will just be the list, C, where C is the list of positive numbers in B.

```
1 start:  pos([-1,44,97,-300,10],    X )

2 rule 3: pos([-1|44,97,-300,10], _C1 ) :- pos([44,97,-300,10],_C1).

3 rule 2: pos([44|97,-300,10],[44|_C2]) :- pos([97,-300,10],    _C2).

4 rule 2: pos([97|-300,10],    [97|_C3]) :- pos([-300,10],        _C3).

5 rule 3: pos([-300|10],             _C4 ) :- pos([10],           _C4).

6 rule 2: pos([10 | ],               _C5 ) :- pos([],             _C5).

7 rule 1: pos([],                    [] )}
```

Figure 4.3: A trace of how Prolog interprets the call: pos([-1,44,97,-300,10],X).

The English description of pos is quite neat, but to get a better idea of what is happening in the process we will look at a handmade trace of the process. An actual trace provided by a Prolog interpreter will probably not look like this. Also, the use of variables in this description is not quite the way a Prolog interpreter would go about creating and naming them. The sequence of steps is shown in Figure 4.3. In this description line 1 is the initial problem. Prolog tries to find a rule to apply. It finds that rules 1 and 2 fail. At line 2 in the figure, Prolog examines rule 3 and finds a way to break up the initial list. The "_C1" term is meant to represent the uninstantiated variable named C, in rule 3. Each recursive call will have a C variable with a different number. Still in line 2, rule 3 can be satisfied if its right-hand-side can be satisfied. The right-hand-side consists of the problem:

```
pos([44,97,-300,10],_C1).
```

This is passed to Prolog and in line 3 of the figure, Prolog comes upon rule 2. It creates another uninstantiated variable, "_C2" to represent the instance of C in this rule. Notice how in this case the second argument will be: [A|C] or [44 | _C2]. What this notation says is that the second argument will become a list with 44 at the front of it and there will be something on the tail of the list. That something will be whatever _C2 eventually becomes. Of course, we know that _C2 will eventually become [97,10], so [A|C] will become: [44,97,10], but at this time, Prolog does not know this. Prolog now goes on to notice that rule 2 can be satisfied if the right-hand-side of the rule can be satisfied. This right side now becomes the problem:

```
pos([97,-300,10],_C2).
```

These calls continue in this manner with Prolog attempting to verify either rule 2 or rule 3 until Prolog gets the problem in line 6:

```
pos([],_C5).
```

This problem matches rule 1. Prolog now knows that _C5 is [ ]. With all the necessary matching completed, Prolog can look up the value of the variable, X, in the original question. It works like this:

```
X matched with _C1,
_C1 matched with [44 | _C2],
_C2 matched with [97 | _C3],
_C3 matched with _C4,
_C4 matched with [10 | _C5],
_C5 matched with [],
giving X = [44, 97, 10].
```

After printing out the value for X, Prolog waits for input from the user. If the user types a ";" Prolog continues the search from this current point. If the user types a carriage return, Prolog returns from all the recursive calls and prompts the user for a new question.

Both member and pos follow the typical pattern found in recursive list processing algorithms: you break off the first item, deal with it, and then deal recursively with the rest of the list. At each level where you broke off an item, you combine your result with the results from the recursive call that processed the rest of the list. Recursion is really quite nice in that if you can simply list all the possible cases that can come up, together with their answers, the problem is solved. Recursion is not normally very efficient but most AI researchers have never really been concerned with efficiency.

4.2.7   Other Predicates

Some other Prolog predicates will prove useful. Two of these are assert and retract. Assert adds facts to the database while retract removes them. The following line adds the fact that Fred likes the Cubs to the database at the end of the likes predicates:

```
assert(likes(fred,cubs)).
```

The statement:

```
retract(likes(X,cubs)).
```

will cause Prolog to start looking at the top of the database for a fact that matches: likes(X,cubs) and if it finds one, then that one fact will be removed from the database.

Two other useful predicates are write, that writes out a single argument without an end-of-line marker, and nl, that writes the end-of-line marker. Here is an example of write and nl:

```
?- write(hello),nl,write(world),nl.
```

This produces:

```
hello
world
```

Another important Prolog predicate to mention is the not predicate.[1] Naturally, it takes a true value and changes it to false and takes a false value and changes it to true. For example, if it is given that Matt likes the Mets, then giving Prolog the question:

```
?- not(likes(matt,mets)).
```

will produce: no. If nothing at all is stated about Bob liking or disliking the Cardinals, then:

```
?- not(likes(bob,cardinals)).
```

will produce: yes. This is a somewhat peculiar result for people well versed in predicate-calculus-based theorem proving from which Prolog was derived. In predicate calculus, without having the appropriate information about Bob and his likes and dislikes, it is impossible to prove the proposition that Bob does not like the Cardinals. Therefore, the Prolog not predicate is not really the same as the logical not found in predicate calculus.

4.3   Rules and Basic Rule Interpretation Methods

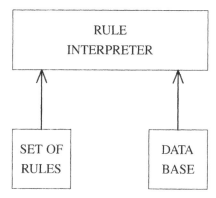

Figure 4.4: Rules and facts are submitted to a rule interpreter or inference engine.

When using rules, in general, the rule base and the facts are fed into a rule interpreter, sometimes also called an *inference engine*, because it works through the rules and facts and attempts to reach new conclusions. A simple diagram of this kind of system is shown in Figure 4.4. In general, the rules can be placed in any order and the facts can also be listed in any order and the system will still reach the same conclusions. Actually, however, there are conditions when the order of the rules and facts may make a difference, but we will

---

[1] Some Prologs use the predicate, \\+, rather than 'not.'

save that problem for a later section. Programming the system focuses on creating a set of rules. The people that create the rules are known as *knowledge engineers*. The set of rules is referred to as the *knowledge base* and the systems themselves are called *knowledge-based systems*. Rule-based systems are also sometimes referred to as *production systems* and *pattern directed inference systems*. To a certain extent, a rule-based system can be made to handle new problems just by having new sets of rules plugged in, a very flexible arrangement. Sometimes, however, a given rule interpreter will be inadequate and will need some reprogramming or even need to be replaced by one more suited to the particular problem.

There are a number of ways for a rule interpreter to deal with a system of rules. In this section we will program the two simplest methods, *forward chaining* and *backward chaining*, and then in later sections look at more complex methods. In this section the methods will be illustrated with a small animal identifying system.

4.3.1   A Small Rule-Based System

		albatross	penguin	ostrich	giraffe	cheetah	zebra	tiger
1	has hair	0	0	0	1	1	1	1
2	has claws	0	0	0	0	1	0	1
3	gives milk	0	0	0	1	1	1	1
4	eyes forward	0	0	0	0	1	0	1
5	has feathers	1	1	1	0	0	0	0
6	has hoofs	0	0	0	1	0	1	0
7	flies	1	0	0	0	0	0	0
8	chews cud	0	0	0	1	0	1	0
9	lays eggs	1	1	1	0	0	0	0
10	tawny color	0	0	0	1	1	0	1
11	pointed teeth	0	0	0	0	1	0	1
12	dark spots	0	0	0	1	1	0	0
13	black stripes	0	0	0	0	0	1	1
14	long legs	0	0	1	1	0	0	0
15	long neck	0	0	1	1	0	0	0
16	black & white	0	0	1	0	0	1	0
17	swims	0	1	0	0	0	0	0
18	flies well	1	0	0	0	0	0	0
19	eats meat	0	0	0	0	1	0	1

Figure 4.5: A set of animals with their characteristics listed as vectors.

The classic small example of a rule-based system has become the animal recognition system of Winston [262], where given some characteristics of an animal, the goal is to determine which of seven possible animals it is. Figure 4.5 lists the seven animals and their characteristics. It would be simple to come up with a set of Prolog language rules for this problem. For example, we could have a rule that an animal is an albatross if it is a bird and it flies well:

```
albatross :- bird, flieswell.
```

Instead of doing it this way, the identification rules will be listed as a series of facts. We will then write our own simple interpreters to work on these rules. For an example of this notation, here is how the rule for identifying an albatross will be coded:

```
idrule([bird,flieswell],albatross).
```

The first item in a rule will be the list of antecedents and the second item will be the consequent. As it happens in all the rules we will use, there will only be a single consequent, but, in general, there may be more than one. The top part of Figure 4.6 shows the set of rules we will use and below them some data for an unknown animal. Those rules with the predicate, 'rule,' are rules designed to "light up" intermediate conclusions while the rules labeled 'idrule' are used to give the identity of the unknown animal.

```
    /*  Rules for intermediate conclusions */

rule([hashair],mammal).
rule([givesmilk],mammal).
rule([flies,layseggs],bird).
rule([hasfeathers],bird).
rule([eatsmeat,mammal],carnivore).
rule([mammal,pointedteeth,hasclaws,eyesforward],carnivore).
rule([mammal,hashoofs],ungulate).
rule([mammal,chewscud],ungulate).

    /*  Rules to identify the animal */

idrule([carnivore,tawnycolor,darkspots],cheetah).
idrule([carnivore,tawnycolor,blackstripes],tiger).
idrule([ungulate,longlegs,longneck,tawnycolor,darkspots],giraffe).
idrule([ungulate,blackandwhite,blackstripes],zebra).
idrule([bird,longlegs,longneck,blackandwhite],ostrich).
idrule([bird,swims,blackandwhite],penguin).
idrule([bird,flieswell],albatross).

    /* some sample data */

flies.
layseggs.
flieswell.
```

Figure 4.6: A set of rules to identify the animals and an example of what the data for an unknown animal will look like.

4.3.2 Forward Chaining

The first major way of interpreting rules is called *forward chaining* and it works much the same way as a feed-forward neural network in that processing begins with the input data and lower-level rules. This method is also known as *bottom-up reasoning*, *data driven reasoning* and *antecedent reasoning*. It works by searching through the rule base in a systematic way for rules whose conditions are true. In general, the search may find more than one rule that could apply, but for now we will simply search through the database and apply the first rule whose conditions are true. In the next section we will deal with the problem of having more than one rule that could be applied.

First, we will illustrate the principle of forward chaining using the following data for the unknown animal:

```
flies.
layseggs.
flieswell.
```

We begin searching through the rules to see if any rule has all its conditions true. At this point we will be using only the rules labeled as 'rule,' and not any of the rules labeled 'idrule.' We come across the rule that if an animal flies and lays eggs then it must be a bird. This rule fires and adds to the database the fact that the animal is a bird by using the Prolog assert clause. The next step will be to remove the conditions that produced this conclusion. If these conditions were not removed, the interpreter would find the same rule again and an infinite loop would occur. After removing the conditions that caused the rule to fire, we begin again and look for another rule to fire. All these operations are easily stated in Prolog as shown below where the predicate f, is for forward chaining:

```
f  :- rule(Conditions,Conclusion),
      fcheck(Conditions),   /* check for the antecedents */
      assert(Conclusion),
      remove(Conditions),
      f.
```

For example, with the data given above for an albatross, we call the predicate, f. First, f tries to match the 'rule' predicate and this rule about mammals is found:

```
rule([hashair],mammal).
```

The fcheck predicate looks to see if all the conditions in the list are true, but they are not. Prolog backs up to the rule predicate and finds another rule about mammals. This fails and Prolog finds a rule about birds:

```
rule([flies,layseggs],bird).
```

Fcheck finds that flies and layseggs are true, so bird is asserted. The facts, flies and layseggs are removed from the database and f is called again from inside f. Notice that this will put a long chain of calls to f on the stack. The process could be done more plainly and less wastefully without recursion if Prolog supported iteration. At any rate, in the second call, f finds the first rule about mammals fails, the second rule about mammals fails, the third rule about birds fails, and in fact, every rule fails because there are no more conclusions that can be reached.

When all the possible rules labeled 'rule' have been fired, we have the interpreter move on to a second definition of f:

```
f :- idrule(Conditions,Animal),
     fcheck(Conditions),
     remove(Conditions),
     writeln(Animal).
```

Here, the program looks for the characteristics of a particular animal and if it finds the answer, it prints out the identity of the animal. With the given data, the idrule about albatross will have its conditions met, and the answer will be printed. If the animal could not be identified, we have the Prolog interpreter fall through to this third definition of f:

```
f :- writeln('unknown animal').
```

The definitions of fcheck, remove, and writeln are given below:

```
/* Fcheck checks the first condition in the list of
   antecedents and then recursively checks the rest
   of them if the first one is true.  */

fcheck([]).
fcheck([First | Rest]) :- First, fcheck(Rest).

/* Remove removes the first fact in the list and goes on
   to remove the rest of them.  */

remove([]).
remove([First | Rest]) :- retract(First), remove(Rest).

/* Writeln, a handy statement to write answers with.  */

writeln(X) :- write(X), nl.
```

4.3.3 Backward Chaining

The second fundamental rule interpretation method is to use the Prolog method of backward chaining. It is also known as *top-down reasoning*, *goal-based reasoning*, and *consequent reasoning*, and it is the way Prolog itself operates. The Prolog predicate, b, defined below, finds a rule that identifies a particular animal:

```
b :- idrule(Conditions,Animal),
     bcheck(Conditions),
     remove(Conditions),
     writeln(Animal).
```

This definition works by selecting an identification rule, a rule that will identify an animal, and then the bcheck predicate checks if the conditions of the rule are true. If the conditions are true, they are removed and the answer is printed, otherwise, the failure causes Prolog to back up and find another identification rule to test. To cover the case when the animal cannot be identified because all the identification rules have failed, we need this second definition for b placed after the above one:

```
b :- writeln('unknown animal').
```

The bcheck procedure is more complex than fcheck. Bcheck first looks to see if the First condition in the list is already true. If it is, the rest of the conditions in the list are checked recursively. If the First condition is not in the database, Prolog moves on to try to find a way to prove the condition is true by selecting a rule that will show that the First condition is true. The definition of bcheck is as follows:

```
bcheck([]).
bcheck([First | Rest]) :- First, bcheck(Rest).
bcheck([First | Rest]) :- rule(Conditions,First),
                          bcheck(Conditions),
                          remove(Conditions),
                          assert(First),
                          bcheck(Rest).
```

As an example, suppose the data is once again that the animal in question flies, lays eggs, and flies well. We call the predicate, b and b will end up selecting the rule for a cheetah as its first candidate:

```
idrule([carnivore,tawnycolor,darkspots],cheetah).
```

When b calls bcheck, it passes on the list of requirements for an animal to be a cheetah. The first rule of bcheck will fail. The second rule will break off the carnivore condition, and ask if carnivore is true. It will not be true, so Prolog will move on to the third bcheck rule. This rule looks for a way to prove carnivore true. The rule predicate in bcheck finds this rule about carnivores:

```
rule([eatsmeat,mammal],carnivore).
```

Another call is made to bcheck with this new list of requirements. The first two rules for bcheck will fail. In the third rule, Prolog will try to find a way to prove the First condition, eatsmeat. This fails and Prolog backtracks.... Eventually, every possible way to prove that the animal is a cheetah fails. After this, more idrules fail. Eventually, albatross is selected as the goal to pursue and this proof finally succeeds.

Forward and backward chaining constitute the two major methods of searching through a rule base. Clearly, backward chaining can be very time consuming if there are many possible final conclusions that can be reached, however, forward chaining can also be time consuming because the interpreter may spend a great deal of time reaching a large number of conclusions that have no bearing whatsoever on solving a particular problem.

4.4 Conflict Resolution

The animal identifying expert system we have been using so far has avoided a problem that occurs in larger expert systems, that of *conflict resolution*. In the forward chaining version of the animal identifying expert system we assumed, and reasonably so, that whenever we found a rule whose antecedents were true we could simply apply the rule. In general, however, in searching through a rule base there could easily be several rules that have their antecedents satisfied. These rules that could be applied are said to be *triggered*. Whichever rule is finally applied is said to *fire* but it becomes a problem to decide which of them should fire. For example, we may have the following set of rules:

IF A and B and C THEN X	(1)
IF A and B THEN Y	(2)
IF A THEN Z	(3)

and all the conditions A, B, and C are true. In this case, rule (2)'s conditions are a superset of rule (3)'s conditions and rule (1)'s conditions are a superset of both rule (2)'s and rule (3)'s conditions. The programmer must tell the interpreter how to choose which rule to fire. The final answer that the expert system produces may well be different depending on which of these rules actually fires.

1	1	1	1	1
1	0	0	0	0
1	1	1	1	1
1	0	0	0	0
1	1	1	1	1

1	1	1	1	1
1	0	0	0	0
1	1	1	1	1
1	0	0	0	0
1	0	0	0	0

1	2	3	4	5
6	7	8	9	10
11	12	13	14	15
16	17	18	19	20
21	22	23	24	25

Figure 4.7: The E and F patterns in 5×5 matrices and the numbering of the units.

Notice how this problem is the same as the problem discussed in Section 2.1 where we developed a simple matrix multiplication technique to use to discriminate between the letters E, F, and H. Instead of using the larger representations of the letters as in Section 2.1, we will use the two smaller 5×5 versions of the letters E and F shown in Figure 4.7. To identify E and F we could have the following rules, where the number indicates the bit is a 1:

IF	1 & 2 & 3 & 4 & 5 & 6 & 11 & 12 & 13 & 14 & 15 & 16 & 21 & 22 & 23 & 24 & 25
THEN	the letter is E
IF	1 & 2 & 3 & 4 & 5 & 6 & 11 & 12 & 13 & 14 & 15 & 16 & 21
THEN	the letter is F

Again, using this rule-based plan, if the unknown letter we want to identify is an F, there is no problem, but if the unknown letter is an E, both these rules are triggered and we are left with the problem of which one should fire. In Section 2.1 we were counting the number of votes for each letter. If we follow that example here, we will declare the letter to be an E. Choosing the rule to fire that has the most specific set of antecedents true is called *specificity ordering*.

Another solution to rule conflicts is to simply take the first rule that is encountered as the one to fire. Of course, this is easily done without looking for other rules that may also be triggered, but this leaves the programmer with the problem of placing the rules in the knowledge base in just the right order.

Other solutions to rule conflicts involve keeping track of how often and when each of the triggered rules has fired. Then a tie is broken by using either the most commonly used rule or the most recently used rule, the least commonly used rule or the least recently used rule. As an example of using the most commonly used rule, suppose you go to your doctor and complain of flulike symptoms. If the flu is quite common at the time, it is likely that the doctor will decide you have just the ordinary flu because (1) it is what everyone else has at the moment and (2), statistically speaking, it is the most likely disease that fits the symptoms. Of course there are occasions where this will not be correct. You could have just come back from a trip up the Amazon river and during the trip you could have acquired a rare tropical disease with symptoms just like those of the flu.

Another method of conflict resolution is to prevent it from happening altogether by adding some additional conditions to the rules. For instance, we may have the following pair of rules where all the antecedents are true and where one set of antecedents is not a superset of the other:

IF A and B and C THEN W

IF D and E and F THEN X and Y and Z

It may be that we need to do all these things, W, X, Y, and Z, but it is possible that these things can be done in a specific order. A concrete example of this might be the problem of decorating a Christmas tree. One expert has this set of rules for decorating Christmas trees:

IF	there is a string of lights available
THEN	put on the string of lights
IF	there is a large ornament available
THEN	put the large ornament at the bottom of the tree
IF	there is a medium ornament available
THEN	put the medium ornament in the middle of the tree
IF	there is a small ornament available
THEN	put the small ornament at the top of the tree
IF	there is some tinsel available
THEN	put the tinsel on

At the beginning of the process there are lots of strings of lights, lots of every size ornament, and lots of tinsel available so every rule is triggered. This problem can be avoided by sequentializing the process. We can break the problem into a series of steps, the "putting on lights phase," the "putting on large ornaments phase," the "putting on medium ornaments phase," the "putting on small ornaments phase," and the "putting on tinsel phase." For each phase, there will be a variable associated with it, such as "lights_phase" for the putting on lights phase. It will be 1 during this phase and 0 otherwise. When one phase is finished, a rule can be used to change the state to the next phase:

IF	lights_phase = 1 and
	there is a string of lights available
THEN	put on the string of lights

IF   lights_phase = 1
THEN   lights_phase ← 0 and
       large_ornaments_phase ← 1

IF   large_ornaments_phase = 1 and
       there is a large ornament available
THEN   put the large ornament at the bottom of the tree

IF   large_ornaments_phase = 1
THEN   large_ornaments_phase ← 0 and
       medium_ornaments_phase ← 1

IF   medium_ornaments_phase = 1 and
       there is a medium ornament available
THEN   put the medium ornament in the middle of the tree

IF   medium_ornaments_phase = 1
THEN   medium_ornaments_phase ← 0 and
       small_ornaments_phase ← 1

IF   small_ornaments_phase = 1 and
       there is a small ornament available
THEN   put the small ornament at the top of the tree

IF   small_ornaments_phase = 1
THEN   small_ornaments_phase ← 0 and
       tinsel_phase ← 1

IF   tinsel_phase = 1 and
       there is some tinsel available
THEN   put the tinsel on the tree

The process has to start with lights_phase = 1. Using this plan, there will be only two rules to choose from at a time, the one that adds something to the tree and the one that changes the state of the problem. A tie can be broken by taking the rule with the longer list of antecedents.

It is also important to note that, whereas we can break the one set of rules for decorating a Christmas tree into five different phases, we can also declare each phase to be a separate little expert system by itself. This has the nice advantage that fewer rules have to be checked in each part of the process. XCON is a famous expert system that configures VAX computer systems and it breaks the process into different phases that are handled sequentially.

Still another alternative for dealing with rule conflicts is to try to find more specific conditions for each rule so that the right consequences are produced without any conflicts. One way to do this using the first example in this section is to transform the rules into these:

IF A and B and C THEN X
IF A and B and not C THEN Y
IF A and not B and not C THEN Z

These rules reflect the kind of data that would have to be submitted to a back-propagation network where every input must have a value.

The final alternative is to admit that you do not know enough to say for sure which rule to use, so choose one at random in the hope that it works out. If this choice fails, back up to the point where you made the arbitrary choice and try another rule.

4.5   More Sophisticated Rule Interpretation

Problems sometimes occur with the simple forward and backward chaining methods we have described, but these can be dealt with using more complex techniques. There are also other capabilities that rule interpreters need and we will mention some of them in this section.

4.5.1   Dealing with Incomplete Data by Asking Questions

One problem is that an expert system may need to operate with an incomplete set of data. This could be because some data is unobtainable or because the person supplying the data has just been careless by not giving all the facts that are available. For instance, a person may walk up to the animal identifying expert system and ask it to identify some animal with a tawny color, long legs, a long neck, and dark spots. Obviously the animal is a giraffe, but without the program having the additional information that the animal has hair and has hoofs, the system will not be able to identify the animal.

One solution to this problem is to make the rule interpreter a little more sophisticated by having it ask the user about facts that the user has not given. This can be done while the system is searching through the list of rules. Let us suppose we had the interpreter using forward chaining and the data that it has on the animal is that it has a tawny color, long legs, a long neck, and dark spots. When the interpreter picks a rule to try to apply, it first looks in the database to see if the antecedents of the rule are true. If they are, there is no problem and the rule fires, but if one or more antecedents are not there, we let the program ask the user if the particular missing facts in the antecedent about the animal are true. With the data for this problem arranged in the order given for rules in Section 4.3, the first rule we happen upon is to try to determine if the animal is a mammal by checking if the animal has hair. The interpreter would then ask the user if the animal has hair. In this case it happens to be a sensible question since this rule will contribute directly to the problem of determining that the animal is a giraffe. In most cases we can expect that the questions asked by systems will not be very sensible. The system might just as easily have found the rule about proving an animal is a bird as the first rule. So, while we can say that this method of getting additional information from the user will eventually work, the interpreter will usually display a considerable lack of "good sense." In addition, the sheer number of questions it might ask could be very annoying to the user.

Of course, asking questions can also be done in a backward chaining expert system and here the questions can be a little more focused. The program can start with the assumption that the animal is an albatross and ask relevant questions for that hypothesis rather than jumping around between rules at random.

4.5.2 Other Activation Functions

When a human expert is given incomplete data the person will not ask a lot of irrelevant questions the way a simple forward or backward chaining system would. A better system is to use another activation function to pick good questions to ask, an analog function that will let you rate the possible answers. In the animal identification problem, one solution that immediately comes to mind is to keep a list of each animal and its characteristics and then search through the list looking for the highest percentage of matches, take the most likely candidates, and do backward chaining, asking questions as necessary about characteristics that the user has not given. So, given that the animal has a tawny color, long legs, a long neck, and dark spots, this should activate the giraffe candidate more than any other.

   If your data is incomplete and you cannot ask about the missing features, you may still want a program that will give you some indication of how likely some conclusion is, and again for that you need some real-valued activation function. Some other techniques and functions that can help are described below.

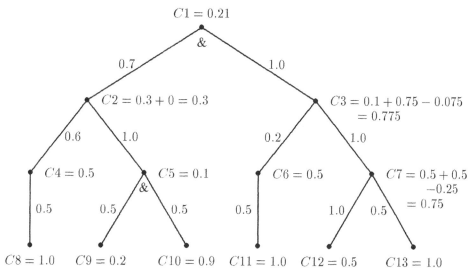

Figure 4.8: An example illustrating MYCIN's method of propagating certainty factors through its network. The nodes marked with & are AND nodes while the rest are OR nodes.

   One of the most commonly used activation functions comes from an early experimental expert system called MYCIN [24], a program designed for analyzing bacterial infections. MYCIN uses rules that are used by an inference engine to construct and evaluate a network. In MYCIN the network is an AND/OR tree like that shown in Figure 4.8. Each node receives values from its children in the same way that nodes in neural networks do by taking the weight on the link between nodes and multiplying it by the value of the child's node. In MYCIN the weights are called attenuation factors because they are in the range from 0 to 1.0. The activation value of a node is called the *certainty factor* and it also runs from 0 to 1 and corresponds to a rating factor like a probability. At the AND nodes the value of the node is calculated by taking the smallest incoming value as the value of the node. At the OR nodes the value of the node is calculated by taking two incoming values, a and b, adding them together and subtracting their product:

$$certainty factor = a + b - a \times b.$$

This formula can be generalized to more than two incoming values. Values for nodes less than some arbitrary threshold, say 0.2, are set to zero. This is a very ad hoc method but it works for many applications. Other, more sophisticated methods have been proposed by more mathematically inclined people who want to make the certainty factors as close as possible to the precise mathematical definition of probability.

Another different method of rating rules and conclusions was developed for the PROS-PECTOR expert system [30]. This system was designed to evaluate the prospects for mineral deposits. It gives all its possible conclusions an initial rating of 1. In PROSPECTOR, the rating is called a *likelihood ratio*. As each conclusion gains evidence, this ratio is multiplied by a factor greater than one. If key evidence for the conclusion is missing, the likelihood factor gets multiplied by a value less than 1. One rule from PROSPECTOR is:

> IF   there is hornblende pervasively altered to biotite
> THEN   there is strong evidence (320,0.001) for potassic
>   zone alteration

In this rule the 320 is the factor to multiply the likelihood ratio by if the antecedent is true and 0.001 is the factor to multiply the likelihood ratio by if the antecedent is not true. When the likelihood ratio gets large enough, the conclusion is considered true. The likelihood ratio is also used by the program to select promising rules to investigate. PROSPECTOR selects likely scenarios to investigate using these ratings and then goes on to investigate them via backward chaining, however, PROSPECTOR is also willing to investigate any theory that the user has.

4.5.3   Uncertain Input

In PROSPECTOR and MYCIN, the answers to questions do not have to be a plain yes (1.0) or no (0.0). Instead, users can enter values that represent their degree of confidence in the answer. In PROSPECTOR, the confidence intervals run in integral values from –5 for definitely not, to 0 for unknown, and to +5 for absolutely certain. In PROSPECTOR, the rules can also have confidence intervals set so that certain rules should not even be considered by the program unless the response from the user is within a certain range, say, perhaps, from +2 to +5.

To cope with the fuzziness of values, several mathematical methods have been proposed. One of them is based on the theory of fuzzy sets proposed by Zadeh. (For a short introduction see [268].)

4.5.4   Extra Facilities for Rule Interpreters

The expert system capabilities described so far have really been only those associated with the pattern recognition and control aspects of the problem. In practical inference engines, you typically need many of the capabilities found in general purpose computer languages, such as having variables that you can do arithmetic on. For instance, we might need an expert system for packing Christmas tree ornaments in boxes. Suppose six big ornaments and four small ones can fit in a box. To do this problem, rules like these will be necessary:

    IF   the large ornament count in box B < 6 and
             there is a large ornament L that needs to be packed
    THEN put the large ornament L in box B and
             increment the large ornament count in box B by 1.

    IF   box B has 6 large ornaments and
             there are < 4 small ornaments in box B and
             there is a small ornament S that needs to be packed
    THEN put the small ornament in box B and
             increment the small ornament count in box B by 1.

    IF   box B has 6 large ornaments and
             box B has 4 small ornaments and
             there are more ornaments that need to be packed
    THEN get another empty box B and
             set the number of small ornaments in this box B to 0 and
             set the number of large ornaments in this box B to 0

Therefore, in this program there will need to be variables, one for each box, that keep track of how many large and small ornaments are in the box. For this, a record structure comes to mind. To "get another empty box" may require creating a dynamic variable. Looping, conventional if statements, arithmetic calculations, and I/O operations are also required. As additions like these are made to the inference engine, the resulting language begins to look very much like a general purpose programming language except it has some new pattern recognition oriented control structures in it in addition to the ordinary ones.

4.6　The Famous Expert Systems

In this section we want to look at the kinds of problems that have been handled by traditional rule-based expert systems by looking at some of the more well-known systems. Two of these early expert systems are DENDRAL for analyzing mass spectrograms and MYCIN for analyzing bacterial infections. These were early systems that demonstrated the principles involved, but they were never good enough to have a technical or economic impact. The systems that made headlines because of their economic benefit were PROSPECTOR, used for mineral exploration, and XCON (alias R1), used to configure VAX computer systems. We also look briefly at the ACE system from AT&T that is used to diagnose faults in telephone cables. For the most part, expert systems are developed using specialized languages and commercial shells that have facilities to make the job easier. Two such languages mentioned below are OPS4 and OPS5. Expert systems are rarely done in the traditional AI languages such as Lisp and Prolog.

4.6.1　DENDRAL

The first famous expert system we will look at is DENDRAL [17]. DENDRAL was designed to analyze mass spectrograms of certain classes of organic compounds. The goal of DENDRAL is to deduce the molecular structure of the compounds. For instance, given some compound like: $C_8H_{16}O$, the program can take in information from a mass spectrogram and conclude that the structure is:

$$CH_3 - CH_2 - \overset{\displaystyle \overset{O}{\|}}{C} - CH_2 - CH_2 - CH_2 - CH_2 - CH_3$$

The compound is called 3-octanone.

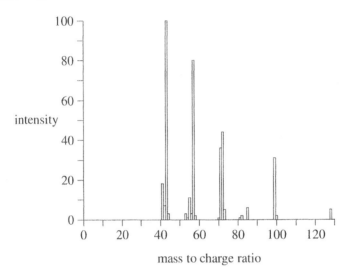

Figure 4.9: A plot of intensity as a function of the mass/charge ratio for the compound 3-octanone. The pattern of peaks allows chemists (and DENDRAL) to guess what substructures are present.

The process begins by taking the compound in question and heating it in an oven so that the molecules break into pieces. The pieces are charged ions that are accelerated by an electric field and then deflected by a magnetic field onto a photographic plate. Naturally, heavy ions are deflected less than light ones, and ions with more charge are deflected more than ions with little charge. The apparatus to do this is much like the system in a CRT except a CRT only uses electrons. The darkness at a particular spot on the photographic plate indicates how many of one particular kind of ion is hitting the plate at that spot. A plot of some results is shown in Figure 4.9.

With the information available from the mass spectrogram, it is possible for an expert to determine what ions are present. DENDRAL uses rules for this like:

> IF   there is a high peak at X, Y, and Z and any size peak at W
> THEN   the substructure U is present

This information about substructures that are present is used by another part of the program to generate possible ways of putting together molecules that contain the necessary constituents. In the next phase, the list of possible candidates is passed on to another part of the program that knows how these kinds of compounds will break up when they are heated. For each of these candidates, the program computes the mass spectrogram that will result, then these simulated spectrograms are compared with the one from the unknown compound. The simulated mass spectrogram that most closely matches the actual spectrogram is chosen as the answer. DENDRAL was actually only capable of dealing with a small number of compounds and it has never been used commercially. Another program called CONGEN is widely used, however, Dreyfus and Dreyfus [28] report that it does not use any techniques that would be regarded as artificial intelligence techniques and so they say it should not be regarded as an expert system.

4.6.2   MYCIN

The MYCIN expert system was designed to analyze information on bacterial infections and give advice on the best possible treatments. It was a backward chaining system and had about 500 rules. The following is part of a sample session with MYCIN from [24]. Comments within curly brackets are not part of the actual dialog. User responses come after a pair of asterisks:[2]

```
- - - - - - - PATIENT-1 - - - - - - -
(1) Patient's name:  (first-last)
**FRED SMITH
                                              {background data}
(2) Sex:
**MALE
(3) Age:
**55
(4) Have you been able to obtain positive cultures from a site at
    which Fred Smith has an infection?
**YES
   - - - - - - - INFECTION-1 - - - - - - -
  (5) What is the infection?
  **PRIMARY-BACTEREMIA
  (6) Please give the date and approximate time when signs or symptoms
      of the primary-bacteremia (INFECTION-1) first appeared.  (mo/da/yr)
   **MAY 5, 1975
  The most recent positive culture associated with the primary-bacteremia
  (INFECTION-1) will be referred to as:
     - - - - - - - CULTURE-1 - - - - - - - -
    (7) From what site was the specimen for CULTURE-1 taken?
    **BLOOD
    (8) Please give the date and time when this blood culture (CULTURE-1)
        was obtained (mo/da/yr)
    **MAY 9, 1975
    The first significant organism from this blood culture (CULTURE-1)
    will be referred to as:
       - - - - - - - - ORGANISM-1 - - - - - - - -
      (9) Enter the identity of ORGANISM-1:
      **UNKNOWN
            {typically the identity is not known and must be inferred}
      (10) is ORGANISM-1 a rod or coccus (etc.):
```

[2] Reprinted from "Production Rules as a Representation for a Knowledge-Based Consultation Program," by Randall Davis, Bruce Buchanan, and Edward Shortliffe, in *Artificial Intelligence*, Volume 8, 1977 with permission from the publisher, Elsevier Science Publishers B. V.

```
**ROD
(11) The gram stain of ORGANISM-1
**GRAMNEG
(12) Have there been POSITIVE cultures from sites which are NOT
     associated with infections about which you are seeking advice?
**NO
(13) Did ORGANISM-1 grow aerobically?
**YES (8)
(14) Did ORGANISM-1 grow anaerobically?
**YES
(15) What is the suspected portal of entry of ORGANISM-1
     into this sterile site?
**GI (6)
         .   .   .
```

My therapy recommendation will be based on the following probable
infections(s) and potential causative organism(s):

INFECTION-1 IS PRIMARY-BACTEREMIA
                                    {possible identities}
<item 1> The identity of ORGANISM-1 may be PSEUDOMONAS-AERUGINOSA
<item 2> The identity of ORGANISM-1 may be KLEBSIELLA-PNEUMONIAE
<item 3> The identity of ORGANISM-1 may be E. COLI
<item 4> The identity of ORGANISM-1 may be BACTEROIDES-FRAGILIS
<item 5> The identity of ORGANISM-1 may be ENTEROBACTER
<item 6> The identity of ORGANISM-1 may be PROTEUS-NON-MIRABILIS

```
  (46) On a scale of 0 to 4, where higher numbers indicate increasing
       severity how would you rate Fred Smith's degree of sickness?
  **3
  (47) Does Fred Smith have a clinically significant allergic reaction
       to any antimicrobial agent?
  **NO
  (48) Fred Smith's weight in kilograms (or <number> POUNDS):
  **70
                       {note that 2 drugs cover for 6 potential organisms}
[Rec 1] My preferred therapy recommendation is as follows:
  In order to cover for Items <1 2 3 5 6>:
    Give:  GENTAMICIN
    Dose:  119 mg (1.7 mg/kg) q8h IV [or IM] for 10 days
    Comments:  Modify dose in renal failure
  In order to cover for ITEM <4>:
    Give:  CLINDAMYCIN
    Dose:  595 mg (8.5 mg/kg) q6h IV [or IM] for 14 days
    Comments:  If diarrhea or other GI symptoms develop, patient
      should be evaluated for possible pseudomembranous colitis.

Do you wish to see the next choice therapy?
                     {alternative choices are available}
**Y
Which, if any, of the above drugs do you wish eliminated from
consideration?
**CLINDAMYCIN

[Rec 2] Next best therapy recommendation:
  In order to cover for items <2 3 4 5 6>:
    Give:  CHLORAMPHENICOL
```

```
   Dose:  1 gm (15 mg/kg) q6h IV for 14 days
   Comments:  Monitor patient's white count
In order to cover for Item <1>:
   Give:  GENTAMYCIN
   Dose:  119 mg (1.7 mg/kg) q8h IV [or IM] for 10 days
   Comments:  Modify dose in renal failure
```

MYCIN was only an experimental system and it has never been used in practical situations. Davis [25] has experimented with an automated knowledge acquisition system for MYCIN called TEIRESIAS.

4.6.3 PROSPECTOR

The PROSPECTOR system was developed by Stanford Research Institute with the help of grants from the Office of Resource Analysis of the US Geological Survey and the National Science Foundation. PROSPECTOR's main claim to fame is its possible discovery of a molybdenum ore deposit in Washington state. There are actually many versions of this story. One version is that the system predicted that there would be a molybdenum deposit at a certain spot near Mt. Tolman and that no one had ever suspected that there was such a deposit there. Another version is that people strongly suspected that there was a deposit in the area, but the program predicted where it was and where other such deposits could also be found. Another version is that the deposit was known to be there in the first place. The concentration of molybdenum ore found by PROSPECTOR was not enough to justify a mining operation at the time. Below are two excerpts from a conversation with PROSPECTOR. The first excerpt is from an early version where the confidence interval for answers given by the user ran from -2 to +2 in integral amounts. In a later version of the program the interval runs from -5 to +5. Here is the sample conversation with PROSPECTOR:[3]

```
-------------Program execution is now starting -------------------
   Do you want to volunteer any evidence?  YES

   A. Space name of evidence:    SPACE-25L
       New likelihood of (Widespread Igneous Rocks):  2
   B. Space name of evidence:  NIL

   Proceeding to establish the likelihood of (Massive Sulfide
       Deposit):

   1.  Do you have anything to say about (Volcanic province and
major fault zone)?  1
   2.  Do you have anything to say about (Mineralization)?  0
   3.  Do you have to say about (Near shore depositional sequences
of andesites, rhyolites, or dacites)?  0
   4.  Do you have anything to say about (Pillow structures) 2
   5.  Do you have anything to say about (Breccia)?  NO
   6.  Do you have anything to say about (Rhyolite or dacite plug)?
YES
```

[3] Reprinted from [30] "Semantic Network Representations in Rule-Based Inference Systems," by Richard Duda, Peter Hart, Nils Nilsson, and Georgia Sutherland, in *Pattern-Directed Inference Systems*, 1978, with permission from the publisher, Academic Press.

6a. Have you anything to say about an entity with the composition of (OR RHYOLITE DACITE)? 1

This entity will hereafter be referred to as ENTITY-1

6b. Have you anything to say about whether ENTITY-1 has the form of PLUG? 2
   7. Do you have anything to say about (Galena, sphalerite, or chalcopyrite filled cracks in rhyolite or dacite)? YES
   7a. Have you anything to say about an entity with the composition of (OR GALENA SPHALERITE CHALCOPYRITE)? 2

This entity will hereafter be referred to as ENTITY-2

7b. Have you anything to say about whether ENTITY-1 has the property of CONTAINING-CRACKS? 2
   7c. Have you anything to say about whether ENTITY-2 is CONTAINED-IN ENTITY-1? 1
           .
           .
           .
   10.   Do you have anything to say about (Olivine or alkaline andesite)? 2
   11.   Do you have anything to say about (Calc-alkaline andesite)? NO
           .
           .
           .
   14. Do you have anything to say about (Clay Minerals)? 0
   15. Do you have anything to say about (Reduction process)? 0
   16. Do you have anything to say about (Bleaching of rocks)? YES
   16a.   Have you anything to say about an entity with the composition of ROCKS? 2

This entity will hereafter be referred to as ENTITY-4

16b.   Have you anything to say about whether ENTITY-4 has the property of BLEACHED? 1
           .
           .
           .
   22.   Do you have anything to say about (Prospect within a few miles of known MDS)? NO

   In summary, the following have been observed:

   The most likely deposit type is (Massive Sulfide Deposit) Its current probability is 0.01465
   The most encouraging observations were: (Galena, sphalerite, or chalcopyrite filled cracks in rhyolite or dacite) and (Rhyolite or dacite plug)
   The only discouraging observation was: (Olivine or alkaline andesite).
   The most useful additional information would be: (Mineralization) and (Breccia)

--------------------------------DONE---------------------------------

In the following example (from [31]), note how the user can get a geological education by asking WHY?

```
13-- Do you know the age of the continental-margin mobile belt?  WHY

The vast majority of porphyry copper deposits are younger than 200
million years.    Major age groups include:

   1.    An older group of 200 to 150 m.y.  (million year) porphyries in
         British Columbia;

   2.    An important younger group of 70 to 50 m.y.  deposits in the
         Mideast, British Columbia (Skeena Arch), and Arizona-New Mexico-
         Sonora (Laramide) copper province;

   3.    The major belt of 60 to 4 m.y.  deposits in the Andes of Argentina,
         Chile, Peru, Ecuador, and Colombia.
            .   .   .
```

4.6.4 ACE

The next system to look at is ACE, short for automated cable expertise, and it may have been the first expert system to be used commercially. It was developed at AT&T Bell Laboratories and it uses forward chaining and has about 300 rules. It is written in OPS4, Lisp, C, and Unix shell. Its job is to look at a database that contains reports of subscriber telephone problems and analyze the causes of the problems. In scanning the reports, ACE does nightly what used to require a human being a month to do. Being able to work this fast is an important ability. Previously, it could take a human expert a month to analyze problems. This meant that the repair staff was involved in doing a large number of short-term fixes that actually resulted from a single larger problem that was undetected.[4]

4.6.5 XCON

The last system to look at is XCON used by Digital Equipment Corporation to properly configure VAX computer systems. XCON stands for eXpert CONfigurer and it is also sometimes known as R1. It is perhaps the most important program that can be credited with fueling interest in expert systems because it reportedly saves DEC millions of dollars a year while doing a better job of configuring VAX computer systems than human beings can do. Besides the cost savings involved, there is also the increased customer satisfaction that comes from having systems correctly configured on the first try. Our descriptions of XCON come from [200].[5]

The history of XCON began in 1974 when a DEC engineer suggested that a program be written to check all customer orders for PDP-11 computers. Some early parts of the

---

[4] This information comes from a talk given by Dr. George V. Otto, AT&T Bell Laboratories.

[5] Reprinted from *The Artificial Intelligence Experience: An Introduction*, by Susan J. Scown, 1985, with permission from the publisher, Digital Press.

system were done in Fortran and Basic, but ultimately, after consulting with members of the Computer Science Department at Carnegie-Mellon University, it was decided to try an artificial intelligence approach. One important consideration involved in the decision was the fact that the number of components used in VAX computers and the specifications for those components were constantly changing. A program that used traditional programming methodology would have to be constantly changed as new and improved components were added. With the AI methodology, these changes could be made very simply by just adding new rules. Scown reports that "Most of Digital's development team agree that the AI solution has been proven and is ultimately the best approach for this problem." In late 1985, XCON had about 4,200 rules and by late 1988 this was up to over 10,000. This large number of rules makes for an interesting new problem: there are very few people who are qualified to modify XCON. Any modifications must be checked to insure that the new fixes do not ruin the parts of the system that are known to work correctly. Researchers are currently looking into this problem and have come up with a knowledge acquisition system for XCON called RIME (see [109] and [7]).

The major researcher and developer of XCON has been Professor John McDermott. Below is a description excerpted from the book, *The Artificial Intelligence Experience* written by Scown (a DEC employee) and published by Digital Press, that describes some of how the system works and the effectiveness of it.

* How XCON Works

XCON accepts as input a list of items on a customer order, configures them into a system, notes any additions, deletions, or changes needed in the order to make the system complete and functional, and prints out a set of detailed diagrams showing the spatial relationships among the components as they should be assembled in the factory.

The users are

* Technical editors who are responsible for seeing that only configurable orders are committed to the manufacturing flow.

* The assemblers and technicians in Digital's manufacturing organization who assemble the systems on the plant floor.

* Sales people, who use XCON in conjunction with XSEL, an expert system that helps them prepare accurate quotes for customers. This can be done on a dialup basis from the customer's site.

* Scheduling personnel who use information from XCON to decide how to combine options for the most efficient configurations.

* Technicians who assemble systems at the customer's site.

Configuration tasks like that of XCON can be thought of as heuristic searches for an acceptable configuration through a search space of possible configurations. Mc-Dermott used a "match search" pattern-matching method that does not deviate from the solution path; backtracking is rarely ever required because XCON's knowledge is usually sufficient to determine an acceptable next step.

Elements of the system include

* OPS5's production memory (knowledge in condition/action rule form about how to configure the systems), the embodiment of the heuristic knowledge base.

* The working memory, which starts with the set of customer-ordered components but by the end of processing has accumulated descriptions of partial configurations that will be used to complete the full configuration.

* The inference engine, which is the OPS5 interpreter. The interpreter selects and applies rules.

* An additional component database (descriptions of each of the components that may be configured in systems).

* User interface software that allows the user to interactively enter and modify orders, review XCON output for those orders, and enter problem reports.

* Traditional software for database access, the collection of statistics on hardware resource utilization and functional accuracy, and for automatically routing problem reports entered by users to the support organization.

The major subtasks within XCON are

* Checking the order for gross errors, such as missing prerequisites, wrong voltage or frequency, no central processing unit, etc. The first subtask is also concerned with unbundling line items to the configurable level, assigning devices to controllers, and distributing modules among multiple secondary buses.

* Placing the components in the central processing unit cabinets and then finding an acceptable configuration of the secondary bus by placing modules in backplanes, backplanes in boxes, and boxes in cabinets.

* Configuring the rest of the components on the secondary bus-panels, continuity cards, and unused backplanes. Also, computing vector and address locations for all modules.

* Laying out the system on the floor and determining how to cable it together.

The production rules in XCON that describe the different subtasks are grouped together. The rules are separated into subtasks both for easier maintenance and to increase efficiency because the interpreter need only consider the rules in a single subtask at any given time.

An example of an XCON rule translated from OPS5 into normal English follows:

R1-Panel-Space

If: The most current active context is panel-space for module-x And module-x has a line-type and requires cabling And the cabling that module-x requires is to a panel And there is space available for a panel in the current cabinet And there is no panel already assigned to the current cabinet with space for module-x And there is no partial configuration relating module-x to available panel space

Then: Mark the panel space in the cabinet as used And assign it the same line type as module-x And create a partial configuration relating module-x to the panel space.

XCON runs in batch mode, processing an order every minute or two. XCON looks into its database for an order to configure and if it finds an order, XCON configures it, updates the database with its output, and looks for another order.

* Testing

At the beginning, Digital used to create a test set of orders, either 25 customer orders or sample orders generated internally. The idea was to test the vast majority of rules on the most difficult orders. Once XCON could run all of these, the developers were confident that the rules functioned well. As new orders came through from customers, the developers would then see where XCON had failed and add those problems to the test cases, constantly making the tests tougher and tougher. These test cases were run against each rule change and/or each formal release. Now, as developers change rules in XCON, they run only those tests they believe are necessary. But before each new release of updated software to the production environment, a complete set of regression tests is run....

* XCON's Performance

Success for XCON has always been difficult to define. At the beginning, the development team had long and heated debates about the defining criteria. They decided that XCON would have to examine all orders, including the most difficult ones. The degree of XCON's accuracy, as judged by human experts was initially 75 percent and rose toward a goal of 95 percent over a period of about a year and a half. To increase accuracy was quite a difficult task because the development team was constantly adding new products and finding more and more "hidden" details about how to properly configure a VAX system. Success also became hard to define because the experts often disagreed on what was correct and what was not.

Another measure of success was acceptance by the technical editors and the engineers in the factory. The technical editors and engineers were at first unwilling to accept XCON as a software product. Later, after the development team had run a large number of orders through XCON, the technical editors and engineers accepted the fact that the system worked.

The average runtime required to configure an order is currently 2.3 minutes. Small PDP-11s sometimes take less than 1 minute to configure, while a 200 line item VAX 8600 cluster may take many minutes. Before XCON and before the complications of clusters, technical editors required 25-35 minutes for an average order and got 70 percent of them correct. XCON provides several hundred pieces of information per order, and creates a usable configuration 98 percent of the time. XCON performs about six times as many functions as the technical editors used to perform.

* XCON's Impact

XCON has allowed Digital to avoid costs that it would have incurred if Digital had been forced to hire more technical editors as the volume of systems sold increased.

XCON has also made possible another significant cost-avoidance and efficiency measure in the manufacturing process. Before XCON's accurate configuration plans were available, systems were sometimes assembled up to a point at which a problem was discovered, and then the system had to be dismantled and reconfigured. This wasted floor space in the assembly plants as well as time. With XCON, the configurations are more dependable, so the manufacturing process can, in turn, be more efficient.

A sample report generated by XCON is shown in Figures 4.10 and 4.11.

```
COMPONENTS ORDERED
LINE  QTY   NAME       DESCRIPTION              COMMENT

 1     1    861CB-AJ   8600 QK001-UZ 12MB 240/50HOS
             KA86-AD       PROCESSOR
             12288 KILOBYTES OF MEMORY
 2     1    TU81-AB    1600/6250 BPI 25/75 IPS 240V
 3     2    CI780-AB   780 INTERPROC BUS ADAPTOR 24   1 OF THESE WERE NOT
                                                      CONFIGURED
 4     2    DR780-FB   DMA CHANNEL, VAL-11/780,120V   1 OF THESE WERE NOT
                                                      CONFIGURED
 5     1    DB86-AA    8600 SECOND SBI ADAPTER
 6     1    H9652-FB   8600 UNI EXP CAB 1 BA11A 240
 7     1    RUA60-CD   RA60-CD, UDA50 CTL, W/CAB
 8     3    RA60-CD    RA60-AA, H9642-AR, 50HZ
 9     5    RA60-AA    205 MB DISK, 50/60HZ, NO CAB
10     1    DD11-DK    DD11-D 2-SU FOR BA11-K
11     1    LA100-BA   KSR TERM W/TRACTOR US/120V
12     1    QK001-HM   VAX/VMS UPD 16MT9

COMPONENTS ADDED
LINE  QTY   NAME       DESCRIPTION              COMMENT

13     1    DW780-MB   8600 SECOND UNIBUS ADAPTER    NEEDED BY THE UNIBUS
                                                     MODULES
14     1    H9652-CB   8600 SBI EXP CAB SWHB 240V3P  NEEDED TO PROVIDE SPACE
                                                     FOR ADAPTORS
15     1    HSC5X-BA   DISK DATA CHANNEL SUP 4 DISK  NEEDED FOR A RA60-AA*

ERROR WARNING

**** THIS NON-STD ORDER REQUIRES MGMT APPROVAL.  THERE ARE MISSING MENU ITEM(S)
FROM THE FOLLOWING MENU(S): LOAD-DEVICE
```

Figure 4.10: Part of the report generated for an order by XCON. The rest is shown in the next figure.

```
CABINET LAYOUT

|------------|
|CONSOLE     |
|LA100-BA 0  |
|            |
|            |
|            |
|            |
|------------|


|------------|--------------------|------------|------------|
|70-19218-01 |70-19219-01         |H9652-CB 1  |H9652-FB 1  |
|FEC CAB # 0 |CPU/KA86            |SBI CAB # 1 |UEC CAB # 1 |
||----------||                    |            |            |
||RL02-FK 0 ||                    |            |            |
||          ||                    |            |            |
||----------||                    |            |            |
|           |                     |            |            |
|           |                     |            |            |
||----------||                    |            ||----------||
||BA11-AL 1 ||                    |            ||BA11-AM 1 ||
||UBA 0     ||                    |            ||UBA 1     ||
||----------||                    |            ||----------||
|           |                     |            |            |
|           |                     |            |            |
|------------|--------------------|------------|------------|


|------------|  |------------|  |------------|  |------------|  |------------|
|H9642-AR 1  |  |H9642-AR 2  |  |H9642-AR 3  |  |H9642-AR 4  |  |TU81-AB 1   |
||----------||  ||----------||  ||----------||  ||----------||  |            |
||RA60-AA 1 ||  ||RA60-AA 1 ||  ||RA60-AA 1 ||  ||RA60-AA 1 ||  |            |
||          ||  ||          ||  ||          ||  ||          ||  |            |
||----------||  ||----------||  ||----------||  ||----------||  |            |
||RA60-AA   ||  ||RA60-AA   ||  ||RA60-AA   ||  ||RA60-AA   ||  |            |
||UNIT # 4  ||  ||UNIT # 3  ||  ||UNIT # 2  ||  ||UNIT # 1  ||  |            |
||----------||  ||----------||  ||----------||  ||----------||  |            |
|           |  |           |  |           |  ||RA60-AA   ||  |            |
|           |  |           |  |           |  ||UNIT # 0  ||  |            |
|------------|  |------------|  |------------|  ||----------||  |------------|
```

Figure 4.11: The rest of the report generated by XCON.

4.7   Learning Rules in SOAR

Learning has always been a weak point of symbolic rule-based AI programs. A notable exception to this has been a program called SOAR.[6] SOAR was originally designed to be a heuristic search system with the initial goal to design as general purpose a program as possible to do as many problems as possible. Normally then, SOAR is discussed in terms of its searching ability, however, in this section we will look at the ability of newer versions of SOAR to learn rules as it searches. Some experiments with SOAR show that its performance gives the Power Law of Practice results first discovered in studies of human learning.

4.7.1   A Searching Example

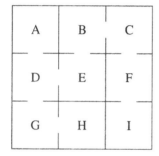

Figure 4.12: This figure shows the layout of some rooms and the task will be to find a way to go from room A to room H.

For a simple example of searching, suppose we have the problem of trying to get from one particular room to some other particular room.[7] Figure 4.12 shows the layout of the rooms. Suppose the goal is to move from room A to room H. There are very few choices here and the tree in Figure 4.13 shows the possible moves a searching program could make. At room A the only room you could move to is D. At D you have two choices and since you do not have any heuristics to lead you in the right direction you could choose to go to either room E or G. Suppose you move to E. From there on there are not any choices and you have to move along to B, C, F, and finally I where there are no alternatives left. This is a good time to back up and try moving from D to G. At G the only move you can make is to go to H and that solves the problem.

The amount of learning you can do in this example is extremely small. The only thing worth learning is:

> IF   you are in room D and the goal is H
> THEN   go to room G

---

[6] The SOAR homepage at: http://www.cs.cmu.edu/afs/cs/project/soar/public/www/home-page.html contains information on SOAR including binaries for Intel and Macintosh systems.
[7] This example is adapted from an article by Smith and Johnson [216].

Figure 4.13: The search for a way to get from room A to room H is straightforward.

Notice that there are various other things you could remember from this experience like when you were in A you can go to D or when you were in B you can go to C, however, unlike people, SOAR attempts to find and keep only the relevant memories, not memories that are unnecessary. In SOAR, the relevant memories are typically called rules, but they might easily be called relevant memories. These rules are also called *chunks* and the process of forming them is called *chunking*.[8]

SOAR has been able to solve an impressive array of problems from simple puzzles up to real world type tasks. One system learned the MYCIN rules [251][9] while another learned some of the hardest parts of the R1 VAX computer configuration task [95].

4.7.2   The Power Law of Practice

In a work published in 1926, Snoddy [219] showed that the time, t, it takes for a human being to trace geometric figures in a mirror depends on the amount of time, N, used to practice the task. The exact relationship was given by the formula:

$$t = bN^{-\alpha}$$

where b and α are constants that have to be determined experimentally. This formula says that the time it takes to do the task will initially be quite long with little practice (a small value for N), but as the task is practiced more, the time to do the task becomes shorter.

---

[8] The terms chunk and chunking come from the research of Miller [122], however, it is not clear if these chunks are exactly the same as Miller's chunks.

[9] Code is available at: http://www-cgi.cs.cmu.edu/afs/cs.cmu.edu/project/ai-repository/ai/areas/planning/systems/soar/neomycin/0.html.

Since then, other researchers have found this same power law of practice holds for many other activities as well such as recalling facts, editing with text editors, checking proofs, and playing solitaire [140]. Newell and others have shown that SOAR programs also produce this power law of practice [141]. In essence, what happens in a SOAR program at the start of learning is that searching takes up quite a lot of time, however, as the chunks build up, the searching time decreases because previously saved results minimize the need to search. Because SOAR programs also generate the power law of practice results, Newell believed that searching and chunking were perhaps the most fundamental aspects of intelligent behavior.[10]

4.8   Rules versus Networks

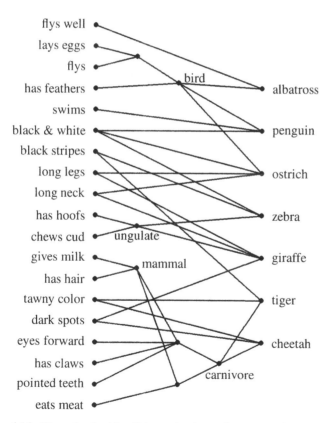

Figure 4.14: The rules for identifying animals can be put together to form a large, sparse, feedforward network. The weights and thresholds for the nodes are not shown. The bird, ungulate, mammal, and carnivore nodes are OR nodes while the rest are AND nodes.

layer	unit	unit value	weight
2	1	0.94134	-3.09056
2	2	0.00147	-0.38760
2	3	0.42901	-4.29914
2	4	0.66736	-0.28191
2	5	0.02098	-0.96560
2	6	0.00156	-2.02424
2	7	0.61722	-0.62367
2	8	0.01811	7.90404
2	9	0.05367	0.51293
3	b	1.00000	-0.78887

Figure 4.15: This figure shows some of the weights in a 19-9-7 backprop network trained to do the animal identification problem. In particular, these are the weights leading into the first output unit. Unlike a rule-based network, there is no easy way to determine the meaning of the weights. For that matter, the "hidden units" in the rule-based network also mean something. The unit value column above lists the hidden unit values in this network and there is no simple interpretation for this collection of values.

As noted in Chapter 1, rules can easily be viewed as small neural networks. If we take the rules in the animal identifying example they can be assembled into the one larger network shown in Figure 4.14 that, of course, looks much like a conventional back-propagation network but with far fewer connections between the nodes. The introduction of hidden units to recognize birds, mammals, ungulates, and carnivores can cut down on the overall amount of processing and of course, they also serve to divide animals into groups with similar characteristics. Another important difference to note is that the units and their connections can easily be stated in English. This is in marked contrast to the type of network you would get from training a back-propagation network on the animal data where you end up with only a collection of incomprehensible numbers as in Figure 4.15. Clark [18] has said that the symbolic rule-based systems have *semantic transparency* because the meaning of each component is easy to read out, whereas the back-propagation type of network does not have semantic transparency. Rule-based systems can therefore quote the rules they used in reaching their decision. Neural systems cannot do this, but if the database of training data is available, they can cite cases to back up their decision. In some applications like, say, a loan approval system, both the rule-based and neural systems can give reasons for rejecting an application in another way. The process can be done by varying the inputs and showing what kind of characteristics a person would have to possess in order to have their loan approved. Then too, it must not be forgotten that the rules in a rule-based system ultimately came from specific cases. Thus, if you ask a rule-based system why it made the decision it did and it quotes you a rule, you should go ahead and ask it where the rule came from!

It is hard to do a fair assessment of rules and networks for use on a given application. For one thing, there are very few expert system projects that have been implemented both

---

[10] See [141], page 96, however, Newell also noted that an argument could be made that pattern match processes were equally fundamental.

ways. For another, it is also possible to do a poor job with one version and a better job with the other and then the results do not mean much. To get fair results from a humanly built network of rules, the expert system creation process should be done independently by a number of different teams and this would be a very expensive experiment to do. Also, the results you get from backprop networks can often vary by quite a lot and there are many techniques that can be applied to improve the generalization of a network. To get fair results for backprop networks you really have to try all these options.

In one report, Saito and Nakano [185] compared a symbolic medical diagnostic expert system with a back-propagation-based one. The network was trained on only 300 cases involving headaches as the only symptom and with 23 possible diagnoses. On a set of test data the network was correct 67 percent of the time versus 70 percent for the symbolic system.

In another experiment, Bradshaw, Fozzard, and Ceci [13] report on two systems to predict solar flares. Both the symbolic and back-propagation versions predict solar flares equally as well as human forecasters. One big difference was that the symbolic system required over one man year of work (over 700 rules) while the network was developed in less than a week. A second big difference is that the symbolic system takes five minutes to do a single prediction while the network takes only a few milliseconds.

There have also been experiments in extracting rules from networks. In the Saito and Nakano experiment mentioned above they did this as well. Thrun has also experimented with extracting rules [234].[11] Another approach by Shavlik and Towell [237, 204][12] is to start with rules, transform the rules into a backprop network, train the network, and then extract rules from the network. In this experiment they found that the final rule set performed better than the network and the network performed better than the initial rules.

4.9 Exercises

4.1. Given data such as:

```
father(john,mary).
father(john,ted).
father(ted,larry).
father(ted,bill).
mother(mary,alice).
mother(alice,carol).
```

Define a rule that will be able to answer questions about who is the grandfather of whom, such as this one:

```
?- grandfather(john,X).
```

Also give rules that will define the grandmother, great grandmother, aunt, uncle, brother, sister, and sibling relationships.

4.2. When the factorial predicate is given the problem: fact(3,A), it reports back that A = 6. What happens if you type a ";" now, instead of a carriage return?

---

[11] For related publications see: http://www.informatik.uni-bonn.de/~thrun/publications.html.

[12] For related publications see: http://www.cs.wisc.edu/trs.html.

4.3. Write a Prolog program to find the nth Fibonacci number.

4.4. For the following structures in list notation, give the structure in the notation that uses
. as a functor that composes trees. Then diagram the trees.

```
[a, [b, c]]
[[a, b], [c, d]]
[[c, [d, [e, f]]], a, b]
```

4.5. Write a Prolog function that will determine if all the members of a list, L, are also
members of a list, X. For example, are all the members of the list, [b,1], present in the list,
[a,1,b,c,x,4,z]?

4.6. Write a Prolog function, vowel, that will take a list of letters and return a list of all the
vowels in the original list. For example:

```
?- vowel([c,h,a,p,t,e,r],X).
```

should give:

```
X = [a,e].
```

Also write a function, consonant, that will return all the letters that are not vowels.

4.7. Write a Prolog function, nodupl, that will remove duplicate entries from a list. For
example:

```
?- nodupl([h,e,l,l,o,w,o,r,l,d],X).
```

should give:

```
X = [h,e,l,o,w,r,d].
```

4.8. Here is a function, append, that runs its first two arguments (lists) together to form a
new list in the third argument:

```
append([],L,L).
append([A|L1],L2,[A|L3]) :- append(L1,L2,L3).
```

So, for instance:

```
?- append([a,b],[c,d],X).
```

gives X = [a,b,c,d]. Trace through the above function call giving all the intermediate results.

4.9. In the text we programmed simple forward and backward chaining interpreters to handle the rule base for the animal identification program. It is a little simpler to let Prolog do
the interpretation by giving it the rules about animals directly, such as in:

```
albatross(X) :- bird(X), flieswell(X).
bird(X) :- hasfeathers(X).
bird(X) :- layseggs(X), flies(X).
```

In the first case, this says that X is an albatross if X is a bird and X flieswell. Write a Prolog program that will do the animal identification problem using data in this format using backward chaining. Write a second program that will use forward chaining.

4.10. Program the forward and backward chaining versions of the animal identifying problem using whatever language is available and convenient. For the backward chaining version, have the program ask questions about whether or not an animal has a certain characteristic if there is no rule available to deduce that the characteristic is true. Make your program ask a question about a given characteristic only once. For instance, if early in the identification process the program asks if the animal has feathers and the user says no, then have the program remember this fact so that it will not ask it again.

4.11. Suppose you are doing the animal identification problem and you have to deal with incomplete data about the unknown animal and you want to ask the user more questions about the animal. How well will the MYCIN activation function work on this problem?

4.12. Suppose you have the data about the seven animals in the animal identification problem as vectors. Find out how well a simple nearest neighbor approach using Euclidean distance will perform at identifying unknown animals when one or two characteristics of each animal are omitted from the description of the unknown animal. Compare this with the results you get from a back-propagation network.

4.13. Hilary Putnam, a skeptic with regards to the accomplishments of AI, has said that expert systems are "just high-speed data-base searchers [165]." Is this true? If it is true, is this bad?

4.14. The rules at the end of Section 4.5 for packing Christmas tree ornaments do not consider what happens if:

    1) you run out of large ornaments to be packed,

    2) you run out of small ornaments to be packed.

Write a more complete set of rules to take these possibilities into account. Do not neglect the case where a box may end up with less than six large ornaments so that there is more room for small ornaments. Assume three small ornaments occupy the same space as one large ornament. You can invent new procedures and variables as necessary. If you know Prolog or Lisp then you may want to program this system with one of these languages. Another possible way to do this problem would be to train a number of cases into a back-propagation network and let the network choose which move to make. How well would this work?

4.15. Instead of having a rule interpreter, why not just write IF-THEN statements in a general purpose program in a language such as C? Consider the impact this would have in doing problems in:

    1) a forward chaining manner,

    2) a backward chaining manner, and

    3) what happens with rule conflicts.

4.16. Get a copy of a book for identifying wildflowers, such as: *A Field Guide to Wildflowers of Northeastern and North-Central North America* by Peterson and McKenny. Produce an expert system to identify *as many* wildflowers as is possible and convenient. Produce a convenient interface for the system so that users can type in characteristics of the flowers as words. In case the user gives an insufficient number of details about an unknown flower, arrange to have the program ask questions about other possible characteristics. If a person submits an unknown flower that does not match any flowers that the program knows about, have the program say it is not sure and have it indicate which known flower the unknown is closest to. There are many ways to do this problem. Choose one or more of the following methods to implement this system. In a class, you may want to assign different students to program different methods and then the results can be compared. Some of the factors that should be used to compare the methods are the ease of programming, size of the code, and execution time. Here are the methods:

a) Use one large back-propagation network.

b) Try using more than one network to learn all the flowers. This gives you the problem of dividing up the flowers among the networks in some appropriate way.

c) Use the Euclidean nearest neighbor scheme or a nearest neighbor scheme based on the number of characteristics the unknown has in common with flowers in the database.

d) Find out how well this variation on a nearest neighbor scheme works: use a nearest neighbor scheme to locate the n nearest neighbors to the unknown flower, where n is perhaps 5 or 10. If there is a perfect match, let this be the answer, otherwise train these n candidates into a back-propagation network and submit the unknown to this network to find out the most likely answer.

e) Use forward and backward chaining in Lisp, Prolog, or some other convenient language.

f) Use any other promising method(s) you can think of.

Again, try to use as many different flowers as is possible and convenient. Note the advantages and limitations of each method as the number of flowers in the system grows. If you do not have time to program any of these methods, you could still try to evaluate the methods without doing any programming. Also, if you do not want to identify wildflowers, you can evaluate the methods for any other application area you can think of.

4.17. One major reason for a company to produce an expert system is to capture the expertise of its human experts who may retire or quit. How else could a human expert transfer his expertise to other people? How was this done before computers?

Chapter 5

Logic

As was stated earlier, Prolog uses a subset of a type of logic called the predicate calculus. The full predicate calculus is much more powerful than Prolog. In this chapter we want to look at the notation and capabilities of predicate calculus and compare them with Prolog. In Prolog, the user asks questions about the database of facts and rules. In predicate calculus the questions are regarded as theorems to be proved so the subject is often called automatic theorem proving. The subject is also often referred to as automated reasoning. The notation we will use comes from a public domain automated reasoning program called Otter[1] with Prolog style variables. Some examples are taken from a collection of problems called the "Thousands of Problems for Theorem Provers" (TPTP) collection of Geoff Suttcliffe and Christian Suttner.[2]

5.1 Standard Form and Clausal Form

Predicate calculus is a form of logic that can express facts about a problem and that uses specific rules to reach valid new conclusions. Of course some of the standard relations in this logic are the familiar "and," "or," and "not," but it also includes the relations "implies" ($\rightarrow$), "is equivalent to" ($\leftrightarrow$), and the quantifiers for variables, "for all" ($\forall$) and "there exists" ($\exists$). One notation in predicate calculus called *standard form* uses all these symbols, however, statements written in this form can be translated into another form called *clausal form* where these latter four symbols are not used. It turns out that clausal form is much more convenient for computer manipulation than standard form so that is the form we will be using here, except we will begin by showing how "implies," "is equivalent to," "for all," and "there exists" can be eliminated from predicate calculus statements.

In the way of notation, we will be using '&' for 'and,' ' | ' for 'or,'[3] and '−' for 'not.'

---

[1] Otter is from Argonne National Laboratory. It includes C code, a 32-bit DOS binary, a Macintosh version, user manuals in various formats, and some sample problems. It is available at http://www.mcs.anl.gov/home/-mccune/ar/otter/index.html and ftp://info.mcs.anl.gov/pub/Otter.

[2] This collection is rather large, a gzipped file of 1M which, when gunzipped comes to 18M. In all likelihood the only problems you will ever want are the ones in the puzzles directory. This package is available from ftp://flop.informatik.tu-muenchen.de/pub/tptp-library, http://wwwjessen.informatik.tu-muenchen.de/~suttner/tptp.html, ftp://coral.cs.jcu.edu.au/pub/research/tptp-library, and http://coral.cs-jcu.edu.au/-users/GSutcliffe/WWW/TPTP.HTML.

[3] If predicate calculus used lists as in Prolog you might get confused between the two different uses for | , however the only use for | in this chapter will be as the 'or' symbol. It is quite common to use ∨ for 'or,' however Otter uses | not ∨.

Predicates and constants will be written as in Prolog. Variables, as in Prolog, will still start with an uppercase character. Predicate calculus also includes functions. In terms of notation, functions will look just like predicates and whenever we use a function it will be identified as a function and not a predicate. In many cases, functions can be replaced by predicates, with no ill effect.

A predicate is a symbol that denotes a relation and its arguments may be variables, constants, or functions. A predicate, with or without a not sign in front of it, is a *literal*. Some literals are then:

$$likes(X, Y)$$
$$-likes(john, mets)$$
$$mammal$$
$$-zebra$$

Literals with a not sign are called *negative literals* and literals without a not sign are called *positive literals*. A clause is a series of one or more literals or-ed together, such as in:

$$b$$
$$-a$$
$$-a \mid -b \mid c$$
$$likes(matt, mets) \mid likes(nancy, yankees)$$

A clause with one literal, either positive or negative, is called a *unit* clause; a clause with more than one literal is a *nonunit clause*; a clause with all positive literals is a *positive clause*; a clause with all negative literals is a *negative clause*; and a clause containing both negative and positive literals is called a *mixed clause*.

Clausal form is a series of clauses and-ed together, such as:

$$likes(matt, mets) \ \& \ likes(nancy, yankees) \ \& \ likes(carol, cubs)$$

Normally, clausal form assumes the presence of the & symbol between clauses so the & operators are not written, instead, clauses are written out one per line and the &s are implied:

$$likes(matt, mets)$$
$$likes(nancy, yankees)$$
$$likes(carol, cubs).$$

Again, as in Prolog, all the variables in a logic statement will be universally quantified, so there will be no need for the "for all x" type of quantifier used in standard form. The other type of quantification in predicate calculus is existential quantification that says that "there exists" a certain particular quantity. One example of this would be the statement that, "Every state has a capital city." In predicate calculus this would be:

$$\forall(S, state(S) \rightarrow \exists(C, capitalof(C, S))).$$

This reads, "for all S, if S is a state then there exists a city, C, such that C is the capital of S." Clausal form eliminates the need for existential quantification by assuming there is a function that finds the quantity that exists. In this case, there could be a function, call it $capitalf$, such that, when it is given a state, S, it gives the capital of the state. The above predicate calculus statement can then be translated to:

$$\forall(S, state(S) \rightarrow capitalof(capitalf(S), S)).$$

Of course, if all variables are assumed to be universally quantified, this can be written as:

$$state(S) \rightarrow capitalof(capital f(S), S).$$

One of the easiest parts of the translation from standard form to clausal form is to eliminate the "is equivalent to" operator as in:

$$a \leftrightarrow b.$$

This is read as "a is equivalent to b" or "a implies b and b implies a." This can be rewritten as:

$$a \rightarrow b \ \& \ b \rightarrow a.$$

Implies is the same thing as an IF statement, so this statement can also be written as:

IF a THEN b
IF b THEN a.

The most important part of translating from standard form to clausal form is to realize that an IF statement of the form:

IF a THEN b

can be translated to:

$$-a \mid b.$$

In English, this latter expression is *"not a or b."* To justify this translation you need to realize that the whole statement, "IF a THEN b," is a statement that by itself is either true or false. This is different from natural language where normally people give IF statements where both a and b are true. English statements are normally meant to be informative in a positive sort of way and not confusing. Below we list the possible cases of "IF a THEN b" in order from obvious to strange:

case	a	b	the whole statement
1	*true*	*true*	*true*
2	*true*	*false*	*false*
3	*false*	*true*	*true*
4	*false*	*false*	*true*

Case 1 is not painful. This is the way that people normally communicate important relationships. It is the normal use for IF statements. Case 2 is slightly stranger in that the statement is, in effect, a lie, but this is still easy to relate to. But now cases 3 and 4 are peculiar. A couple of IF statements to illustrate cases 3 and 4 might be:

IF an elephant is pink THEN the elephant cannot fly.
IF an elephant is pink THEN the elephant flies.

In the real world under ordinary circumstances elephants are not pink, so it does not make any sense to even say these things. Mathematically inclined people are not always so sensible and instead of saying that statements like these are silly, they declare them to be "vacuously true" on the grounds that you cannot find a counter-example to prove them false. In practical situations, letting them be true does no harm and does some good because then a statement of the form, "IF a THEN b," has the same Boolean value as "*not a or b*," as shown below:

$$a \qquad -a \qquad b \qquad -a \mid b$$

$$
\begin{array}{llll}
true & false & true & true \\
true & false & false & false \\
false & true & true & true \\
false & true & false & true \\
\end{array}
$$

The translation from IF statements is therefore exceedingly simple: replace the IF with − and the THEN with | . For a larger example, we will take these rules about identifying a zebra:

> IF *hashair* THEN *mammal*
> IF *mammal* AND *chewscud* THEN *ungulate*
> IF *ungulate* AND *blackandwhite* AND *blackstripes* THEN *zebra*

Translating the rules about identifying a zebra, we get:

> −*hashair* | *mammal*
> −*mammal* | −*chewscud* | *ungulate*
> −*ungulate* | −*blackandwhite* | −*blackstripes* | *zebra*

In doing this translation we made use of the fact that $-(a \mathbin{\&} b)$ is the same as $-a \mid -b$.

A feature present in predicate calculus that is missing in Prolog is the ability to reach negative conclusions. For instance, take the statement:

$$\text{IF } male(X) \text{ THEN } -female(X),$$

This says that if X is male then X is not female. Prolog has no way to say this, since it does not have the capability of stating a negative conclusion. (Of course, there is a not operator that will invert a Boolean value, but it cannot be used to find negative consequences.) The closest you can come in Prolog in this regard is to create a predicate, notfemale, and use it like so:

$$notfemale(X) \; :- male(X).$$

Another important feature missing from Prolog is the ability to accept translations from statements of the form:

$$\text{IF } a \text{ THEN } (b \mid c).$$

In clausal form this corresponds to:

$$-a \mid b \mid c.$$

To be able to accept this translation, Prolog would have to allow multiple predicates to the left of the ": −" operator such as this where the ";" means "or":

$$b; c : -a.$$

This type of statement has just simply been banned from Prolog so that it can be efficient in trying to reach conclusions. Clauses, such as those in Prolog where there is only one possible positive conclusion, that is, one positive literal, are called *Horn clauses*. In Prolog, the one positive literal goes to the left of the ': −' while all the negative literals go on the right without any not signs.

5.2   Basic Inference Rules

When it comes to manipulating clauses to look for conclusions, there are some nice simple *inference rules* for doing so. Inference rules allow valid conclusions to be drawn. For the time being, to keep things simple, we will work with clauses where the predicates do not have any arguments. After that we will look at the more complicated situation.

5.2.1   Inference Rules

The first of these inference rules we look at will be *binary resolution*. In binary resolution, we select two clauses, one of which has a particular positive literal in it, and the other has the same literal but with a not sign in front of it. An example of binary resolution can be done with the following two clauses:

$$-a \mid b \qquad \text{(IF } a \text{ THEN } b)$$
$$-b \mid c \qquad \text{(IF } b \text{ THEN } c)$$

The obvious conclusion to draw from these two statements is, "IF a THEN c." To draw the conclusion in clausal form using binary resolution, all that is necessary is to run the two clauses together into a new statement, but canceling out the b and $-b$, giving:

$$-a \mid c \qquad \text{(IF } a \text{ THEN } c)$$

This new conclusion, $-a \mid c$, is called a *resolvent* of $-a \mid b$ and $-b \mid c$. "Running statements together" and canceling the matching literals of opposite sign is allowed only if a single pair of literals is canceled. Notice that this definition prohibits resolving together the following two clauses and canceling *two* pairs of literals (but not one):

$$-a \mid b$$
$$-b \mid a.$$

Some other valid examples of binary resolution can be done with the following pairs of clauses:

$$-a \mid b \text{ and } a \text{ gives } b$$

$$-a \text{ and } a \text{ gives}$$

$$-a \mid -b \mid -c \text{ and } c \text{ gives } -a \mid -b$$

The middle example produces the *empty clause*.

A second inference rule is *UR-resolution*, for Unit Resulting resolution. UR-resolution takes one nonunit clause and one or more unit clauses and produces a unit clause. For instance, UR-resolution can be performed on the following set of clauses:

$$-a \mid -b \mid c$$
$$-c$$
$$b$$

In this case, c and $-c$ cancel and b and $-b$ cancel, giving a new clause, $-a$.

A third inference rule is *hyperresolution*. Hyperresolution is used to produce a positive (but not necessarily unit) clause by starting with a negative or mixed clause and some positive clauses. Again, literals with opposite signs cancel, but each of the positive clauses can only use *one* of its positive literals to cancel with a negative literal in the negative or mixed clause. An example of hyperresolution can be done with the clauses:

$$-a \mid -b \mid c$$
$$a \mid d$$
$$b$$

and produces:

$$c \mid d$$

There are many other inference rules that can be used but we will not attempt to cover them.

5.2.2 Clauses with Variables

So far we have been applying inference rules using only predicates without arguments. We now look at the more complicated case of using inference rules with structures and variables. Resolution can be applied to two clauses when the two clauses can be "unified." Unification is just a form of pattern matching with variable instantiation, like that done in Prolog programs. In a really simple case there are these two clauses:

  1   $-likes(X, lemonade) \mid likes(X, yankees)$
  2   $likes(nancy, lemonade).$

In trying to match the two clauses, 1 and 2, the first literal of 1 is the negative of 2, if X is instantiated to the value, *nancy*:

  1′ $-likes(nancy, lemonade) \mid likes(nancy, yankees).$

The two clauses, 1′ and 2 can be resolved using binary resolution giving:

  3   $likes(nancy, yankees).$

In a case using hyperresolution we have:

  4   $-father(X, Y) \mid -father(Y, Z) \mid grandfather(X, Z)$
  5   $father(adam, seth)$
  6   $father(seth, enosh)$

and this gives:

$$grandfather(adam, enosh).$$

Notice that if we had the following clause, 5′, instead of 5, the resolution could not proceed because it would not be possible to do a consistent set of instantiations:

5′ $father(adam, cain)$.

As in Prolog, variables can match variables and each variable has meaning only within its own clause. When you do the matching, rename all the variables in the relevant clauses so that each variable is used only once in the set of relevant clauses. For instance, trying to keep everything straight in this set of clauses for hyperresolution can be confusing:

$$-b(Z, X) \mid -c(Y, X) \mid d(X, Y, Y, Z)$$
$$b(X, X)$$
$$c(X, Y).$$

These clauses can be changed into:

$$-b(Z1, X1) \mid -c(Y1, X1) \mid d(X1, Y1, Y1, Z1)$$
$$b(X2, X2)$$
$$c(X3, Y3).$$

It is easiest to do this problem slowly as two binary resolutions. Using the first two clauses you get $Z1$ matching $X2$, and $X1$ matching $X2$, giving:

7   $-b(X2, X2) \mid -c(Y1, X2) \mid d(X2, Y1, Y1, X2)$
8   $b(X2, X2)$
9   $c(X3, Y3).$

Clauses 7 and 8 can be resolved to give:

7′ $-c(Y1, X2) \mid d(X2, Y1, Y1, X2).$

Clauses 7′ and 9 can then be transformed to:

7″ $-c(X3, Y3) \mid d(Y3, X3, X3, Y3)$
9   $c(X3, Y3)$

and resolved to:

10   $d(Y3, X3, X3, Y3).$

In addition, variables can match predicates and functions as in:

$$-harms(Y1, X1) \mid -do(Y1)$$
$$-command(Z2, shoot(X2)) \mid harms(shoot(X2), Y2).$$

Resolution produces:

$$-command(Z3, shoot(X3)) \mid -do(shoot(X3)).$$

A serious problem occurs, however, with these two clauses:

$$a(X, X)$$
$$-a(f(Y), Y).$$

Here, X will match $f(Y)$, but $Y = X$, so $f(Y) = f(X) = f(f(X))$, and so on. This represents an unacceptable match and a test must be made to avoid problems like this. The test is called an *occur check*. Many Prologs do not implement this test since it requires a fairly large amount of time to constantly check for this rare situation. Other Prologs allow the user to turn this checking on or off.

5.3   Controlling Search

Prolog has a specific search algorithm for trying to reach conclusions, but in resolution, there is no specific search algorithm other than to try to resolve together everything you can think of. This leaves a person or a program free to resolve together some set of clauses that the person or program feels will be beneficial in finding the solution to a problem. Naturally, people have a better feel than computers for choosing important clauses to resolve together. A major topic of automated reasoning is to try to find ways of speeding up the search process. In this section we will look at some of the search algorithms that have been developed so far. None of these search strategies (or any others) comes with a guarantee that it will speed up the search process. Furthermore, only experienced human beings can make a reasonable guess as to what method is best for a particular problem.

5.3.1   The Problem with Blind Searching

The problem we will be working with will be to identify a zebra given the following clauses:

A1	$-hashair \mid mammal$
A2	$-givesmilk \mid mammal$
A3	$-mammal \mid -hashoofs \mid ungulate$
A4	$-mammal \mid -chewscud \mid ungulate$
A5	$-ungulate \mid -blackandwhite \mid -blackstripes \mid zebra$
A6	$hashair$
A7	$chewscud$
A8	$blackandwhite$
A9	$blackstripes$

A normal human solution would be to try to derive the clause, "zebra," all by itself. One way to do this is the following. First, resolve clauses A1 and A6 together to give:

A10 $mammal.$

The next step is to take A4, A7, and A10 together to give:

A11 $ungulate.$

The final step is to take A5, A8, A9, and A11 together to give:

A12   *zebra*

and the proof is complete.

Proofs by human beings are seldom as simple as this. Proofs completely by computer are much harder because computers do not have the ability to look ahead. The computer has no reason to prefer the above proof finding steps. It could just as easily choose pairs of clauses that give these useless, but valid, conclusions:

$-hashair \mid -hashoofs \mid ungulate,$
$-hashair \mid -hashoofs \mid -blackandwhite \mid -blackstripes \mid zebra,$
$-givesmilk \mid -hashoofs \mid ungulate,$
and so on.

The computer program is effectively trying to cross a stream on rocks while lost in a fog and without having *any* heuristics to guide it. Notice too, how in the resolution-based method of theorem proving, *whole new clauses are added* to the database of facts and rules. The difficulty involved in finding proofs is compounded by the fact that, the more clauses there are in the database, the more time it takes to search the database and derive new clauses. Sometimes so many new clauses are derived that the program (or a human user controlling the reasoning program) has to dispose of clauses that seem to be useless. Needless to say, when disposing of any clauses, you run the risk of disposing of something important and you may make the whole proof impossible. The Prolog searching strategy was designed to, among other things, make it unnecessary to store large numbers of clauses.

5.3.2   Proof by Contradiction

One of the most basic methods used by automated reasoning programs is the method of proof by contradiction. It is often used in combination with other methods. It works as follows. If we start with a set of clauses that, by themselves, are consistent (meaning there are no contradictions within them), then if we want to prove c, add $-c$ to the list of clauses, start resolving clauses, and look for a contradiction to develop. Another way to state this proof by the contradiction approach is to say that if c is true, adding $-c$ to the list of clauses will eventually produce the empty clause. For an example of this, done again in a human fashion, here is the zebra identification problem with a 10th clause, $-zebra$:

B1 $-hashair \mid mammal$
B2 $-givesmilk \mid mammal$
B3 $-mammal \mid -hashoofs \mid ungulate$
B4 $-mammal \mid -chewscud \mid ungulate$
B5 $-ungulate \mid -blackandwhite \mid -blackstripes \mid zebra$
B6 $hashair$
B7 $chewscud$
B8 $blackandwhite$
B9 $blackstripes$
B10 $-zebra$

We can use B5, B8, B9, and B10 together to give B11:

B11 $-ungulate.$

We can take B1 and B6 to give:

B12 $mammal.$

Now take B4 and cancel its terms with B7 and B11 to give:

B13 $-mammal.$

With this contradiction the proof is finished. Of course, we could always go a step farther and resolve B12 and B13 to give the empty clause. Notice that with this strategy it seems like there should be many ways to find a contradiction and generate the empty clause.

5.3.3 The Set-of-Support Strategy

This proof by contradiction strategy can be used effectively with other strategies. The first such strategy we will show is the *set-of-support* strategy. The zebra problem is shown below with the clauses renamed:

	C1	$-hashair \mid mammal$
	C2	$-givesmilk \mid mammal$
	C3	$-mammal \mid -hashoofs \mid ungulate$
	C4	$-mammal \mid -chewscud \mid ungulate$
	C5	$-ungulate \mid -blackandwhite \mid -blackstripes \mid zebra$
*	C6	$hashair$
*	C7	$chewscud$
*	C8	$blackandwhite$
*	C9	$blackstripes$
*	C10	$-zebra$

Clauses C6 through C10 have been marked with asterisks. This set of marked clauses can also be regarded as a separate list, the set-of-support list. The idea behind the set-of-support strategy is that certain specific facts (and maybe rules) about a problem are more important to concentrate on. These get marked with an asterisk. When it comes time to resolve two different clauses, the set-of-support strategy requires that at least one of the clauses must be marked with an asterisk. The clause that results will also get an asterisk. For instance, we can start with the first marked clause, C6, and resolve it with C1 to give C11:

*C11 $mammal$

On the other hand, we will not resolve, for instance C1 and C3, because they are not marked. In more detail, the set-of-support strategy works as follows. First choose a clause from the set-of-support list of clauses. This clause is called a "given clause" and you try to resolve it with all the other clauses in the list. After it has been resolved with all the other possible clauses, it is marked as having been used (and perhaps placed on a separate list to indicate this), so we will not attempt this procedure with it again. Some later clause may, however, be resolved with this clause so this clause is not completely eliminated from the problem. Continuing with the problem, we notice that C6 has been resolved with as many other clauses as possible. We now move on to the next clause in the set-of-support group, clause C7. It can be resolved with C4 giving:

*C12 $-mammal \mid ungulate$

Continuing in this way gives:

C8 + C5 → *C13 $-ungulate \mid -blackstripes \mid zebra$
C9 + C5 → *C14 $-ungulate \mid -blackandwhite \mid zebra$
C9 + C13 → *C15 $-ungulate \mid zebra$
C10 + C5 → *C16 $-ungulate \mid -blackandwhite \mid -blackstripes$
C10 + C13 → *C17 $-ungulate \mid -blackstripes$
C10 + C14 → *C18 $-ungulate \mid -blackandwhite$
C10 + C15 → *C19 $-ungulate$

At this point, we have run out of clauses from the original set-of-support list, but we continue with the new clauses that have been added. Continuing on with C11 gives:

C11 + C12 → *C20 $ungulate$

A program will note that C19 and C20 give the empty clause and declare the proof finished.

Notice how this procedure had some value in moving the search in the right direction. It did not bother to generate clauses such as:

$-hashair \mid -hashoofs \mid ungulate,$
$-hashair \mid -hashoofs \mid -blackandwhite \mid -blackstripes \mid zebra,$
$-givesmilk \mid -hashoofs \mid ungulate,$
and so on,

that have no value in solving the problem.

5.3.4 Weighting

There is another scheme called *weighting* that can also help to make a program concentrate on relevant clauses. One method of weighting clauses in the set-of-support group is to rate how large they are by counting the number of symbols in them, not counting commas and parentheses (as it happens, the example that follows does not use any commas or parentheses anyway). Then the clause from the set-of-support group with the lowest weight is resolved with all the other clauses in the set. This amounts to the idea that "if it is getting simpler, it is getting better." To show how this works, we will start in on the zebra problem again:

D1 $-hashair \mid mammal$
D2 $-givesmilk \mid mammal$
D3 $-mammal \mid -hashoofs \mid ungulate$

D4	$-mammal \mid -chewscud \mid ungulate$
D5	$-ungulate \mid -blackandwhite \mid -blackstripes \mid zebra$
* D6 (w=1)	$hashair$
* D7 (w=1)	$chewscud$
* D8 (w=1)	$blackandwhite$
* D9 (w=1)	$blackstripes$
*D10 (w=2)	$-zebra$

Weights for set-of-support clauses are shown in parentheses. There are four candidates in the set-of-support group rated 1. We arbitrarily break this tie by taking the first one, D6, and try to resolve it against all the other clauses, D1 through D10. This gives only:

D6 + D1 → *D11 (w=1) $mammal$

D6 is marked as used to disqualify it from being chosen again. Again, a clause needs to be chosen. There are five clauses rated 1. We arbitrarily break the tie somehow and suppose that clause D11 is the one selected (another tie-breaking consideration might be to work with the most recent clause generated). D11 can be resolved with D3 and D4:

D11 + D3 → *D12 (w=4) $-hashoofs \mid ungulate$
D12 + D4 → *D13 (w=4) $-chewscud \mid ungulate$

Clauses keep being generated in this manner, where the lowest weighted clause is selected to resolve against the other clauses. Finishing this problem is left as an exercise.

5.3.5 Prolog's Strategy

Finally, there is the Prolog method of choosing clauses to resolve. In this method we have one clause (the question) that we start with. We negate it and scan sequentially through the list of clauses until we find a clause where a resolution is possible. We mark this place and start over at the top of the list with this new statement, looking to match some other clause, and so on. The goal is to derive the empty clause. For example, here is the zebra problem again:

E1	$-hashair \mid mammal$
E2	$-givesmilk \mid mammal$
E3	$-mammal \mid -hashoofs \mid ungulate$
E4	$-mammal \mid -chewscud \mid ungulate$
E5	$-ungulate \mid -blackandwhite \mid -blackstripes \mid zebra$
E6	$hashair$
E7	$chewscud$
E8	$blackandwhite$
E9	$blackstripes$

We add the clause $-zebra$ and start searching for a clause this can be resolved with. We find one at E5 and mark this place. Our new question becomes:

$$-ungulate \mid -blackandwhite \mid -blackstripes.$$

Starting with the first of these terms, $-ungulate$, we go to the top of the list and try to find a clause that contains ungulate. We find E3, mark this place and after resolution, we have the problem:

$$-mammal \mid -hashoofs \mid -blackandwhite \mid -blackstripes.$$

This will be resolved with E1 giving:

$$-hashair \mid -hashoofs \mid -blackandwhite \mid -blackstripes.$$

This will be resolved with E6 giving:

$$-hashoofs \mid -blackandwhite \mid -blackstripes.$$

Trying to resolve a clause with this one will fail because there is no clause containing hashoofs. The program will go back up the search tree and try to prove mammal using E4, and so on. The main advantage of the Prolog searching plan is that it saves quite a lot on the storage of clauses and thus the processing will be much faster as well.

5.4   An Example Using Otter

This section gives an additional example of using predicate calculus to solve a little puzzle and shows how the Otter program handles it.

5.4.1   The Problem

The problem is a little example taken from Lewis Carroll.[4] It is:

```
1)  The only animals in this house are cats.
2)  Every animal is suitable for a pet, that loves to gaze
    at the moon.
3)  When I detest an animal, I avoid it.
4)  No animals are carnivorous, unless they prowl at night.
5)  No cat fails to kill mice.
6)  No animals ever take to me, except what are in this house.
7)  Kangaroos are not suitable for pets.
8)  None but carnivora kill mice.
9)  I detest animals that do not take to me.
10) Animals that prowl at night always love to gaze at the moon.
The problem is to prove that ''I always avoid a kangaroo''.
```

Setting up the clauses for problems can often be difficult but in this case the 10 rules translate quite easily to the following clauses:

```
-inhouse(A)  | cat(A).
-gazer(A) | suitablepet(A).
```

---

[4]By way of the TPTP collection. The source given there for the problem is: L. Carroll, *Lewis Carroll's Symbolic Logic: Part I, Elementary, Part II, Advanced*, C.N. Potter, New York, 1986.

```
set(prolog_style_variables).
set(auto).
set(print_kept).
clear(detailed_history).
list(usable).
-inhouse(A) | cat(A).
-gazer(A) | suitablepet(A).
-detested(A) | avoid(A).
-carnivore(A) | prowler(A).
-cat(A) | mousekiller(A).
-takestome(A) | inhouse(A).
-kangaroo(A) | -suitablepet(A).
-mousekiller(A) | carnivore(A).
takestome(A) | detested(A).
-prowler(A) | gazer(A).
kangaroo(b).
-avoid(b).
end_of_list.
```

Figure 5.1: The text submitted to the Otter program.

```
-detested(A) | avoid(A).
-carnivore(A) | prowler(A).
-cat(A) | mousekiller(A).
-takestome(A) | inhouse(A).
-kangaroo(A) | -suitablepet(A).
-mousekiller(A) | carnivore(A).
takestome(A) | detested(A).
-prowler(A) | gazer(A).
```

So for instance, the first clause translates to: "if any animal, A, is in the house then that animal, A, is a cat." The next step is to try to prove that "I always avoid a kangaroo." This can be done by giving the clauses:

```
kangaroo(b).
-avoid(b).
```

The first clause says that b is a specific kangaroo (remember that we are using the Prolog standard for variables where strings that start with uppercase letters are variables and strings that start with lowercase variables are predicates or constants). Now we want to prove that "I always avoid a kangaroo" so you would like to be able to derive avoid(b) but instead, we will deny the goal and look for a contradiction. Strictly speaking, this problem needs a statement to the effect that a cat is not a kangaroo, however, Carroll did not state the problem this way.

All the above clauses were given to Otter in the form shown in Figure 5.1. The first statement specifies Prolog style variables (the default is something else). The second statement gives the program the authority to guess what techniques will be most useful and which clauses will go in the set-of-support. In order to get the program to output all

We have a non-Horn set without equality. Let's try
ordered hyper_res, unit deletion, and factoring, with
satellites in sos with and nuclei in usable.

```
    dependent: set(hyper_res).
    dependent: set(factor).
    dependent: set(unit_deletion).

-----------> process usable:
** KEPT (wt=4): 1 [] -inhouse(A)|cat(A).
** KEPT (wt=4): 2 [] -gazer(A)|suitablepet(A).
** KEPT (wt=4): 3 [] -detested(A)|avoid(A).
** KEPT (wt=4): 4 [] -carnivore(A)|prowler(A).
** KEPT (wt=4): 5 [] -cat(A)|mousekiller(A).
** KEPT (wt=4): 6 [] -takestome(A)|inhouse(A).
** KEPT (wt=4): 7 [] -kangaroo(A)| -suitablepet(A).
** KEPT (wt=4): 8 [] -mousekiller(A)|carnivore(A).
** KEPT (wt=4): 9 [] -prowler(A)|gazer(A).
** KEPT (wt=2): 10 [] -avoid(b).

-----------> process sos:
** KEPT (wt=4): 11 [] takestome(A)|detested(A).
** KEPT (wt=2): 12 [] kangaroo(b).
```

Figure 5.2: The "auto" feature of the Otter program has scanned the input and decided that hyper-resolution may be a good strategy plus factoring and unit deletion, two other strategies we have not mentioned. It has also chosen two clauses, numbered 11 and 12, to go into the set-of-support list.

```
=========== start of search ===========

given clause #1: (wt=4) 11 [] takestome(A)|detested(A).
** KEPT (wt=4): 13 [hyper,11,6] detested(A)|inhouse(A).


given clause #2: (wt=2) 12 [] kangaroo(b).

given clause #3: (wt=4) 13 [hyper,11,6] detested(A)|inhouse(A).
** KEPT (wt=4): 14 [hyper,13,1] detested(A)|cat(A).


given clause #4: (wt=4) 14 [hyper,13,1] detested(A)|cat(A).
** KEPT (wt=4): 15 [hyper,14,3] cat(A)|avoid(A).


given clause #5: (wt=4) 15 [hyper,14,3] cat(A)|avoid(A).
** KEPT (wt=4): 16 [hyper,15,5] avoid(A)|mousekiller(A).


given clause #6: (wt=4) 16 [hyper,15,5] avoid(A)|mousekiller(A).
** KEPT (wt=4): 17 [hyper,16,8] avoid(A)|carnivore(A).


given clause #7: (wt=4) 17 [hyper,16,8] avoid(A)|carnivore(A).
** KEPT (wt=4): 18 [hyper,17,4] avoid(A)|prowler(A).


given clause #8: (wt=4) 18 [hyper,17,4] avoid(A)|prowler(A).
** KEPT (wt=4): 19 [hyper,18,9] avoid(A)|gazer(A).


given clause #9: (wt=4) 19 [hyper,18,9] avoid(A)|gazer(A).
** KEPT (wt=4): 20 [hyper,19,2] avoid(A)|suitablepet(A).


given clause #10: (wt=4) 20 [hyper,19,2] avoid(A)|suitablepet(A).
** KEPT (wt=2): 21 [hyper,20,7,12] avoid(b).

----> UNIT CONFLICT at   0.66 sec ----> 22 [binary,21,10] $F.
```

Figure 5.3: Here is the sequence of steps the program went through to find the proof.

the clauses it keeps for every given clause, the flag, print_kept has to be set. The clear_history flag also concerns the format of the output from the program and is not important here. Figure 5.2 shows that the program decided to use hyperresolution, factoring, and unit deletion. These last two rules were not discussed here and they do not come up in this example either. It also decided to put two clauses into the set-of-support list.

The trace shown in Figure 5.3 lists the given clauses as they are taken out of the set-of-support. Beneath each given clause is the list of clauses that were generated using the given clause. For instance, given clause 1, clause 11 in the list produces clause 13 listed just below. That line shows that clause 13 was derived from clauses 11 and 6 via hyperresolution and has a weight of 4. Clause 13 then goes into the set-of-support list. The program continues taking clauses from the set-of-support list until it finds a contradiction or runs out of clauses (or memory, or time). In this case it was quite simple and the program derives clause 22, `avoid(b)` and this contradicts `-avoid(b)`. This check for contradictions is an extra step that is done after every given clause has generated all the new clauses that it can, and so it is not really part of the set-of-support strategy.

5.5   The Usefulness of Predicate Calculus

Wos et al. [267] give some good examples of how automated reasoning can be used to do problems in logic circuit design and verification, research in mathematics, systems control, and program debugging and verification. They point out that programming a logic-based solution for just one problem normally requires so much effort that it is usually easier to do the problem by hand. On the other hand, they say that if you are going to do a lot of closely related problems, automated reasoning can be worthwhile because you will build up a database of facts and relations that can be used over and over.

5.6   Other Reasoning Methods

Predicate calculus theorem proving has multiple liabilities when it comes to trying to produce human level reasoning capabilities. In the first place, the logic is true or false and human reasoning often comes in shades of gray, that is, people think certain conclusions are likely or unlikely, but not certain. For one approach to using continuous-valued logic and variables see [226].

A second failing of predicate calculus theorem proving is that it is *monotonic logic*. In monotonic logic whatever conclusions are derived are correct and their truth cannot be changed if more assertions are added. For example, suppose you are given that John needs to drive to town. To drive to town John needs a car. John has a car and therefore it follows that he can drive to town. In monotonic logic this conclusion stands, but suppose that there was a terrible snowstorm and there is now three feet of snow on the roads. The conclusion that John can drive to town should be withdrawn. A system of logic where conclusions can be withdrawn is called *nonmonotonic* logic and some researchers are pursuing this alternative (for instance, see [107]).

Other researchers believe even nonmonotonic logic is not the way people do logic and they are looking for other explanations for how people reason. For instance, Johnson-Laird and Byrne [76] propose that people use model-based reasoning. They take in the facts and construct a mental model of the situation and then look for counter-examples to refute their conclusion.

5.7 Exercises

5.1. Translate the following IF-THEN statements to clausal form:

$$\text{IF } a \text{ THEN } b$$
$$\text{IF } a \text{ \& } b \text{ THEN } c$$
$$\text{IF } a \mid b \text{ THEN } c$$
$$\text{IF } a \text{ THEN } (b \text{ \& } c)$$
$$\text{IF } a \mid b \text{ THEN } (c \text{ \& } d)$$
$$\text{IF } a \text{ \& } b \text{ THEN } (c \mid d)$$

5.2. Repeat Exercise 4.1 using the predicate calculus notation used in this chapter.

5.3. Given the following database of information about oceans, countries bordering oceans, and the fact that the countries are South American, Asian, or African, show that there is an ocean that borders on Asia and Africa.

```
ocean(atlantic).
ocean(indian).
borders(atlantic,brazil).
borders(atlantic,uruguay).
borders(atlantic,venezuela).
borders(atlantic,zaire).
borders(atlantic,nigeria).
borders(atlantic,angola).
borders(indian,india).
borders(indian,pakistan).
borders(indian,iran).
borders(indian,somalia).
borders(indian,kenya).
borders(indian,tanzania).
sa(brazil).
sa(uruguay).
sa(venezuela).
african(zaire).
african(nigeria).
african(angola).
african(somalia).
african(kenya).
african(tanzania).
asian(india).
asian(pakistan).
asian(iran).
```

Do this problem with the techniques of this chapter. Can this problem be done in Prolog?[5]

---

[5] From the TPTP collection. The problem itself originated with: D.A. Plaisted, "A Simplified Problem Reduction Format," in *Artificial Intelligence*, 18(X), 1982, pages 227–261.

5.4. In Figure 5.3 the program had the clause `kangaroo(b)` as a given clause, but the program did not generate any new clauses from it. Why not?

5.5. All the dated letters in this room are written on blue paper. None of them are in black ink, except those that are written in the third person. I have not filed any of them that I can read. None of them, that are written on one sheet, are undated. All of them, that are not crossed, are in black ink. All of them, written by Brown, begin with "Dear Sir." All of them, written on blue paper, are filed. None of them, written on more than one sheet, are crossed. None of them, that begin with "Dear Sir," are written in third person. If there is a letter by Brown, can it be read?[6]

5.6. Suppose everyone either likes the Cubs, likes the Mets, or likes tennis. This means that the person who likes the Cubs does not like the Mets or tennis and a person who likes the Mets does not like the Cubs or tennis and the person who likes tennis does not like the Cubs or Mets. Then suppose anyone who likes the Cubs or Mets does not like lemonade. Given that Nancy likes lemonade, prove that Nancy likes tennis. Do this using the techniques of this chapter and then determine if the problem can be done in Prolog as well.

5.7. Someone who lives in Dreadbury Mansion killed Aunt Agatha. Agatha, the butler, and Charles live in Dreadbury Mansion and are the only people who live therein. A killer always hates his victim and is never richer than his victim. Charles hates no one that Aunt Agatha hates. Agatha hates everyone except the butler. The butler hates everyone not richer than Aunt Agatha. The butler hates everyone Aunt Agatha hates. No one hates everyone. Who killed Aunt Agatha?[7]

5.8. Using the clause weighting scheme in Section 5.3 do the proof of zebra. Decide on some method of breaking ties and stick to it. Does weighting work very well for this problem? Can you devise any other strategy that might be useful for picking clauses to resolve together that will solve this problem faster?

5.9. Below is a small map, consisting of the regions, a, b, c, d, and e:

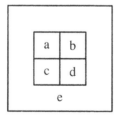

The map can be colored using three colors, say red, green, and blue, such that a region of one color never has a common border with a region of the same color, except possibly, if the regions meet at a point.

    a) Produce a set of clauses that captures these requirements. For instance, if a is red, then b, c, and e cannot be red. Every area is either red or green or blue and if an

---

   [6] Taken from the TPTP collection. It originated with Lewis Carroll, *Lewis Carroll's Symbolic Logic: Part I, Elementary, Part II, Advanced*, C.N. Potter, New York, 1986.

   [7] From the TPTP collection. The problem itself originated with: F.J. Pelletier, "Seventy-Five Problems for Testing Automatic Theorem Provers," in *Journal of Automated Reasoning*, Volume 2, Number 2, pages 191–216, 1986.

area is one color is cannot be either of the other possible colors. Some relationships that you might think to put in your set of clauses are the ideas that if a is some color, d must be that same color or if b is some color, c must be that same color. While this is true, it is not really a requirement of the problem, so do not include these relationships.

b) Using the clauses you developed above, assume a is red and b is green and then find the colors of the areas, c, d, and e.

c) Now, given *only* that a is red, prove that d is red.

d) Can this problem be done in Prolog?

Chapter 6

Complex Architectures

The pattern recognition, neural networking, and rule-based methods we have looked at so far are capable of doing some very simple types of pattern recognition problems, and while these techniques are extremely useful, they are inadequate for more complex problems and they are especially inadequate if the goal is to produce a system with something like the capabilities that people have. A realistic (humanlike) computer architecture is necessary for a program to simulate human thinking, but the architecture must also be designed with the storage of patterns in mind. In this chapter we will look at some of the architectural requirements, as well as more complex methods of structuring data and memories for AI programs, including not just neural networking methods, but also traditional symbol processing methods.

Most of the neural methods to be found in this chapter represent new proposals designed to operate much like the traditional symbolic methods. They have been included here for two reasons. One, it shows just how far backprop can be pushed into doing all sorts of things. Second, there has been a great deal of criticism from the symbol processing camp in AI that networks are too limited to do what people do, yet here are examples that show that networks can do some of these things, although it must be admitted that they work rather poorly compared to their symbolic counterparts and may not be realistic.

6.1 The Basic Human Architecture

With respect to neural networking, we have looked at some fairly simple one-step pattern recognition problems that can be accomplished by training one neural network, however, no one believes that the human mind operates by using just one large *uniform and undifferentiated* network. It evidently consists of a number of networks working together. Minsky has stated that there are at least 400 specialized architectures within the brain [89] and Minsky and Papert speculate that the brain consists of large numbers of parallel distributed processing systems with some serial control mechanisms [125]. At the highest level, research indicates that there are two separate processors in the brain, the left hemisphere and the right hemisphere. In most people, the left brain is the logical, verbal, and symbolic processor, while the right brain is the intuitive, nonverbal, and creative processor, but in some people this arrangement is reversed. Furthermore, there are subsystems concerned with processing each of the five senses. Even the memory system consists of at least three parts, a short-term memory, an intermediate-term memory, and a long-term memory.

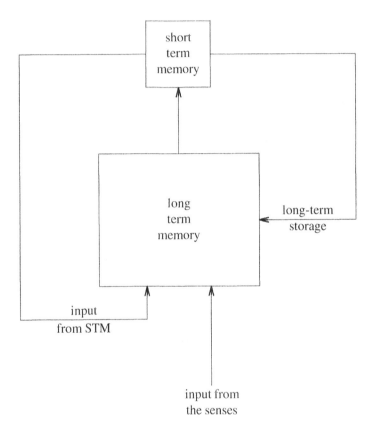

Figure 6.1: A rough and very high-level outline of the architecture of the human mind.

A rough diagram of the the human mind is shown in Figure 6.1 where the emphasis is on the memory systems within the mind. The long-term memory (LTM) represents a system that is a little bit like the kinds of simple one-step pattern neural recognition architectures we have been considering. There is input coming from the senses and it is passed on to a long-term memory. Here, the input brings to mind some thoughts that are closely associated with the input. The output of this box is what comes to mind, it is the thought that the system generates.

In Figure 6.1 we have the output from the long-term memory feed into short-term memory (STM). This memory has a very limited capacity, and in fact, research shows that people can hold about 7 plus or minus 2 different facts in short-term memory at one time [122]. If people try to keep more than this approximate number of facts in their short-term memory, some of the facts will be forgotten. This makes the human mind somewhat like standard computer systems where the CPU has a limited number of registers.

In operation, facts in short-term memory will influence what is retrieved from long-term memory. This influence is indicated by the line labeled "input from STM" that runs from STM to LTM. To show how the contents of STM influences what is retrieved, we can take the sentence, "John shot some bucks." If we only know from long-term memory that John is a hunter and we hear this sentence, the long-term memory will produce the idea that John was hunting. If, on the other hand, we first heard that "John went to Las Vegas," then this fact in short-term memory will take precedence over the fact in long-term memory that John is a hunter and the system will produce the idea that John lost some money gambling.

It is generally thought that the short-term memory is electrical in nature and that patterns of charge in the nerve cells store the short-term memories. The memories stored in short-term memory eventually have to be stored away permanently in long-term memory and this pathway is shown in Figure 6.1. Actually, there is evidence for an intermediate-term memory that takes and holds these memories before they are stored in long-term memory, but this structure is not shown in the figure.

This takes care of a simple overview of the internals of the mind, but a mind also has external components as well. A person is equipped with devices (eyes, ears, and hands) that allow a person to store memories in the external world and retrieve them. The external memory devices that can be plugged in are paper and pen, books, audio and video recordings, and computers.

The arrangement we gave here is very simple but it will be enough for what follows. Hunt [71] gives a more detailed discussion of how the human architecture seems to be arranged based on psychological research. Newell [141] argues that developing complex architectures that approximate the human mind should be a major pursuit in cognitive science and artificial intelligence and he proposes one based on his research on problem solving and learning.

6.2 Flow of Control

The operation of the whole architecture we just outlined is different from the kinds of computer architectures that people are familiar with. Notice that the architecture is very much like the forward chaining architecture used in interpreting rules described in Chapter 4. In that system, the rule interpreter looks for rules to fire. It selects one, updates its environment, and then takes a look at the new environment that now exists. This prompts some other rule to fire and so on. In the architecture we described in the previous section, a new thought comes to mind based on inputs from the physical environment and the contents of the memory systems. This new state of the system contributes to producing the next state of the system. What comes to mind is not always predictable because the mental and physical environment will always be at least a little bit different from anything the system has seen before. Besides that, if the selection of the next thought is being done by a Boltzman machine type of relaxation algorithm, the uncertainty is compounded. This different kind of architecture results in a system that can easily be distracted, look brilliant, or make stupid mistakes.

The first item to look at is how the operation of the architecture can be distracted from a task it is working on. Let us take the task of the recital from memory of the two lines from "Locksley Hall":

> I the heir of all the ages in the foremost files of time,

and

> For I doubt not through the ages one increasing purpose runs.

Now suppose we had a human being trying to recite the verses from "Locksley Hall" and this human being might soon become an heir to a fortune. This is William James' description of what may happen:

> But if some one of these preceding words—'heir,' for example—had an intensely strong association with some brain-tracts entirely disjointed in experience from the poem of 'Locksley Hall'—if the reciter, for instance, were tremendously awaiting the opening of a will which might make him a millionaire—it is probable that the path of discharge through the words of the poem would be suddenly interrupted at the word 'heir.' His *emotional interest in that word* would be such that its *own special associations would prevail* over the combined ones of the other words. He would, as we say, be abruptly reminded of his personal situation, and the poem would lapse altogether from his thoughts.

In terms of the general model described at the beginning of this chapter, this strong association with becoming an heir combines with the current word, 'heir,' and this combination sets the person off on thoughts that have nothing to do with the poem. Of course, after a little while, this thought about becoming a millionaire is going to decay and the person is going to remember that they were reciting the poem and restart the process from where they left off.

One benefit of this control structure is that when a person is working on a problem and something goes wrong, the person will notice this, interrupt the current process, and try to deal with what went wrong. Maybe the whole process will have to be started over again or maybe the process can be restarted from the point where the problem occurred. This is unlike typical computer programs where the system will not be aware that anything has gone wrong. The typical computer program needs to be programmed ahead of time to deal with absolutely every case that can come up. For instance, suppose you tell a robot programmed in the traditional manner to dig a hole. Ordinarily this is a simple task, however, every once in a while something unusual will happen. For instance, these things can happen:

1) the robot hits a rock so large it cannot be removed;

2) the robot finds some bones or a fossil;

3) the robot finds buried treasure;

4) the robot finds a buried cable;

5) it starts to rain and it rains so hard the sides of the hole collapse;

6) the robot is digging a hole in Alaska in the summertime and it hits permafrost;

7) a plane crashes nearby.

The kind of control flow found in people is really necessary for any system that has to deal with a world that is very complex and where everything cannot be anticipated ahead of time.

This architecture also has another benefit. A chance combination of ideas in the memory systems together with something really unique in the physical environment can lead a person to think of things that the person would not otherwise have thought of. For instance, you could have an idea about how to build a better mousetrap or a better artificial intelligence. The bubble chamber used in particle physics experiments was invented by a physicist who was watching bubbles form in a glass of beer. Also, Einstein is reported to have had an insight into special relativity while sitting in a train in a train station. When trains just begin to move, they move so slowly that you cannot feel your train moving. If there is a train on a nearby track it can seem as if it is the other train that is moving. Under these conditions you cannot immediately tell if it is your train or the train on the next track that is moving. Einstein then applied this insight to produce special relativity.

On the negative side, if the conscious architecture is not careful, things can go wrong. For instance, suppose the architecture is confronted with doing algebra. Human short-term memory seems generally incapable of handling problems of this sort so people fall back on using paper to store intermediate results. As a person is manipulating the formulas there may be many transformations that apply. The system will pick one, then another, then another. Suppose that the recognizing of things to do comes from a Boltzman-machine-like algorithm that finds patterns as minimas in an energy landscape. Given this algorithm, it is entirely possible that some things that need to be done will be overlooked either because (1) the random nature of the process happens to miss them, (2) some minima are too shallow (reflecting a lack of practice), or (3) because the system is cooled too quickly. Cooling too quickly means being in a hurry, and of course, it is well known that people in a hurry are more likely to make mistakes. Doing things too fast also makes possible another similar type of error. Before one transformation is completely finished, another necessary operation captures the attention of the system. The system moves to do this operation and because of the limitations of short-term memory, the uncompleted operation is thrown out and forgotten.

6.3 The Virtual Symbol Processing Machine Proposal

There is currently a big division in AI between believers in the traditional symbol processing methods and believers in the nontraditional neural networking methods. Much of the time, this debate between the two approaches has been that the mind must be one or the other, neural *or* symbolic. However, there are some intermediate positions on this debate. One of these intermediate positions is that the mind could be a parallel distributed processing system that has the ability to simulate a symbol processing architecture. In terms of ordinary computer architecture, the mind's parallel distributed processing system is used to run a virtual symbol processing machine.

One argument for how this is accomplished is as follows. The mind's neural pattern recognition abilities are extremely good, but its ability to do symbol processing is extremely limited. But then people also have the ability to manipulate the environment in suitable ways to make up for their lack of internal symbol processing ability. They can use pencil and paper to store memories and to perform symbol processing activities. For example, if we have the problem of multiplying two numbers, there is a series of steps that can be performed with symbols on paper that will do this task. Each step in the process is quite simple and each of these steps is something that a parallel distributed processing architecture is very good at. Suppose the problem is to multiply 79 times 23. The algorithm is to write down the numbers like so:

$$\begin{array}{r} 79 \\ \underline{23} \end{array}$$

Now you start by saying "9 times 3 is 27." Write down the 7 below the 3 and place the 2 up near the seven. Now, "7 times 3 is 21." Add in the 2. Place the result below the line to the left, and so on and so on. Each step is a simple pattern recognition step of the kind that PDP systems are so good at, however, the system consisting of the PDP system, the hand, and the pencil and paper form a symbol processing machine architecture. The next step in the argument is to realize that a mind can imagine doing these steps without actually using the pencil and paper. The result is that the parallel distributed processing architecture is, by itself, running a complete symbol processing computer architecture.

One of the very nice features of this arrangement is that the virtual symbol processing system can call on the underlying PDP system to provide quick answers to those simple problems, like "9 times 3." If the virtual symbol processor had to search through a list of multiplicands, multipliers, and their products like the following ones stated in Prolog:

```
times(1,1,1).
times(1,2,2).
      .
      .
      .
times(9,3,27).
      .
      .
      .
times(10,10,100).
```

it would be quite time consuming, but the PDP architecture can do this step very quickly.

This idea of implementing symbol processing on PDP architectures comes from Rumelhart, Smolensky, McClelland, and Hinton [182]. For more on this viewpoint also see [218] and [18].

6.4 Mental Representation and Computer Representation

Now we need to begin to look at how knowledge and memories are stored in the human mind and how the storage is done in symbolic and neural computer systems. Of course, it is not known how people store memories and knowledge so the methods given below and in the following sections of this chapter are speculative and probably not very realistic. The main methods developed so far have been for symbolic systems where the memories and knowledge are stored as trees of symbols. Before looking into all these methods, however, there is a major new criticism of purely symbolic methods that has arisen, the problem of defining symbols or the *symbol grounding problem*, that we must consider. Finally, in this section there is one other topic related to the issue of memory storage, the need for the system to build up and take apart structured representations.

6.4.1 A Problem with Symbolic Representation

To work on the problem of mental representations, let us start with a specific thought that needs to be stored. Let this thought be expressed by the sentence:

Mary ate cookies.

How this thought is stored in a person's mind would depend on how the thought was entered into a person's mind. If you see Mary eating cookies you will have some visual record of this event. If you only hear about it or read about it you may well store some vague abstract images of this statement that the words conjured up in your mind from your past experiences of seeing people eat cookies. The new memory you store would be some combination of your past experience of cookie eating with Mary put into the image as the person eating the cookies. Ultimately, it seems that whatever it is that you store, the new memory must be just some pattern of neural activity, connection strengths, or chemical compounds, but researchers are far from knowing how to represent and manipulate such things as yet.

The traditional symbol processing approach to this problem is to store some structure of symbols to represent this event. Prolog programmers will probably propose this:

```
ate(mary,cookies).
```

which internally will appear as a linked list of symbols where the first symbol is the predicate. To people it is pretty obvious what all these symbols mean, but to an artificial intelligence program these symbols do not mean anything and many people argue that they should mean something to the program. An extreme symbolicist may actually claim that the symbols used in an artificial intelligence program do not have any meaning and they do not need to have any. They just get used in rules and tree structures and that is all. Thus for instance, suppose a system is given the facts that "Mary ate cookies" and "John ate candy." The system translates these inputs into some suitable internal representation, like Prolog. Now ask the system, "Who ate candy?" The question is translated into a compatible form and the whole problem becomes just a simple pattern matching search to find the answer. Symbols have to be tested for equality and that is all.

The extreme symbolic approach may be adequate for many applications but it seems implausible for many more complex problems. Suppose the symbol processing system sees the following statement:

Mary swallowed the basketball in one gulp.

The oddity of this statement will prompt two reactions in a person: either it is false or there are some strange details that are missing from the story. A simple symbol manipulator will not find anything strange in this statement, because it does not know the *meaning* of the statement. Of course, by adding a great number of facts and rules it is possible for a more complex symbol processing system to also become suspicious of it, but then the extremes you have to go to to do this look suspicious to many people.

The process of producing the more complex symbolic system might work like this. First, enter into a dictionary the fact that Mary is a girl. Second, define the properties of a girl using a *property list*. A property list gives some details of a symbol and it is roughly the symbolic equivalent of a list of microfeatures of an object in the PDP notation. In Lisp there is a specific facility for defining the properties of symbols, but in Prolog programmers must make up their own notation, such as this:

```
property(girl,[[object, human],
          [gender, female],
          [size, medium]]).
```

For some applications, the level of detail in the above list may be sufficient, but it is not sufficient for the "Mary swallowed the basketball in one gulp" problem. More detail can be given by adding more property lists defining the symbols, human, female, and medium with property lists. Let us just follow up on the symbol, human, by defining the properties of type human:

```
property(human,[[texture, soft],
          [shape, humanoid],
          [animate, yes]]).
```

but all these symbols must be defined as well. Let us just single out humanoid for definition giving perhaps:

```
property(humanoid,[[feature, head],
            [feature, body],
            [feature, [leg, [quantity, 2]]],
            [feature, [arm, [quantity, 2]]],
                .
                .
                .
```

and then single out head for definition:

```
property(head,[[feature, [eye, [quantity, 2]]],
              [feature, [ear, [quantity, 2]]],
              [feature, nose]],
              [feature, mouth],
              [[shape, spherical],
                      .
                      .
                      .
```

and finally single out mouth for definition:

```
property(mouth, [[size, 2]   /* say two inches */,
                      .
                      .
                      .
```

After all this, the system will still have a rather poor understanding of what a person is. It knows that a person has these parts, but it still does not know how they are arranged. Is the nose located above or below the mouth? Is there one eye in the front of the head and one eye in the back? Or, are both of them behind the ears? A head has a spherical shape but what is a spherical shape? A sphere could be described by the set of points such that $x^2 + y^2 + z^2 = r^2$, but what are points and what do those symbols, x, y, z, r, 2, +, and = mean? What happens is that trying to define symbols by only making reference to other symbols is a poor way to fully describe a person, or for that matter to describe anything at all.

Of course, we are nearing the point where a complex symbolic system can determine that "Mary swallowed the basketball in one gulp" is suspicious. The following details must be added: add the facts that basketballs are 12 inches or so in diameter, a mouth is about 3 inches across, add the rules that eating something involves moving it into the mouth and that the size of the something involved must be less than or equal to the size of the mouth. These additions make it possible for a complex symbol system to finally "understand" that the sentence is false or there must be some other unusual details that make the sentence make sense. One really improbable aspect of this arrangement is that the programmer has to think up all sorts of rules ahead of time to deal with the most silly possible situations and this is clearly impossible. It seems just as improbable that people's minds somehow automatically generate all these facts and rules.

6.4.2   Symbol Grounding as a Solution

Problems like the above have led some people to propose that the mind is not wholly symbolic or wholly rule based and a good portion of human reasoning must be based on processing images of things and not just symbols. Symbols themselves must be defined in terms of either visual or other sensory images. The problem of defining symbols without making reference to just more symbols has become known as the *symbol grounding problem* [61, 62, 63]. Harnad [63] describes the problem of defining symbols for an artificial intelligence using only symbols as comparable to the following problem:

> Suppose you had to learn Chinese as a *first* language and the only source of information you had was a Chinese/Chinese dictionary! This is more like the

actual task faced by a purely symbolic model of the mind: How can you ever get off the symbol/symbol merry-go-round? How is symbol meaning to be grounded in something other than just more meaningless symbols?

People manage the grounding of symbols by storing visual images and other sensory images of human beings and other objects in the world. The images provide a grounding for symbols that finally defines them without reference to other symbols. Notice that to even describe the idea of symbol processing, it is necessary to ground the idea with images of processing symbols in algebra or logic. People generate a symbol when they find many instances of very similar patterns that cluster together or "form a valley in an energy landscape" as did the 'Cubs fans' in Section 3.6. Using the symbol then calls up the major characteristics of the cluster.

Obviously, neural networks have some capability for grounding some simple symbols, but it is not clear how a network could ground a complex symbol like "human" because, in the case of a human being, it is not just the physical features that ground the concept of human being, examples of human behavior are also needed.

In addition to actually grounding symbols, sometimes symbols can be fairly well defined using only other symbols. Suppose you did not know what a zebra was. You could look it up in a dictionary, where it might say something to the effect that a zebra is a horse with stripes. But this understanding is only possible if you know what a horse is and what stripes are. If you do not know what a horse is, then you have to look up horse where the definition will tell you that a horse is an animal with four legs, but if you do not know what an animal is, then . . . .

Harnad has made the point that a hybrid symbolic/nonsymbolic system may be just what people use and be just what is needed for artificial intelligence systems. Symbols would be defined by a nonsymbolic parallel distributed processing system, while the higher levels of thinking could use the grounded symbols in fairly standard symbol processing algorithms. This plan would also fit in with the idea that one half of the brain does intuitive and nonsymbolic processing while the other half does symbolic processing.

In summary, in order to create a fairly realistic approximation of human memory and thinking, it seems as if symbols will have to be understood the same way people understand them. This will mean giving the artificial system some kind of senses that approximate human senses. In the meantime, computer understanding of the world will be very limited.

6.4.3 Structures and Operations on Structures

Another important constraint on the storage and processing of memories is that the memory records should be made up of decomposable parts that can be recombined as necessary. First, there is the case of dealing with visual memories. Suppose you read this little story:

> Mickey was up on a ladder painting the side of his house and singing a merry tune. Meanwhile, Chip and Dale were happily gathering nuts beneath a tree. Suddenly, Pluto sees the chipmunks and charges after them. Chip and Dale run beneath the ladder and . . . .

In reading this story, people take the familiar images they have seen before and go on to construct a new and unique series of images based on what they read. The ability to cut apart images from old situations and paste them into new ones is something that is needed in a humanlike architecture. Part of what is needed can be seen in the neocognitron. In the neocognitron there is an operation built into it that will look at a picture, "concentrate" on one portion of it, and identify it. Thus, the neocognitron may already have solved part of the problem in that parts can be cut out of a picture. Now someone needs to come up with the additional neural circuitry that will transfer the specific object to another picture. These operations, we have assumed, would give the system an "imagination." Reading and really understanding a story seems to require taking in the words and letting your mind construct images of what is happening in the story, so this capability will be necessary for getting computers to understand natural language.

In symbol processing, the decomposable parts are subtrees or symbols. In symbolic languages there are operators or functions that can be used to build up or break down the structures. Prolog has the | operator to either break a list into a head and a tail or build up a new structure. Lisp has the functions, car, cdr, and cons to do structure modifying operations.

6.5   Storing Sequential Events

The most clearly obvious example of structure in human experience is that one event follows another and another and another. That is, events form a kind of linked list. Given that people are moving forward in time and recording their present experiences as they go along, this simple linked list structure is inevitable. In this section we look at symbolic and neural methods for storing lists of items.

6.5.1   The Symbolic Solution

For a concrete example of storing linear structures let us use the two lines of poetry from "Locksley Hall":

> I the heir of all the ages in the foremost files of time,

and

> For I doubt not through the ages one increasing purpose runs.

The traditional AI programmer has absolutely no doubt about how to represent these structures. Clearly a linked list, like this one in Prolog is called for:

```
[I, the, heir, of, all, the, ages, in, the, foremost, ...  ]
```

It becomes extremely easy for a traditional system to recite these words. Either recursion or iteration can be used to move through the list.

Whether using a simple linked list is a realistic way (the actual human way) of storing a sequence of items is a point that can be argued both ways. On the one hand, it seems really unlikely that people have anything like the capability to create linked lists from unused memory cells the way a Prolog interpreter does. When people have to memorize a sequence, it appears on the surface that, because they have to repeat the sequence in their minds a number of times, they are "wearing down a groove" in memory, or adjusting weights the way a neural network adjusts weights. On the other hand, there is another phenomenon that lends some credibility to the traditional symbolic solution. There is evidence that memory in the human mind comes in *chunks* and only a certain amount of information can be stored per chunk. The chunks are then linked together. This phenomenon can be seen in a simple test. If you go to memorize π out to quite a few decimal places, you will find that you will group the digits in chunks with perhaps two to four digits per chunk. For instance, you might group them like so, where the blanks separate the chunks:

$$3.14 \ 159 \ 265 \ 3589 \ 79 \ 323 \ 846 \ 264 \ 338 \ 3279 \ 50$$

So this may be an indication that people store small amounts of data in one node and then link the nodes together in the way that Lisp and Prolog operate. For more on chunking see [140], [141], [122], and [11].

6.5.2 Neural Solutions

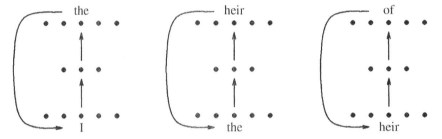

Figure 6.2: A simple network can be wired up to take its output and recirculate it back to the input layer to produce the next word in the sequence.

Now we can turn some simple neural networking solutions to the problem of remembering the lines of poetry. The first simple solution to getting a network to remember a sequence of symbols is to train a network to output the i+1st word of a sequence when it is given the ith word on input. After training, you can input a symbol, say "I," and it will produce "the" on the output units. Then "the" can be recirculated down to the input units and "heir" will appear on the output units, and so on. Networks that recirculate values back to earlier layers are said to be *recurrent*. The situation is shown in Figure 6.2. To a certain extent, this will work, but only when the same word occurs only once in the whole list to be memorized. We still could not have a network store both lines from "Locksley Hall" because "I" occurs in both the first and second lines. There is no way this network can learn when the answer should be "the" and when it should be "doubt." Of course there are also the words, "the" and "ages" which are repeated and which cause problems as well.

These problems can be partially fixed by adding a simple short-term memory to the input layer of the network. This short-term memory will always remember one previous word in the sequence. Figure 6.3 shows the layout of the network for several steps in the recall process. First, when the network begins recalling the sequence, "I" is placed on the input units on the left while the other input units on the right that remember the previous word are empty. The word, "the," appears on the output units. Now "I" is shifted to the

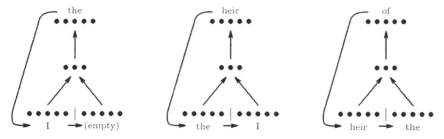

Figure 6.3: By adding a simple short-term memory (the right half of the input units), a network can remember sequences that contain more than one instance of each symbol.

short-term memory units while "the" is copied down to the input units. This produces "heir" on the output units, and so on. With this arrangement, the network still cannot deal with the phrase, "the ages" that occurs in both lines of the poem. Obviously, this could be solved by adding another short-term memory register to the input layer to remember more of the context.

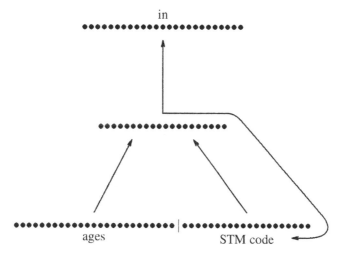

Figure 6.4: This neural network with a simple short-term memory can memorize the two verses from "Locksley Hall." A word is input on input units 1–25 together with a pattern on the short-term memory (or context) units 26–45 to give a pattern on the hidden units and then a word on the output units. Then the values on the *hidden units* are copied down to input units 26–45 for use with the next input pattern. The short-term memory code on the context units depends on *all* the words that came before although, in practice, a network like this will only have traces of previous words up to about five words back.

There is another solution to this problem. The short-term memory can take its values, representing the context, from *the hidden layer* as shown in Figure 6.4. The pattern of activation on the hidden units that is going to be copied down to input units 26–45 (the context units, or short-term memory) is not going to be something that has any obvious meaning. It will be a somewhat arbitrary pattern (like a hash code) that will be a function of the current input plus the previous pattern on the context units. It will then remember something about the whole sequence of patterns that went before, but there will be no obvious way a person can look at that pattern of activation and figure out what it means. When the first "I" is input, it will combine with the pattern on the context units to produce "the" on the output units. By the time the second "I" is input, there will be a new pattern on the context units that will produce "doubt" on the output units. The network will also produce the correct output for each of two different "ages."

To illustrate what happens, we will train a network to actually memorize the two verses from "Locksley Hall." The patterns used for the words will be the ASCII codes of the letters of the words with the first three bits removed and with words longer than five characters shortened to just five characters. The strings submitted to the network will be:

i	the	heir	of	all	the	ages	in	the	frmst	files	of
1	2	3	4	5	6	7	8	9	10	11	12

time	for	i	doubt	not	thru	the	ages	one	incre	purpo	runs
13	14	15	16	17	18	19	20	21	22	23	24

In training this network, the first pattern input was a pattern of 45 values of 0.5. This produces some pattern of activation on the hidden units and then some pattern on the output units. The network should have the code for "i" on the output units. Weights in the network are adjusted in the usual way to try to always produce the correct output values. Now, when the word "i" is placed on the input units, the pattern from the hidden units is copied down to input units 26–45. Now all the input units activate the hidden units and the hidden units activate the output units. The word "the" should appear on the output units. The network is repeatedly trained on all the input and output sequences until the tolerance is met. Training this kind of network is slow and it is usually done slowly because, as the weights are modified, the hash codes also change. Often these networks will move off in a direction that briefly increases the total error over all the patterns instead of decreasing it.

As an example of what the codes look like, here is what happened in one particular training session. The following vector was the pattern that was present on the hidden layer after "ages" at position 7 had been input:

0.00	0.91	0.00	0.00	0.03	0.00	0.00	0.02	0.74	0.00
0.05	0.00	0.00	0.97	1.00	0.97	0.00	0.94	0.98	0.00

while it had this pattern when the "ages" at word 20 was input:

0.00	0.92	0.00	0.00	0.00	0.00	0.00	0.00	0.92	0.00
0.08	0.00	0.00	0.03	0.73	0.79	0.00	0.75	0.60	0.00

The two patterns are much the same, yet different enough to prompt the network to produce two different responses. In the first case, the network produces "in," and in the second case it produces "one."

This technique of taking values from the hidden units and using them as a kind of short-term memory, coding the state of the network, was originated by Elman [35]. Before this, Jordan [77, 78] proposed doing the same thing with the output units. In addition, other variations are possible including circulating values from both the hidden and output layers back to the input layer. Allen [2] has called all these types of networks with short-term memories, *connectionist state machines.*

Needless to say, these types of networks can be used to do simple language processing tasks and there have been experiments where a recurrent network was used to find patterns in stock market activity (for one, see [81]) or gold market activity [106].

One experiment with this type of architecture hints that it may be close to how people store a list of items in memory. When people are given a list of items to remember they are more likely to recall the first and last elements of a list than the middle elements of the list. Nolfi et al. [144] trained a network to recall lists of seven letters and it turned out that the network was also more likely to remember the first and last letters of a list than the middle ones.

6.6   Structuring Individual Thoughts

We have already started out working on the problem of how specific memories can be stored in symbolic systems, but now we will look at more variations on storing symbolic data, as well as neural methods to store the symbolic structures.

6.6.1   The Symbolic Methods

Once again, if we work with "Mary ate cookies," one possible internal representation of this statement will be a linked list:

```
[mary, ate, cookies].
```

In the above scheme, the meaning of each symbol in the list is implicit. Some more sophisticated applications may require a different variation on this simple linked list notation. They may require specifying something about each item in the list that identifies it. Thus, there is also the following representation where the order of the facts is not important because each symbol is given an identification tag:

```
[[action, ate],
 [actor,  mary],
 [object, cookies]]
```

Actually, a program for processing language might be more sophisticated still in that when it reads the word Mary and looks up this word in its dictionary, it finds that Mary is a person and Mary is a female, so the program may turn out a more complex structure such as:

```
[[action, ate],
 [actor,  [person, [name, mary]],
                   [gender, female]],
 [object, cookies]]
```

As another example, the more complex sentence:

The little girl, Mary, ate some chocolate chip cookies.

might produce:

```
[[action, ate],
 [actor, [person, [name, mary],
                   [gender, female],
                   [age, child],
                   [size, little]],
  [object, [cookies, [kindof, chocolate_chip],
                     [reference, indefinite],
                     [quantity, some]]]]]
```

It is also possible to accommodate the thought that John and Mary ate cookies like so:

```
[[action, ate],
 [actor, [and, [person, [name, john]],
               [person, [name, mary]]]],
 [object, cookies]]
```

Or, we could specify the actor as some group with a unique internal symbol and use two lists to contain all the data:

```
[[action, ate],
 [actor, group_001],
 [object, cookies]]

[group_001, [[person, [name, john]],
             [person, [name, mary]]]]
```

Whatever the case, representations in a traditional symbolic system simply become tree structures of symbols with however much detail is needed for a given application.

Another traditional symbolic data structure for the storage of knowledge is the *semantic network*. Semantic networks record the same sort of facts you find in Prolog statements, but the representation consists of one large connected network. Suppose we again have the statement, "The little girl, Mary, ate some chocolate-chip cookies." Figure 6.5 shows a semantic network that stores these facts. In this network, the objects are nodes in the networks and the relationships are on the arcs connecting the nodes. The arrow on an arc indicates how the relationship should be read. For example, the arrow from Mary to person indicates that Mary is the name of a person. Rather than searching a database of facts as in Prolog, in a semantic network you search the nodes and relationships in the network. Most people believe that semantic networks do not contain any capabilities beyond what you get from predicate calculus notation and so they add little or nothing to the problem of representing facts in symbolic systems. Nevertheless, the pictorial notation can be quite appealing and therefore semantic networks are quite commonly used.

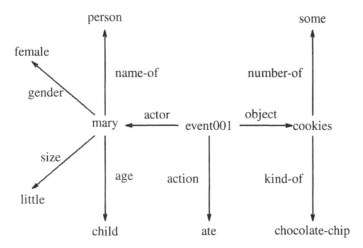

Figure 6.5: A semantic network is a traditional symbolic structure for representing facts. In this example, the network stores the event001: "The little girl, Mary, ate some chocolate chip cookies."

6.6.2   Neural Methods

Representing thoughts such as the ones seen above in a neural system is harder. The possibilities have not yet been completely explored but some representational schemes are available. In the simplest case above, "Mary ate cookies," one obvious scheme is to simply divide up a vector into three parts. Let the first part be the name of the person involved, let the second part be the type of event involved, and let the third part be the object. For a sentence of the form:

<person> <verb> <object>

let us have the person be either John or Mary, the verb be ate or got, and the object be either cookies or candy. Each of these six words can be coded as follows:

100000 for John
010000 for Mary
001000 for ate
000100 for got
000010 for cookies
000001 for candy

Then for the statement, "Mary ate cookies," we have the pattern:

010000 001000 000010

where the first six positions code for the actor, the next six for the action, and the final six for the object.

This plan where you have to allocate a specific number of fields within a vector has a severe limitation. There are those cases cited above like "The little girl, Mary, ate the chocolate chip cookies" where the number of items in an individual thought can vary. Obviously, you can try to plan ahead by using a vector that has a space available for every possible parameter, but planning this far ahead will not work for every case that can come up unless you enforce some limits on the complexity of the thoughts you will represent.

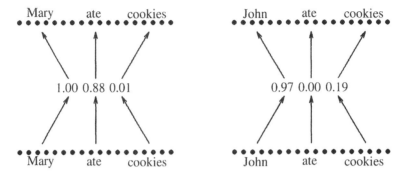

Figure 6.6: The hidden layer of an 18-3-18 auto-associative network contains a compressed representation of the input.

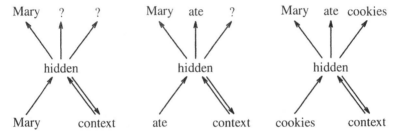

Figure 6.7: As codes for the words of a sentence are input one-at-a-time, the context units accumulate a coded representation of what was input. When the final word is input, the hidden layer has a pattern that has to produce the pattern, "Mary ate cookies" on the output units. This pattern on the hidden layer is a compressed representation of the output layer.

There are other ways of coding thoughts like "Mary ate cookies" into a network that are rather strange. The techniques involve training patterns into a three-layer network and then taking the pattern on the hidden layer units as a compressed and coded representation of its input pattern or patterns. For a simple case of this we can take the eight possible sentences we used above that can be derived from the pattern:

<person> <verb> <object>

and train them into an 18-3-18 auto-associative network. The networks in Figure 6.6 show the coding that arose for the sentences, "Mary ate cookies" and "John ate cookies." In a slightly more complex case, the same type of coding can be produced on the hidden layer units of the recurrent network shown in Figure 6.7 where the words, Mary, ate, and cookies are submitted to the network one-at-a-time. In this arrangement the number of words that can be input is arbitrary, but the output coding has to be fixed by the programmer.

In a still more complex example, Pollack [158, 159] has demonstrated how variable-sized tree structures can be compressed into a fixed size vector using an auto-associative

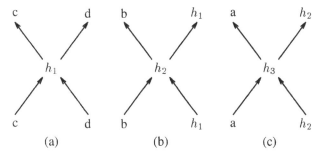

Figure 6.8: To train a network to encode the tree, [a, [b, [c, d]]], as a vector, the three sets of patterns shown above are trained into the network. In (a), c and d produce the code h_1; in (b), b and h_1 produce the code, h_2; and in (c), a and h_2 produce the final code, h_3.

network. For example, suppose the tree is:

$$[a, [b, [c, d]]]$$

The process of coding this tree as a vector is shown in Figure 6.8. First, place the symbols, c and d on the input and output units. There will be a compressed representation, h_1 on the hidden units. Next, place b and h_1 on the input and output units. This produces a compressed representation, h_2 on the hidden units. Finally, place a and h_2 on the input and output units giving a compressed representation, h_3 on the hidden units. These three pattern sets have to be trained into the network in the usual way and the hidden layer codes will change as the learning proceeds. When the learning is finished, the final h_3 represents a fixed size vector that codes the entire tree. To get the tree back, put h_3 on the hidden units, giving a and h_2 on the output units, then put h_2 on the hidden units, and so on. Pollack calls this whole scheme *recursive auto-associative memory* (RAAM). In one test Pollack showed that when a RAAM was loaded with a number of tree structures representing sentences, there were regularities that developed in the vector codes suggesting that the vector forms may be useful for other neural operations such as drawing inferences and answering questions, and in fact, in another experiment Pollack produced a network that could do simple inferences.

Another new method for storing tree structures is *holographic reduced representations* [154].

The next architecture to look at is DUAL, an architecture designed for storing relations in a network (see Dyer, Flowers, and Wang [32] and Dyer [33]). Suppose we have the following data on John and Mary in the form of relations:

likes(john,candy)	likes(mary,cookies)
gender(john,male)	gender(mary,female)
size(john,medium)	size(mary,medium)

Here, John and Mary both have three relations each, but in general, each person could have any number of relations. The DUAL architecture can be used to produce vector encodings for both John and Mary, codings that can be used to retrieve these relations about John and Mary. First, train a network called the short-term memory, STM, with the following input and output patterns for John:

input	output
likes	candy
gender	male
size	medium

This will produce some list of weights, call it J, for John, and we will save this list on the side somewhere. Second, retrain the STM network with the following input and output patterns for Mary:

input	output
likes	cookies
gender	female
size	medium

Let the list of weights for this network be called M, for Mary, and we save it as well. Now take both sets of weights, J and M, and use them *as data* to train an auto-associative network with a hidden layer and call this network LTM, for long-term memory. When you place the pattern J on the input to the LTM network, there is some coded representation of this on the hidden units. Store this away as a code for John. In effect, John gets defined by every relation you have associated with him. To get a code for Mary, put M on the input to LTM and save the pattern on the hidden units. The process is summarized in Figure 6.9.

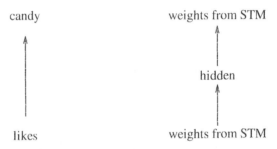

Figure 6.9: The DUAL Architecture: the short-term memory network on the left learns the facts about John and saves these weights as a list, J. The short-term memory is then re-initialized and learns the facts about Mary and saves these weights as a list, M. The lists, J and M are now trained into the long-term memory on the right, which is auto-associative. When the list J is placed on the input units of the LTM, the hidden layer has a code for John. Likewise, when M is placed on the input units of the LTM, the hidden layer has a code for Mary. To restore information about one person, put that person's hidden layer code on the hidden layer of LTM then take the output weights and copy them over to STM. Now with an input pattern on STM, say, likes, you get what the person likes on the output units.

Now, if you need information about John, say, what is it he likes, look up the vector code for John, place it on the hidden units of the LTM network, take the output of LTM and load these weights into the STM memory network, place the code for "likes" on the input units of STM, and the answer "candy" will appear on output. For a more complex use of DUAL to represent semantic networks, see [33].

The above schemes for representing complex structures in a neural network are new and there is no guarantee that they will turn out to be generally useful.

6.7   Frames and Scripts

One of the important characteristics of the human information processing architecture is
its associative memory. When some facts are brought to our attention, a number of other
related facts will come to mind from long-term memory. So, suppose you know that Mary
loves to eat oatmeal raisin cookies. If you hear that Mary ate some cookies, with no mention
of exactly what kind they were, your knowledge of Mary becomes activated and comes
to mind. One item that will come to mind is that Mary likes oatmeal raisin cookies in
particular and so she probably was eating oatmeal raisin cookies. In traditional symbolic
systems, bringing to mind typical expectations is achieved by using a *frame*. The frame
concept in AI was originated by Minsky [123], although the idea has also been known as a
schema in cognitive science for a long time. When the situation involves a series of actions,
the frame becomes more complex and it is known as a *script*. We will start by looking at a
PDP-based schema example and then change it into a more complex symbolic-frame-based
model. We will then look at a typical simple application of a script.

6.7.1   Schemas and Frames

The schema model we will look at is from Rumelhart, Smolensky, McClelland, and Hinton [182]. The model consists of an interactive activation network with 40 units. Each unit
represents an object or characteristic that can be present in five different kinds of rooms.
The five kinds of rooms are: an office, a bedroom, a bathroom, a living room, and a kitchen.
The 40 objects or characteristics that could be present are:

ceiling	walls	door	windows	very-large
large	medium	small	very-small	desk
telephone	bed	typewriter	bookshelf	carpet
books	desk-chair	clock	picture	floor-lamp
sofa	easy-chair	coffee-cup	ashtray	fireplace
drapes	stove	coffee-pot	refrigerator	toaster
cupboard	sink	dresser	television	bathtub
toilet	scale	oven	computer	clothes-hanger

The researchers asked people to imagine a typical example of each kind of room, for instance, a kitchen, and then check off which of these 40 items would typically be found in
such a room. Based on the number of occurrences of each item, a set of weights was calculated. The presence or absence of each object or characteristic then activates or inhibits
the other 39 objects or characteristics. Not surprisingly, if some units are clamped on, then
other units that are often associated with the clamped units also come on. For instance,
clamping on coffee pot and sink will light up refrigerator, cupboard, walls, and so forth.
This is just another example of pattern completion by networks.

Now we will look at how traditional symbol processing methods would do this problem
using frames, but we will also increase the complexity of the model as well. A possible
frame for a kitchen, stated in Prolog might be as follows:

```
frame(kitchen,[[part_of, building],
               [contents, [[walls, [quantity, 4]], ceiling,
                          [electrical_outlets, [quantity, 4]],
                          floor, window, door, table,
                          [chairs, [quantity, 4]], stove,
                          pictures, refrigerator, telephone,
                          microwave_oven, coffee_pot, clock]]]).
```

In this frame, the contents entry has a list of typical items found in a kitchen. If a computer program was reading about a kitchen, the items in this list are the things it would expect to find present in a typical kitchen. In addition, in this frame, we note that a kitchen is part of a building. There would be similar frames for the contents of a typical living room, a typical bedroom, and so on. A lot of this information is repeated in each frame, for instance, all the rooms have four walls, a ceiling, a floor, and some electrical outlets. In situations like this where a lot of information is repeated, the repeated information is often factored out into a new frame. In this case, it would be appropriate to create a new frame called room and label a kitchen, a living room, and a bedroom as special cases of a room. In addition, the specific rooms need to be flagged as instances of the more general type of object, a room. The definitions of a kitchen and a room might go something like this:

```
frame(room,[[part_of, building],
            [contents, [[walls, [quantity, 4]], ceiling, floor,
                       [electrical_outlets, [quantity, 4]],
                       window, door]]]).
```

```
frame(kitchen,[[instance_of, room],
               [size, large],
               [contents, [[chairs, [quantity, 4]], table,
                          pictures, refrigerator, telephone,
                          microwave_oven, coffee_pot, clock,
                          stove]]]).
```

In addition, we note that as rooms go, the size of a kitchen is typically large. Now in operation, if a program needs to know if, say, a typical kitchen has electrical outlets, it starts with kitchen and looks for outlets. It will fail to find any mention of them, but then it will go one level up to the room frame and there it will find that electrical outlets are typically part of a room, so they are typically part of a kitchen.

A program utilizing this frame organization may sometimes need to keep track of the details of a specific kitchen. For a case like this, it can form a frame for the specific kitchen. Let this kitchen be named simply kitchen_001. Suppose the program learns that kitchen_001 is medium size, has three chairs, a table, a stove, and a computer. Furthermore, it is given that it is missing pictures, a window, and a coffee pot. The frame for kitchen_001 could be this:

```
frame(kitchen_001,[[instance_of, kitchen],
                   [size, medium],
                   [contents, [[chairs, [quantity, 3]],
                              table, stove, computer]]]).
                   [missing, [pictures, window, coffee_pot]]]).
```

Because this frame gives the size of the kitchen as medium, this entry will override the [size, large] entry in the kitchen frame. The contents entry shows what the system *knows for certain* is present in the kitchen and the missing entry shows what is known to be missing from this kitchen. Now, if the program needs to know if a particular item is present in this kitchen, it starts by looking at kitchen_001. If necessary, it consults the kitchen frame and then the room frame. For an item like, say, a microwave oven that is not listed as missing or present in kitchen_001, the program will go to the next level up where it will find that a microwave oven will typically be present, but of course, it cannot be absolutely sure that one is present. This whole arrangement of frames has become a *frame hierarchy*, a tree structure as shown in Figure 6.10.

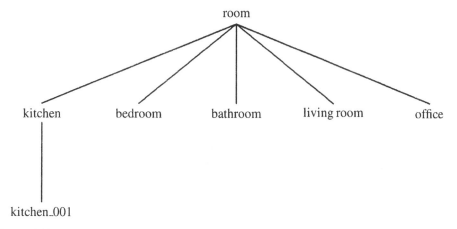

Figure 6.10: A kitchen, bedroom, bathroom, living room, and office are all instances of a room. Kitchen_001 is an instance of a kitchen. This collection of frames is called a frame hierarchy.

6.7.2   Scripts

Frames are intended to represent static information. To represent a sequence of events, the very similar concept of a script is used. Some scripts from actual programs will be discussed in Chapter 10. Scripts can be implemented in symbolic form or even as a collection of neural networks. For now, we will just mention the general concept and how it amounts to another instance of pattern completion. Below is a short and simple script given in just plain English about going to the airport to take a plane:

1) go to airport

2) if you have a ticket go to step 4

3) buy ticket

4) if you only have a small carry-on size piece of baggage, then go to step 6

5) give baggage to baggage handler and obtain baggage claim tickets

6) go to plane boarding area

7) if the plane is ready to board, then get on, else wait

8) board plane

9) leave plane

When a program is given a script like this, it is possible for it to look at a story and infer the steps that are missing. For instance, maybe the program is given the short story:

> John went to the airport. He had a lot of heavy bags because he was going to be gone a long time. He bought a ticket and went to wait for the plane. He arrived at O'Hare three hours later.

With the script and this story the program can conclude that John also gave baggage to the baggage handler and obtained baggage claim tickets, despite the fact that this was not mentioned in the story. This is just a typical pattern completion problem, but the subject matter is at a higher level than in a frame.

People constantly make inferences about stories they hear. In fact, some experiments (see [12]) indicate that this tendency to make inferences is so strong that sometimes people do not even remember exactly what was said. They remember their inferences just as though they were actually part of the story. For instance, again given that Mary likes oatmeal raisin cookies and you hear that Mary was eating cookies, sometime later on you might actually "remember" that someone said that Mary was eating oatmeal raisin cookies, despite the fact that this was not explicitly mentioned. In the airport story given above, you might remember that it was said that John gave his baggage to the baggage handler or that John got on board the plane, despite the fact that this was not explicitly stated. Mistakes such as these give clues as to how memory is organized. Among AI researchers, Schank has been particularly active in researching the use of scriptlike knowledge, both by people and in programs. His conclusions include the idea that memory is extremely important in the human reasoning process and that the storage of ideas in memory has to be a very flexible process [194]. He has also been working on how scripts could be created by building them up from specific instances.

6.8 Exercises

6.1. Write a Prolog (or Lisp) program that will take in statements like the following (say, at least the following, but if you can think up more, please do so!) and decide whether or not they make sense:

```
ate(mary,basketball)
ate(mary,cheese)
ate(mary,crow)
ate(mary,snails)
ate(mary,hat)
```

If you do not want to write a program you can still outline what is needed to handle these statements.

6.2. We have many *visual* images of what a kite is, however, programs do not have them. Write very detailed Prolog or Lisp statements that define what a kite is and how to build one, assuming the program only knows what wood, paper, rags (for the tail), and adhesive or staples are. Assume it knows how to cut these things to whatever size your program calls for. Document in English what your statements mean.

6.3. Train a recurrent neural network to output an ASCII "5" when it is given the ASCII sequence of inputs, "2," "+," and "3," but have it output an ASCII "6" when it is given the sequence, "2," "*," and "3." Output the "5" or the "6" only after all the inputs have been made. Have the network output 0.5 for each of the bit positions otherwise.

6.4. A recurrent network given in the text took the codes for words in short sentences like "Mary ate cookies" and collected them for output. Produce a network that will do the reverse, so for example, given the pattern on the input units for:

Mary ate cookies,

have the network sequentially output the words: Mary, ate, cookies.

6.5. Train a recurrent network to output 1 when a short string of letters contains a letter 'a' and output a 0 otherwise. The letters should be given to the network one letter at a time. (This might take longer than you expect.)

6.6. Take the sentence, "Don grows Reliance seedless grapes." and code this using the various symbolic notations described in Section 6.6.1 that apply.

6.7. Give the process you would go through to code the tree:

```
[[a, b], [c, d]]
```

using recursive auto-associative memory.

6.8. When you go to decode any RAAM representation, how will you know when the output units contain a code for a symbol or a code for a subtree?

6.9. Computers come in many different sizes, many different types of CPUs, with many different operating systems, and various numbers and types of peripherals. Create a frame hierarchy that will catalog all the variations among computers.

Chapter 7

Case-Based and Memory-Based Reasoning

In traditional AI the assumption has always been that people operate by using *knowledge* about the world, that is, a collection of facts that describes the world. In recent years another alternative has been proposed: people store *experiences* and operate by comparing their past experiences with the current problem. This new idea about how people operate has produced the memory-based and case-based methods of solving problems and in this chapter we will look at these new developments and compare them with the traditional solutions.

7.1 Condensed versus Uncondensed Knowledge

An important characteristic of all or almost all of the traditional symbolic methods is that they assume that knowledge must be condensed down to rules, frames, scripts, semantic networks, and so forth. These structures store knowledge about the world, but knowledge that has been stripped away from actual experiences. Complete experiences are not stored. Some examples could be that you might list the facts that apples are usually red, although sometimes they are green or yellow, elephants are gray and wrinkled and very big. Elephants are animals and all animals are alive. Roses are flowers and are alive but then they are not animals, and so on. These facts are entered into programs without the system ever having seen a complete apple, elephant, or rose. In natural language processing programs, rules are entered into the system for creating or recognizing correct sentences, yet the system has no memory of ever having seen or written a complete sentence. Even when a system is designed to do knowledge acquisition by taking in specific examples, the object is still to take the specific set of examples and condense them down to a set of facts and rules. Deciding when to save a new fact or create a new rule is obviously a problem.

The alternative to condensed knowledge is uncondensed knowledge. Here, complete experiences are stored, so that a program would have to store a complete image of an apple, its taste, its smell, and so on. In the case of an elephant, instances of how it moves around would have to be stored in addition to its size, shape, and so forth.

A special case of condensed knowledge is what could be called "compressed knowledge." This represents the compression (think of the Unix compress command) of large amounts of actual data down to rules so that the actual data can be thrown away. Recall that in Sections 3.1 and 3.2, we gave several possible methods of doing pattern recognition. One of the simplest was the linear pattern recognition method. When we fed the data of Figure 3.1 into the linear pattern recognition algorithm it produced an equation that allows anyone to determine whether a point, (x, y), belongs in one class or the other. In effect, the knowledge of eight cases was compressed down to an equation that is independent of the cases used to derive the rule. All the cases that led to this rule can be disposed of. This makes the linear pattern classifier much like the rule-based approach of traditional symbolic AI.

On the other hand, for an example of using uncompressed knowledge there is the nearest neighbor pattern recognition scheme that does not compress knowledge. The separation of the pattern space is achieved by comparing a given point with all its neighbors and finding the closest match. All the cases have to be remembered.

The question then comes up as to whether or not neural networks store compressed or uncompressed knowledge. On the one hand, once a network learns all its cases, the cases can be thrown away, so looking at it this way, the network does not store cases, just as a linear pattern classifier does not store cases. Also, when a network stores the cases it automatically produces a rulelike structure that is much the same as a system of traditional rules. But the question can be argued the other way as well. If you look at a Hopfield-type network there is an energy minimum for each case that the network stores. Now even though a back-propagation feed-forward type of network does not have energy minima associated with each of its cases, this is mostly just because it is convenient not to bother with that exact type of network. In the feed-forward type of network, given a set of certain known features, the goal is to predict the values of another set of unknown features. Another point in favor of calling a network a case-based method is that when a pattern is placed on the input units of a trained network, all the inputs must take on a value. The network looks at all the features of a pattern, not just a few. This is unlike the traditional rule-based arrangement where the rules are designed by hand to take into account only the important features that are used to draw a specific conclusion. That is, suppose you want to determine a Bears fan using a traditional rule-based architecture and the rule:

```
bears_fan(X) :- cubs_fan(X), from_chicago(X).
```

You only need to look at two inputs, cubs_fan and from_chicago to determine who is a bears_fan. In the back-propagation network that does this problem, every one of the 19 input features must be taken into account, not just some subset. In fact, the traditional idea that a whole vector of parameters about a person gets condensed down to a list of relevant symbols is another example of how traditional methods emphasize the condensation of knowledge. Traditional rules, then, condense knowledge even more than the linear pattern classifier. Another point in favor of saying that networks store cases is that networks are derived from specific cases. Generally, traditional methods of knowledge storage involve using handcrafted rules and other structures that are not derived from specific cases. Still another point is that neural networkers usually say that networks simply store cases when there are enough hidden units available to store all the cases. Neural networkers usually want to have too few hidden units in a network just so that the network will generalize and not just memorize specific cases. So neural networks seem to fall in between purely rule-based and purely cased-based methods.

For an example of how an AI problem can be done using condensed and uncondensed knowledge, we only need look at the animal recognition problem in Chapter 4. One way to do the recognition problem is to use traditional rules, another way is to store the data into a back-propagation network where cases are stored but rulelike abilities can develop, and the final way is to use a nearest neighbor approach.

7.1.1 Arguments For Condensed Knowledge

The traditional AI method of using condensed knowledge became the dominant form of pursuing AI for a number of different reasons. One really important reason for the pursuit of AI through condensed knowledge is because this is the way that sciences like physics and mathematics have been done. Many AI researchers have simply migrated from these areas into AI. In mathematics the usual mode of operation is to try to form a formal system that contains the minimum number of necessary assumptions but from which you can still derive, by theorem proving, a large number of interesting conclusions. In physics the approach is to attempt to find the minimum number of formulas, laws, and assumptions that explain how the universe works. The neat results that are possible in these sciences became a model for how intelligence should be researched and modeled.

In Chapter 1 we have already mentioned that natural language has provided a number of examples that have led researchers to believe that people develop rule systems to deal with language and words. For one, children pick up regularities in language and apply what they learn to new situations. In another example of this, studies of the errors children make when forming the past tense of verbs indicated to researchers that children were actually forming a rule base. In addition, it is fairly easy to produce rules of grammar that describe all the correct and only the correct statements that can be made in a natural language, so the assumption has been that using rules is exactly what people are doing when they use language.

Another major reason for the emphasis on the use of condensed knowledge is that it has never seemed very intelligent to solve problems just by storing cases and looking up the answers associated with those cases. This approach seemed too trivial.

The last major reason for the attractiveness of condensed forms of knowledge representation is that by using rules it is extremely easy to program computers to use reasoning methods like those in Prolog to reach conclusions. Such programs can often quickly deduce solutions to simple problems. Since the results on simple problems come so easily, it becomes extremely tempting to pursue this line of research to solve harder problems as well.

Besides the *arguments* for condensed knowledge, there is a natural psychological explanation as to why computer researchers have favored the idea. The typical computer is the von Neumann computer with a single CPU and a large amount of memory. If you need to look for a close match to a piece of data you must spend a lot of time moving lots of data through the CPU. On the other hand, if you have rules or formulas that you can apply to limit the search, the process is much faster. This mindset shows up in typical computer programs as statements like: "if x < 5 then …" or "if ch >= 'a' and ch <= 'z'." Thus, rules make classification and decision problems run faster on a von Neumann-type computer than if a table look-up scheme was used. Of course now, in recent years, other computer architectures have been proposed where the memory locations can all be searched in parallel and this sometimes makes simple searching practical and shows how profoundly the technology we have to work with has influenced the thinking about thinking.

7.1.2   Arguments Against Condensed Knowledge

One criticism of condensed knowledge is that traditional AI programs rarely store any specific experiences. If they work on a problem, once they solve it, they normally have their memories wiped clean of this experience, so the program cannot use the experience it has gained working on one problem and apply it to another. The program has to muddle its way through all the steps it has tried before. If people behaved this way and could not remember or benefit from their own past experiences, they would not be regarded as intelligent. This is a major point made by Schank [194]. Schank has been interested in programming models of natural language processing and story understanding. In his early programs, knowledge was stored in scripts and frames, but it became evident to him that framelike/scriptlike structures of condensed knowledge were not realistic. For one thing, when one of his script-based programs read a story it did not learn anything new about the world that it could use later to help it understand another story. For instance, if a program reads about someone visiting a fast food restaurant, say a Burger King, then that experience should make it easier for the program to understand what happens when it reads about someone visiting a McDonald's. Obviously, people function this way. When a person visits a fast food restaurant for a second time, there is only a single case available to guide behavior. There just is not enough data in a single case to form rules, frames, or scripts. This method of understanding the present situation by comparing it with past cases is one of the main features of case-based reasoning. Schank believes that people use memory rather than formal rules.

Other critics expect that it will be impossible to arrive at a formal theory of the world that is anything like the kind of formal theories that mathematicians and physicists produce. The real world is a much bigger world than the abstract worlds of mathematics and theoretical physics and it is messy. There are more relationships within a set of data than a small set of rules can cover. An intelligent creature has to store all the cases. It is true that when the creature encounters enough relevant cases it can then go about trying to formulate rules that help it understand the world, but it still never loses its cases.

Perhaps the most well-known criticism of traditional symbol processing AI comes from Hubert Dreyfus [27] and Hubert and Stuart Dreyfus [28] and [29]. A major point of theirs is that philosophers have tried for a long time to do this same task of describing the world with symbols and rules, yet in the end, the philosophers had to give up and admit they could not do it.

Some other objections of Dreyfus and Dreyfus are that (1) people are not sequential logical machines like today's digital computers, (2) real human *experts* do not use rules, and (3) no rule-based system will be able to equal the performance of human experts. They also make the following observations about expert behavior. In the learning phase of any endeavor, people are typically given rules about what is the proper response in a given situation. Some examples are how to fly a plane, how to give CPR, and how to play chess. In this mode of behavior, people are "uncertain" about what to do so they fall back on the rules that instructors have given them. True experts, they say, gain enough *actual experience* about the endeavor that they just end up *knowing* what the proper response is. They just give up following rules and go with their instincts. Dreyfus and Dreyfus rate human performance ability on a scale of 1 to 5 with 1 being a novice, 2 an advanced beginner, 3 competent, 4 proficient, and 5 an expert. Competent level practitioners are the last group that uses rules and they therefore maintain that expert systems that use rules should really be referred to as competent systems.

7.1.3   Problems with Condensed Representations

Now we will look at a few actual problems that can arise by using condensed knowledge. First we will look at a frame-based problem and then look at some rule-based problems.

The frame-based problem will be the problem of storing knowledge about a kitchen. In the last chapter we gave a frame-based solution to representing data about a kitchen. If all a program needs to know about are, say, a few styles of modern kitchens, there may not be any problem with using a standard, traditional, frame-based representation. On the other hand, if the program needs to be like a human being and needs to know about a lot of different kinds of kitchens, there are problems that arise. For instance, if we found a certain kitchen in an old house had a hand-operated pump (instead of a faucet), the chances of finding a bathtub in the kitchen are enhanced, as are the chances for kerosene lamps, possibly a wood-burning stove, no refrigerator, no phone, no computer, no electrical outlets, and so on. The frame-based representation of a modern kitchen fails in this case. An obvious solution for the frame-based representation is to add another frame for 'old kitchen' and give this frame a representative set of default values for an old kitchen. Yet, in between a 19th century kitchen and a late 20th century kitchen there is a whole continuum of kitchen styles and moreover, the expected contents of a kitchen depends on the size of the kitchen and the size of the kitchen depends on the size of the house. A large modern house will have a large kitchen and maybe multiple stoves and refrigerators. Given that a kitchen has a certain feature or a certain size, it enhances and inhibits the probability of its having other particular features. One obvious solution to this problem is to increase the complexity of a frame by including entries that state that if a kitchen contains feature A, then there is a certain probability that features B and C are present. But then, given that A and B are present, what is the probability of C? The number of facts that need to be stored can become fairly large and this whole plan seems unrealistic.

A neural networking approach would be more realistic. In the neural network approach, we could store many instances of kitchens in the network and it would end up producing an energy valley for the concept of 'kitchens,' just as storing many instances of Cubs fans ends up forming a general concept of a 'Cubs fan.' Actually, we could claim that this valley for kitchens amounts to a frame for kitchens, but a frame created by neural networking methods. The neural network automatically produces rough probability estimates that certain features of a kitchen will be present, given that certain other features are present.

We will now look at a couple of examples of how condensing knowledge using rules can fail. For the first example, we will repeat one from Section 4.4, the problem of distinguishing between an E and an F pattern in 5 × 5 matrices. The letters and a map to number the bits are shown in Figure 7.1. To identify E and F we could have the following rules, where the number indicates the bit is a 1:

1	2	3	4	5
6	7	8	9	10
11	12	13	14	15
16	17	18	19	20
21	22	23	24	25

Figure 7.1: The E and F patterns in 5×5 matrices and the numbering of the units.

IF   1 & 2 & 3 & 4 & 5 & 6 & 11 & 12 & 13 & 14 & 15 & 16 & 21
        & 22 & 23 & 24 & 25
THEN   the letter is E

IF   1 & 2 & 3 & 4 & 5 & 6 & 11 & 12 & 13 & 14 & 15 & 16 & 21
THEN   the letter is F

One of the key aspects of the rule-based implementation of this problem as compared to a neural networking implementation, is that the rule-based approach has effectively taken the list of microfeatures for letters like F:

$$11111 \quad 10000 \quad 11111 \quad 10000 \quad 10000$$

and replaced it with a list of symbols that is only a list of the features that are normally present. As we noted above, this represents another way of condensing knowledge. Introducing symbols saves one from having to deal with the very long bit strings (composed mostly of zeros) that neural networks need. In many ways this is a welcome innovation, unfortunately, there are disadvantages as well. If you give the system a letter E, then the rules for both an E and an F will be triggered. Some conflict resolution strategy is needed to settle the matter. Another way to deal with this problem is even more obvious. We can redefine the letter F so that all those features that should *not* be present are in fact not present. Doing this, of course, causes the advantage of introducing symbols to disappear and this turns the solution into a nearest neighbor type of approach.

Real-life situations are more complicated than the above pattern recognition problem. One such problem that has been suggested by McCarthy is the problem of whether or not a creature, X, can fly.[1] Here is a simple rule about the matter, stated in Prolog:

```
fly(X) :- bird(X).
```

Of course, this is clearly not valid for flightless birds like penguins and ostriches so we should really write:

```
fly(X) :- bird(X), notflightless(X).
```

This rule can still cause trouble because birds that can normally fly can have broken wings, baby birds cannot fly, and there are many other such conditions that prohibit flight. McCarthy even suggests this unusual scenario: what if the bird has its feet stuck in cement?

---

[1] This example comes from McCarthy and was reported in [249].

When you have to cope with all the possible conditions that can arise, even rather bizarre ones, rule-based reasoning loses a lot of its attractiveness. Rule-based programs can reach many conclusions that are obviously not valid. People often say that such programs do not have any "common sense." The situation here is at least as bad as it is in ordinary programming where programmers have to try to anticipate every strange combination of data that can come up. Maybe the problems are even worse when it comes to rule-based programming because the application areas are even larger than those of conventional programming.

The reaction of some traditional AI researchers to the problems inherent in current rule-based systems, like the problem of when an animal can fly, has been to work on developing a sophisticated form of reasoning called nonmonotonic logic. In [107], McCarthy discusses a simple outline of the project.

The reactions of other AI researchers to the problems with using condensed knowledge have been to develop the memory-based and case-based approaches. Memory-based methods are much the same as a nearest neighbor approach. Case-based methods also use elements of the nearest neighbor approach, but the kinds of problems that case-based systems can do are more sophisticated. The next two sections look at memory-based and case-based alternatives to condensed knowledge.

7.2   Memory-Based Reasoning

The idea behind memory-based reasoning (MBR) is to simply store large numbers of cases in memory, then when a new problem comes up, search memory for the closest cases and apply some statistical analysis to these close cases to derive an answer. The search uses a measure of closeness or a *metric* that could be simple Euclidean distance or some other distance measure. In this section we look at some simple examples of this approach.

7.2.1   A Simple Example

For a very simple example of memory-based reasoning we can use the sports example from Section 3.6. That data is shown again in Figure 7.2. First, we simply store this data away in a conventional computer's memory. Now, suppose we want to take a pattern for a Cubs fan from Chicago and look up what other characteristics are typically associated with such a person. We code our Cubs fan as:

```
1010 ??? ??????????? ?????
```

where ? means the bit could be either 0 or 1. We now search memory for all the entries that match this pattern and we get these four:

```
1010 101 100000000000 01000
1010 010 010000000000 11000
1010 011 000000100000 11000
1010 100 000000000010 01000
```

To determine typical values for the characteristics in the last five columns, we could compute the average values:

```
0.50  1.00  0.00  0.00  0.00
```

```
1010 101 100000000000 01000
1010 010 010000000000 11000
0101 101 001000000000 00001
0100 010 000100000000 00100
1000 101 000010000000 00100
0101 011 000001000000 00001
1010 011 000000100000 11000
0101 010 000000010000 00011
1000 000 000000001000 00100
0100 000 000000000100 00100
1010 100 000000000010 01000
0101 100 000000000001 00001
```

first 4 columns: Chicago, Cubs, New York, Mets
next 3 columns: Democrat, Republican, likes lemonade
next 12 columns: names
last 5 columns: Sox, Bears, tennis, Yankees, Jets

Figure 7.2: The data for the sports fans in Section 3.6.

That is, 50 percent of the time the Cubs fan from Chicago will be a Sox fan, 100 percent of the time, the Cubs fan from Chicago will be a Bears fan, and so on. Notice how neatly this was done *without using any special learning algorithm*. If we did some more complex analysis we would discover that all Republican Cubs fans are also Sox fans, but none of the Democrat Cubs fans are Sox fans.

7.2.2 MBRtalk

One of the earliest experiments in using the memory-based reasoning approach has been MBRtalk, a program used to generate speech that runs on the Connection Machine.[2] A Connection Machine can be configured with up to 65,536 processors and this makes it ideal to do a parallel search of a large database to find close matches and then to compute how to pronounce a given word. When the back-propagation-based NETtalk program was trained with a series of examples and tested on those examples, it chose the correct phoneme 94 percent of the time. For a set of words it was not trained on it was 78 percent correct. MBRtalk used the same database of words but used a 15-letter window consisting of a "central" letter, the four preceding letters, and ten following letters. Also input were the preceding four phonemes and stress fields. The program gets 100 percent of the instances of the training set correct and is 86 percent correct on unknown words.

Looking at how MBRtalk works is interesting, but we will not attempt to do a complete description of how it works. We will look at some of the simpler parts and skip most of the statistical analysis that is done. In the example that follows (from Waltz and Stanfill [248]), they use the smaller-size window used by NETtalk.

---

[2] The Connection Machine is a registered trademark of Thinking Machines Corporation.

Record		Δ	$\frac{1}{\Delta+0.01}$
...gyps	J +	0.0000	100.00
...gesu	g +	0.0196	33.78
...gemi	J +	0.0361	21.69
...geni	J +	0.0369	21.32
...gewg	g +	0.0420	19.32
...gene	J +	0.0690	12.66
...gin.	J +	0.0806	11.04
.regist	J −	0.1127	8.15
.vegeta	J −	0.1835	5.17
ghage..	J −	0.2059	4.63

Figure 7.3: In order to figure out how to pronounce the 'g' in ...gyps, MBRtalk searched its memory first for records that have 'g' in the middle and then used some more complex methods to rate how close each record it turned up was to ...gyps. The second column shows this distance Δ and the third column is $1/(\Delta + 0.01)$. Next, the third column values are added up for each of the candidate pronunciations and the candidate with the largest value is the winner. The table is adapted from [248].

To look at how MBRtalk operates, suppose the program has the problem of determining how to pronounce the 'g' in gypsum. The first step is to do a parallel search of memory to find all the records that contain a 'g' in the central position. This is the pattern that MBRtalk will try to match against all the records in the database:

$$...gyps$$

where the dots stand for empty spaces. Now, some statistical analysis is performed on this subset of records to find the closest matches to this unknown record. The statistical analysis is somewhat ad hoc (but plausible) and represents another metric. The calculations turn up the ten records shown in Figure 7.3. The Δ represents the distance each record is from '...gyps.' The closest record has a distance of 0.0000 and comes from the word, gypsy. In this record, the 'g' in ...gyps is pronounced as 'J' and the stress field is rising. Among the other records, there are some records that predict that the 'g' should be pronounced as 'g' (as in globe). Taking the answer for the closest record is one way to decide on the answer for the unknown. Instead, Waltz and Stanfill resorted to a more complicated plan. First, add 0.01 to each distance and then divide by 1.0, giving the values shown in the column on the right in the figure. Now for each possible pronunciation of 'g,' add up the associated numbers in the column on the right. The winner is the one with the highest total. As it turns out, the 'g' pronunciation totals 53.02 and the 'J' pronunciation totals 184.66.

7.2.3 A HERBIE Solution to Reading

A slightly newer system that reads text has been produced by Wolpert [264] and it has an error rate one third the error rate of NETtalk, uses the memory-based reasoning approach, and is simpler than MBRtalk. The algorithm used is one of a type of generalizing algorithm

Wolpert calls a HERBIE (for HEuRistic BInary Engine). Each database record, i, consists of a vector, $\vec{x}_i$, that is the pattern, and a corresponding answer vector, $\vec{y}_i$, that has a 1 entry for the answer and 0 elsewhere. The unknown, $\vec{q}$, is compared with every $\vec{x}_i$ where the distance, $\vec{d}_i$, between $\vec{q}$ and $\vec{x}_i$ is defined as the number of places where the letters differ and then the four closest matches are used in the formula below. A vector $\vec{r}$ is computed like so:

$$\vec{r} = \frac{\sum_{i=1,4} \vec{y}_i / d_i}{\sum_{i=1,4} 1/d_i}$$

and the answer, j, is the index of the component of $\vec{r}$ that has the largest value. This is called the "center of mass" rating scheme because each $\vec{y}_i$ can be considered a point in space while each $1/d_i$ is the "mass" at $\vec{y}_i$. With this scheme, the closer a record is to the unknown, the more the entry contributes and a perfect match gives an infinite contribution.

7.2.4 JOHNNY

Another memory-based system developed on the Connection Machine is JOHNNY [222]. JOHNNY learns how to read using a set of rules about how to pronounce words and by consulting its memory of pronunciations for words. JOHNNY knows how to pronounce words but it has to learn the associations between the printed words and their pronunciations. JOHNNY does this by reading a word and then using the rules to try to "sound out" possible pronunciations. When this set of candidate pronunciations is found, the stored list of pronunciations is searched for the closest match to these candidates. When the best match is found, the correct phonemes for the word are stored in the format used by MBRtalk. This knowledge then becomes available for future words so as time goes on, JOHNNY's ability to correctly pronounce words steadily increases.

Category	Number	Examples
Long and short vowels	12	?a? → @
Terminal e is silent	1	?e. → −
Vowels + w	6	?aw → c
Second vowel of pair is silent	25	ea? → −
Simple consonant rules	26	?c? → s
		?c? → k
		?b? → b
Consonant clusters	13	?th → T
		?th → D
		th? → −
Doubled consonants are silent	21	bb? → −

Figure 7.4: Some examples of the simple rules that JOHNNY starts with adapted from [222].

Some of the 104 rules used by JOHNNY are shown in Figure 7.4. The left side of a rule represents a three-character window and the symbol on the right side stands for the correct pronunciation. In this notation, '?' stands for do not care, '.' stands for an empty space, and '−' stands for silent. The set of rules used in JOHNNY is not complete and also contains ambiguities, such as the fact that the letter 'c' may give the sound 's' or 'k' and that 'th' may give the sound 'T' or 'D.'

When a word such as 'the' is encountered, it is broken up into records. For 'the' these records are: '.th,' 'the,' and 'he.'. The memory of rules is searched to try to find the closest matches to these records. The closeness is determined by a simple rating scheme. Letters that mismatch give a large penalty for the rule while letters that match do not cares give a smaller penalty. For example, the rule, ?ch → C mismatches '.th' on one letter and receives a large penalty for this and a small penalty for matching on do not care. The rule, ?t? → t mismatches on two do not cares. Ultimately, matching on the pattern, '.th,' JOHNNY retrieves the pronunciations, 'T' and 'D' as the most likely pronunciations. Continuing on, 'the' matches silent (–) and 'he.' also matches silent.

With these candidate pronunciations, JOHNNY searches its memory of phonetic encodings of words for the closest match. The searching finds that the closest pronunciation in memory is 'D–i' (the), with one mismatch on the third phoneme. Stanfill reports that this process comes up with the wrong word 7 to 12 percent of the time. At this point, several possibilities exist. In unsupervised mode, JOHNNY simply associates the pronunciation with the printed word, whether it is right or wrong. A second option is to allow a human teacher to correct it. Naturally, the supervised learning produces better results, however, we will just continue on looking at the unsupervised learning mode. Having come up with a word, the word and its pronunciation are stored in memory in a format suitable for using MBRtalk. This format uses a nine-letter window, and 'the' is stored as the three records:

$$....the.. \quad \rightarrow \quad D$$
$$...the... \quad \rightarrow \quad -$$
$$..the.... \quad \rightarrow \quad i$$

JOHNNY was tested on several training sets. In these tests, on the first pass through the training set, JOHNNY uses only rules. In the second and third passes, JOHNNY uses MBRtalk and the knowledge it has accumulated about pronunciations together with the rules. Using a set of 8,192 words selected at random from a dictionary, the error rate on the first pass (using only rules) was 7 percent. At the end of the second pass the error rate was 3 percent. In another experiment using different training sets, some of the rules, such as rules for 't' and 's' were deleted. In this case, the initial error rate was 20 percent but it fell to about 8 percent after the third pass.

This experiment with JOHNNY was made to illustrate the utility of the memory-based reasoning approach and that it could use rules and learn from the words it encounters. Learning successes and failures from experiencing specific cases is a central tenet of the memory-based and case-based reasoning approaches. Stanfill notes that this use of rules is the exact opposite of how rules are typically treated in AI programs. Typically, the goal is to deduce rules from data. In this case, JOHNNY starts with rules and supplements its knowledge by experience. Stanfill says that in this model, "rules are better viewed as a means of communication: a teacher instructs a student in a set of rules; the student then uses the rules to get started in an activity, and proceeds to learn as it solves an increasing number of problems."

Some nice features of memory-based reasoning are first, that a system only needs examples to learn about its problem domain, second, it can explain its decisions by citing precedents, and a third, really humanlike capability is that because it can tell a good match from a poor match, it "knows when it knows." Rule-based or other systems using condensed knowledge do not know when they do not know. Even back-propagation networks are not able to spot inputs that are unusual, so like ordinary programs and rule-based systems, networks simply fumble on ahead even when it is perfectly ridiculous to do so.

7.2.5   PACE

A new memory-based program, called PACE, for Parallel Automatic Coding Expert [23], has been used to classify a person's responses on the 1990 US census, again using a Connection Machine. The goal of the program is to take natural language input from a form and classify the person's industry and occupational categories. PACE can process about 60 percent of the returns accurately using 132,247 previously classified returns as its database. This performance was better than the 47 percent that a rule-based expert system called AIOCS, for Automated Industry and Occupation Coding System, developed by the Census Bureau gave. PACE was developed in four person-months whereas AIOCS took 192 person-months.

7.3   Case-Based Reasoning

Case-based reasoning (CBR) closely resembles memory-based reasoning. Making an exact distinction between the two methods is somewhat difficult. Both methods are new and both MBR and CBR make extensive use of memory by storing solutions to problems that were previously encountered. Unlike the memory-based methods that run on the Connection Machine, case-based methods have been implemented on standard hardware using standard symbolic processing languages, especially Lisp. Case-based methods concentrate on finding a close case and then modifying it to suit some new circumstances, while the emphasis in MBR has been to search a very large number of cases and use a rating scheme to find a likely answer. Because part of a case-based program requires finding the closest match, possibly MBR could be described as one component of CBR, or ultimately, the two may come to mean exactly the same thing. While both MBR and CBR are "new," elements of both approaches can be found in some early AI programs.

7.3.1   Case-Based Reasoning in People

A good example of the philosophy of the case-based reasoning approach can be shown by looking at how people learn to program computers. The problem to look at will be that of developing a new program in Snobol given only two small examples of Snobol programs. If you happen to know Snobol, try to pretend that you do not. The two Snobol programs are shown in Figure 7.5. The first program reads in numbers, finds their sum, and prints it out, while the second program reads in a number representing the radius of a circle, calculates the area of the circle, and prints out the area of the circle.

Our new problem now is to write a program that will find the sales tax on the following purchase of three items given that the sales tax is 7 percent:

```
* * * * * * * * * * * * * * * * * * * * * * * * * * * * * * * * * * * * * * * * * * * * * * * * * * * * * * * *
*  SNOBOL PROGRAM NUMBER 1
*
*  THIS PROGRAM READS THE NUMBERS AFTER THE END STATEMENT AND
*  PRINTS THEIR SUM.
*
      SUM = 0
LOOP  CARD = INPUT                                      :F(PRINT)
      SUM = SUM + CARD                                  :(LOOP)
PRINT OUTPUT = 'SUM WAS ' SUM
END
      1
      2
      3


* * * * * * * * * * * * * * * * * * * * * * * * * * * * * * * * * * * * * * * * * * * * * * * * * * * * * * * *
*  SNOBOL PROGRAM NUMBER 2
*
*  THIS PROGRAM READS IN THE RADIUS OF A CIRCLE AND PRINTS
*  THE AREA OF THE CIRCLE
*
      R = INPUT
      AREA = 3.14 * R * R
      OUTPUT = 'AREA OF CIRCLE OF RADIUS ' R '= ' AREA
END
      5
```

Figure 7.5: Two small Snobol programs that can be broken apart to create new programs.

```
* * * * * * * * * * * * * * * * * * * * * * * * * * * * * * * * * * * * * * * * * * * * * * * * * * * * * * * *
*  THIS PROGRAM READS IN PRICES OF ITEMS AND FINDS THE TAX ON
*  THOSE ITEMS.
*
      SUM = 0
LOOP  CARD = INPUT                                      :F(PRINT)
      SUM = SUM + 0.07*CARD                             :(LOOP)
PRINT OUTPUT = 'SUM WAS ' SUM
END
      12.99
      9.95
      7.50
```

Figure 7.6: A first attempt at writing a program to find the tax on some items.

floppy disks	$12.99
paper	$ 9.95
ribbon	$ 7.50

As experienced computer programmers, we know some of the requirements. We need to write a loop that will take each value one at a time, find the tax on the item by multiplying it by 0.07, and then add that value in to some variable. You continue this process until you run out of data, then you print the answer. If you do not know Snobol, you are going to have a problem, but a good strategy is to look around for similar programs and try to adapt them. The program that fits most closely is program number 1. To a person with programming experience, it is certain that the line:

```
SUM = SUM + CARD                                    :(LOOP)
```

is the part of the program that is adding up the numbers. The only part that does not fit the program you need to write is the part about the quantity that needs to be added up. Program 1 is adding up 'CARD,' but we are also interested in adding up the tax and the tax needs to be computed somehow. We need to figure out how to go about multiplying the tax rate times CARD and replacing the original CARD with this expression. Fortunately, program 2 shows how to calculate the area of a circle. Knowing that multiplication is required, we can guess that the * is the multiplication operator in Snobol, so we can tentatively conclude what must be put into the gap in the first program. We produce the program shown in Figure 7.6. Unfortunately, however, when the program is submitted to the Snobol interpreter, it fails with a syntax error in the line that was changed. Without knowing Snobol (and the true meaning of Snobol error messages), we are in a bit of trouble, however, a very careful comparison of the two Snobol program shows that there is always a blank around every operator. In almost every other computer language the blanks around the operators are optional, however, some people do intentionally put blanks around operators just to make the expressions look nice. In desperation, and for lack of anything else to try, we change that line to:

```
SUM = SUM + 0.07 * CARD                             :(LOOP)
```

Now the program works correctly, since Snobol does require blanks around its operators. Note how experience with other computer languages made it possible to make good guesses about how to put together this program in a strange language. Notice too, how the experience from those other languages also increased the odds of making the mistake of not putting blanks around the multiplication operator.

Actually, just learning computer programming for the first time amounts to building up an inventory of cases that can be used for problems that come up later. Notice too, whether a person is learning programming for the first time or just learning a new language, it is the *examples* that are the most useful. The manuals that provide the precise descriptive rules are a lot less easy to use than the manuals that are filled with examples.

We could summarize the case-based reasoning approach as follows:

1) memory is searched for a close match;

2) the close match is examined for parts that do not belong;

3) the parts that do not belong are replaced by parts that might work; and

4) if the proposed solution fails, try to identify the reason why and fix it.

5) Return to step 1 or 3 and continue as long as you can think of a possible 'fix.'

6) While going through steps 1 to 5, keep a record of things that worked so you can use them again in a new problem and keep track of things that did not work so that you do not use them again.

Most case-based systems have been produced using symbol processing methods on conventional computers. In these systems, the storage of episodes in memory is important because when a new case comes up, it is important to be able to quickly find all the relevant old cases. Thus, to some extent, CBR becomes a database problem.

7.3.2 CHEF

The one actual case-based system we will look at is CHEF, a program written in Lisp to design recipes in the area of Szechwan cooking [59]. The program starts with a set of 10 initial recipes for such dishes as stir-fried beef with green beans, broccoli with tofu, ginger chicken, vanilla souffle, and others. CHEF does not rely completely on cases to do its reasoning; it also uses rules about how to handle some parts of the problem.

BEEF-WITH-GREEN-BEANS

A half pound of beef
Two tablespoons of soy sauce
One teaspoon of rice wine
A half tablespoon of corn starch
One teaspoon of sugar
A half pound of green beans
One teaspoon of salt
One chunk of garlic

Chop the garlic into pieces the size of matchheads.
Shred the beef.
Marinate the beef in the garlic, sugar, corn starch, rice wine, and soy sauce.
Stir fry the spices, rice wine, and beef for one minute.
Add the green beans to the spices, rice wine, and beef.
Stir fry the spices, rice wine, green beans, and beef for three minutes.
Add the salt to the spices, rice wine, green beans, and beef.

Figure 7.7: When searching for a stir-fried beef with broccoli recipe, CHEF finds the closest match is this stir-fried beef with green beans recipe.

Suppose the first problem presented to CHEF is to create a stir fry dish using beef and broccoli. CHEF first searches its memory to find the closest recipe to apply. Naturally, it finds the recipe for stir-fried beef with green beans is the closest match. The recipe is shown in Figure 7.7. Now CHEF will make a copy of this old recipe and begin to modify it. The places in the recipe that mention green beans are replaced with broccoli. At this time, the program looks through a set of what are called, "ingredient critics." One ingredient critic finds that broccoli should be chopped into pieces the size of chunks. This step is added to the new recipe.

Now, this new recipe is turned over to another program that will do a simulated job of cooking the meal. CHEF knows what the results of cooking the meal ought to be. Among them, the beef should be tender, it should taste sweet, salty, savory, like garlic, and the broccoli should be crisp. While CHEF was only given the requirements to make a stir fry dish using beef and broccoli, these other requirements were stored with the definition of stir frying. They are added to the requirements for the entire recipe. Unfortunately, the cooking program came back and reported that the broccoli was soggy instead of crisp.

This failure requires that CHEF look into the cause of the problem. When it does, it finds a rule that when meat is cooked, some liquid comes out of the meat and CHEF decides that it must be this extra liquid that makes the broccoli soggy. This was not a problem with the green beans because green beans are not as delicate as broccoli. Having uncovered the problem, CHEF consults a list of possible strategies to correct the matter. Three possible strategies come to mind. One is to try to find a plan that meets all the requirements yet does not produce the liquid in the pan that makes the broccoli soggy. CHEF does not find such a plan. A second strategy is to check to see if something can be done to remove the liquid in the pan produced by the beef. Nothing is found. A third strategy is to alter the steps in the cooking process by splitting the process into separate steps and then bringing the results of the two steps together at the end. CHEF chooses this strategy and then the new steps are the following:

> Stir fry the broccoli for three minutes.
> Remove the broccoli from the pan.
> Stir fry the spices, rice wine, and beef for three minutes.
> Add the broccoli to the spices, rice wine, and beef.
> Stir fry the spices, rice wine, broccoli, and beef for a half minute.

Now CHEF creates a "demon" to warn of possible future problems with meat and vegetables that can become soggy. The demon is just a memory record that is added to a list of records containing planning failures. In doing this, CHEF does not just create a warning about beef and broccoli being the problem. CHEF notices that it found out that the problem occurred because meat gives off liquid and certain kinds of vegetables are susceptible to becoming soggy. The demon is therefore designed to warn of failure when meat and a crisp vegetable are stir-fried together. In this way, CHEF generalizes on its experience. A memo of failure is also attached to the new beef and broccoli recipe. This can be important when CHEF has to create other recipes that may produce the same problem.

Now suppose CHEF is given the problem of trying to create a stir fry dish using chicken and snow peas. It has to search its memory of plans to find one that satisfies these requirements, but first it consults the list of failures. Here it finds the demon that warns that if the dish is stir-fried and there is meat and a crisp vegetable required, then it should avoid the problem that occurred with the beef and broccoli recipe. Since snow peas are classified as a delicate vegetable, this demon's warning must be taken into account. Now when CHEF looks for a plan that meets the requirements that are called for, there will be the extra requirement that the recipe avoids the failure that occurred in the original beef and broccoli recipe. Even if there is a recipe on file for, say, stir-fried chicken and green beans, CHEF will end up preferring the plan used in beef and broccoli, even though the stir-fried chicken and beans seems closer on the surface to stir-fried chicken and snow peas.

Starting from 10 initial recipes, CHEF created 21 new recipes, but it could have created many more. There are a lot of details about how CHEF operates that we will not look into. Interested readers should see [59]. The matter of how the case memory is organized and how CHEF finds the closest matches is especially interesting, however, and we will now look into this aspect.

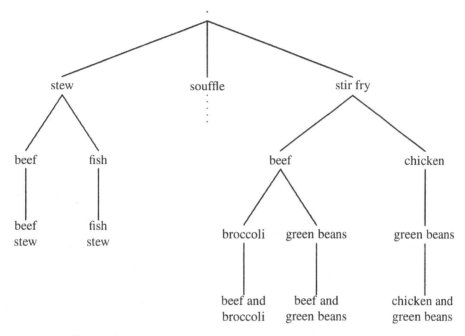

Figure 7.8: A discrimination net for recipes like the one used in CHEF.

CHEF finds the closest matching recipe by taking the requirements for the new recipe and consulting a discrimination net. A portion of a hypothetical such network that stores recipes is shown in Figure 7.8. The network is organized and searched in such a way that the most important features of the required recipe are taken into account first. For instance, suppose a recipe calls for making stir-fried fish. Suppose the system does not have an exact plan for this, but there are plans available for fish stew and stir-fried chicken. Converting a fish stew to stir-fried fish is likely to be much harder than converting stir-fried chicken to stir-fried fish. Therefore, the stir-fried chicken recipe must be regarded as a better plan to work from than the fish stew. Thus, the type of dish is more important than the exact ingredients used in the dish. The discrimination net is arranged so that the type of dish is the first criteria that the system uses to match on. CHEF's searching algorithm always looks for the recipe that will require the fewest modifications.

For another example of how CHEF looks for the nearest match, suppose that the problem is to produce a stir-fried fish with green beans recipe. Looking down the discrimination net, two recipes are possible, beef and green beans and chicken and green beans. This looks like a tie, but CHEF would go on to look up the tastes of chicken, beef, and fish in a semantic network. There, it finds that the taste of beef is strong while the taste of both chicken and fish are both weaker and more similar. Because of this similarity of taste factor, CHEF would take the chicken and green beans recipe as closer because the spices that go into the chicken dish should work better than the spices that go into the beef dish.

Another example of how the CHEF pattern look-up scheme behaves is the following. Suppose the goal is to find a stir fry dish that is hot and includes shrimp. CHEF would go down the stir fry branch of the network. Suppose it does not find any reference to shrimp. In this case CHEF realizes that shrimp is a seafood. The search is performed again, this time looking for any sort of seafood. Now in doing the search it might find, for instance, a recipe for spicy red fish. It would then have to modify the red fish recipe to use shrimp.

Finally, the demons that CHEF uses to warn of failures are also used in the search through the discrimination net. Remember that when the program gets a problem it looks through its memories of failures for possible problems that may come up. Suppose the program has gone through the soggy beef and broccoli problem. It now has the demon that warns of this problem and it has also a memo to this effect attached to the final beef and broccoli plan. Suppose the system now gets the problem of making a stir fry dish with chicken and snow peas. First it consults the memory of failures and finds that the goal of avoiding soggy vegetables applies. Suppose the available plans are the following: first, the beef and green beans, second, the new beef and broccoli with its memory of the failure, and third, a recipe for stir-fried chicken and green beans. This third one does not have the splitting of cooking steps because green beans will not get soggy in the extra liquid from the chicken. Fourth, there is a stir-fried shrimp and snow peas recipe. This recipe does not have the splitting of cooking steps in it either, this time because shrimp does not give off any liquid when cooked. Now, if the choice of plan was based on getting the plan with the closest set of ingredients, the shrimp and snow peas plan would be the plan of choice, however, the way CHEF works, the warning of failure that was generated before the search of the discrimination net is a more important goal to achieve. CHEF will therefore pick the beef and broccoli recipe and work to modify it.

Some other comments about CHEF are in order. At the highest level of operation, the creation of recipes, a case-oriented solution is used. However, at lower levels, CHEF does not work from previous cases and grab solutions from those cases. For instance, when broccoli is called for in a recipe, CHEF actually looks up this ingredient in a table and finds out that the broccoli should be cut up into chunks. A completely case-based system would have a case on file that included cutting up broccoli into chunks. That step would be extracted from the previous recipe and spliced into the current one. Another example of this comes with the set of general strategies used to fix the beef and broccoli recipe. CHEF consulted a list of these strategies to apply. A completely case-based system would recall cases where ingredients are cooked separately and then combined in the last step. A plan from one of these cases would then be adapted to the beef and broccoli case. There are also many places where CHEF uses rules. So, at the lower levels, CHEF still uses the traditional AI methods and a lot of handcrafting is necessary. Designing a program that relies only on cases would be interesting and challenging. It would be especially interesting, not to mention useful, if most of the handcrafting could be eliminated from the process.

7.4 Other Case-Based Programs

Some of the other CBR systems that have been designed are HYPO by Ashley and Rissland [5, 6, 175], that performs legal reasoning in the domain of trade secret law. While it is common to think of laws as rules and therefore that legal reasoning is rulelike, legal reasoning actually relies heavily on cases. The JULIA program [87] functions as a caterer's assistant. PARADYME [88] is an implementation of JULIA on the Connection Machine. CASEY [90] diagnoses heart disease. TRUCKER [58] plans truck routes under constantly changing requirements for delivery and pickup. CBR has also been suggested for use in military planning.

Most research into automatic programming to date has concentrated on trying to produce programs that contain a lot of rules about how to produce correct programs from scratch on the first try. Recently, however, Williams [260] has been working on a case-based approach to programming like that discussed in the Snobol example, except Williams is using very small Lisp programs.

7.5 Exercises

7.1. If you did not know back-propagation or other pattern recognition techniques, and you had to program an ordinary digital computer to tell the difference between a dog and a cat, how would you go about doing it? If you had to make up rules to make explicit the difference between a dog and a cat, what would they be? One easy way to tell the difference would be by the sound they make, but what criteria would a deaf person use or a person viewing pictures of dogs and cats use? If the animal in question was not a dog or a cat, consider how rules could determine that. Remember that dogs and cats can come in all shapes and sizes.

7.2. Dreyfus and Dreyfus argue that whereas human beings may start to learn about a subject area by using rules they have been given, the real experts move on to a stage where they no longer use rules because they cannot give rules for their decisions. Is there a reasonable counterargument for their observation?

7.3. In Section 7.1 there was the problem of deciding whether or not a bird can fly. In the text we pointed out how rule-based systems would have trouble coping with this problem. Suppose a bird does have its feet stuck in cement. How would people determine whether or not the bird can fly?

7.4. Read and summarize the following article in the Winter 1988 issue of *Daedalus*: "Mathematical Logic in Artificial Intelligence," by John McCarthy. Then comment on how similar his nonmonotonic logic proposal is to case-based reasoning.

7.5. Read and summarize Chapter 2, "Reminding and Memory" of Schank's book, *Dynamic Memory* [194].

7.6. Read and summarize the article, "Trading MIPS and Memory for Knowledge Engineering," *Communications of the ACM*, August 1992.

7.7. Consider how well case-based reasoning methods would work at solving the third and fourth grade arithmetic word problems exercise in Chapter 1.

7.8. Is there any way to tell by looking at the outputs of a system whether internally it is using rules or just using lots of cases that cover the whole space?

Chapter 8

Problem Solving and Heuristic Search

On many occasions people will not know the answer to a particular problem because they have never seen the problem before. However, given that they have seen similar problems in the past, they can use their knowledge to attempt likely solutions to the new problem. This trial and error problem solving behavior has been an important aspect of traditional AI research and many studies made of how people solve problems indicate that they do exhibit the trial and error sort of behavior found in heuristic programs.

On the other hand, there are some people who suspect that human problem solving is not *always* this mechanical. Many famous scientists have said that when they were working on exceptionally difficult problems they received a "flash of insight" apparently from out of nowhere and saw the solution to a large complex problem as a single thought. With this flash of insight the need to search for a solution seems to vanish, thus, there is the possibility that sometimes people go beyond using heuristic search techniques. For some accounts of this, see [55] and [148].

In this chapter we will look at the fundamental heuristic search techniques and how they were applied in some early and very well-known programs, the 8-Puzzle, the Geometry Theorem Prover, and Symbolic Integration programs. Besides the methods presented here, there are many other heuristic search techniques.

8.1   The 8-Puzzle

The standard example used for an introduction to heuristic search has become the problem of solving the 8-puzzle. The 8-puzzle is a smaller version of the well-known 15-puzzle. The 15-puzzle consists of a 4×4 array of small numbered tiles and an empty space. The goal of the puzzle is to take a given initial configuration and change it through a series of moves into a given final configuration. For instance, there is the problem of turning:

	2	3	4
1	5	7	8
9	6	10	11
13	14	15	12

into

1	2	3	4
5	6	7	8
9	10	11	12
13	14	15	

An 8-puzzle problem might be to turn

1		3
7	2	4
6	8	5

into

1	2	3
8		4
7	6	5

There are two basic ways to program these problems, searching "blindly" without looking at the patterns that occur in the problem and searching by looking at the patterns that develop as the problem is solved. First, we will do two blind search methods and then look at heuristic search methods.

8.1.1   The Blind Search Methods

In the first method we simply have the program exchange the empty square with each of its 2, 3, or 4 neighbors and continue to do so until it stumbles upon the goal state. Needless to say, we need to select a method for generating all the possible moves in an orderly fashion and without repeating configurations that have already occurred before. Two methods come to mind. First, with a *depth-first* search we start moving down the tree to some maximum depth before we stop the downward movement and go back up the tree of possibilities. Second, there is a *breadth-first* search where we generate the tree level by level until we find the answer or give up. First we will try a depth-first search. We will start with the following pattern:

1		3
7	2	4
6	8	5

and try to turn it into

1	2	3
8		4
7	6	5

With the initial board configuration, we have the option of moving the blank left, right, or down. Doing all three of these yields three new configurations:

1

	1	3
7	2	4
6	8	5

2

1	2	3
7		4
6	8	5

3

1	3	
7	2	4
6	8	5

The number preceding the board will be a unique number for the particular configuration that also indicates the order in which it was generated. The initial matrix will be numbered 0. We have not yet stumbled upon the final answer, so we will have to continue moving the empty square around. But first, the initial matrix, 0, will be put on a list of matrices that have already been used to generate all their children. We name that list, USED. The newly generated matrices, 1, 2, and 3 will be put, in order, on a list named UNUSED. The contents of the two lists are shown below:

USED LIST: 0

1		3
7	2	4
6	8	5

UNUSED LIST: 1

	1	3
7	2	4
6	8	5

2

1	2	3
7		4
6	8	5

3

1	3	
7	2	4
6	8	5

To continue the search, the first matrix on the UNUSED list is taken and used to generate more matrices. Matrix 1 is therefore taken from the UNUSED list and all its possible children are generated. The new matrices are:

Matrix 1 is moved to the USED list. Checking both these new matrices against matrices in the USED and UNUSED lists, we find that the second of these two is the same as matrix 0, so we discard it. Matrix 4 is put on the front of the UNUSED list. The two lists are now:

USED LIST:

UNUSED LIST:

We have not reached the goal, so we select the first matrix from the UNUSED list, matrix 4 and create new matrices from it giving these 3:

Only two of these are new. Matrices 5 and 6 go on the UNUSED list. Matrix 4 goes on the USED list. We continue taking matrices from the UNUSED list and generating new matrices until we stumble on the answer, run out of memory, or reach some maximum acceptable depth. Suppose we choose this last alternative and let the maximum depth of search be three. In this case, matrices 5 and 6 will not now go on the front of the UNUSED list, they will just be discarded. Now we generate alternatives from the first matrix on the UNUSED list, matrix 2. Figure 8.1 shows the tree of matrices generated when the maximum depth of search is three. All this searching has yet to turn up the correct answer, so if a depth of three was the maximum depth allowed, the program would have to report a failure.

The second possible blind search method to use is a breadth-first search, where each level of the tree is generated, one layer at a time, until we hit upon the goal node. With this organization, node 0 produces the same nodes 1, 2, and 3. Node 0 goes on the USED list. Nodes 1, 2, and 3 go on the UNUSED list. We now take the first matrix on the UNUSED list, matrix 1, and use it to generate the following matrices:

Again the second of the two is found on the USED list so it is disposed of. Matrix 4, however, is added to the *end* of the UNUSED list and matrix 1 goes on the USED list. The

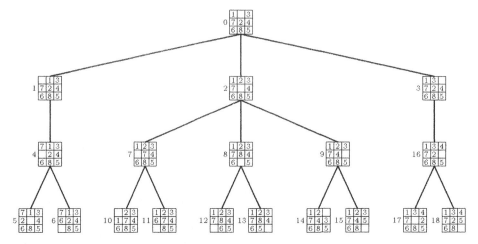

Figure 8.1: A depth-first search to a depth of three fails to find the goal. The number to the left of each board indicates the order in which the configuration was generated.

USED and UNUSED lists are now:

USED LIST:

UNUSED LIST:

We now take the next matrix off the UNUSED list, matrix 2, and generate its children giving:

These then go at the end of the UNUSED list while matrix 2 goes on the USED list. We continue in this manner, always taking matrices from the beginning of the UNUSED list and adding the matrices' children to the end of the UNUSED list, until we find the answer. In Figure 8.2 we show the tree that is generated after three layers have been created, but again, the goal has not yet been reached.

8.1.2 Heuristic Searches

The disappointing results so far could encourage us to look for a better searching strategy. If we take a look at nodes 13 and 18 in Figure 8.2, it seems as if node 13 is closer to the goal than node 18. In fact, from node 13 the goal is just two moves away. It would clearly be useful to have a formula that will neatly quantify the idea of "being closer to the solution." For a start, let us try the following plan. If we let a "square" mean either a tile or the blank space, we notice that matrix 18 has eight squares out of place, whereas we started with

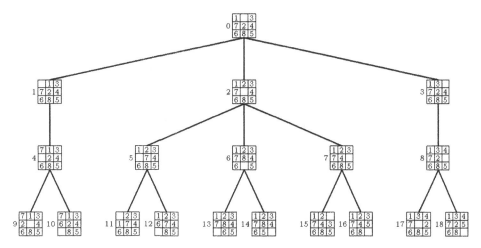

Figure 8.2: A breadth-first search to a depth of three fails to find the goal.

node 0 with only five out of place. On the other hand, node 13 has only three squares out of place so it is a good bet that node 13 is closer to the solution than nodes 0 or 18. If we use the criteria that we always work with the matrix that is closest to the goal, where closest is defined as fewest number of squares out of place, the search should go rather more quickly than either method tried so far. We will refer to the number of squares out of place as the "disorder" of the matrix. Using disorder as an indication of which matrix to use next will make the search we conduct a heuristic search. (Notice how similar this is to trying to minimize energy in a Hopfield network.) A heuristic search uses some patterns in the problem to guide the search in the most likely direction. Heuristic searches do not always guarantee that a solution will be found, however, if the heuristics are any good, a solution will almost certainly be found and it will be found much faster than if you use a blind search approach. Starting from the beginning, we take the initial node, 0, find all its children, and put them on the UNUSED list in ascending order according to the number of squares out of place. Node 0 goes on the USED list. The lists are now:

USED LIST:

UNUSED LIST:

where the disorder ratings are written to the right of each matrix. The tree showing the search is shown in Figure 8.3. Node 2, with a disorder of 4 is selected as the node to work with. Doing so gives the third layer of Figure 8.3. The new nodes are inserted into places in the UNUSED list according to increasing values of their disorder. Nodes with disorder equal to existing nodes will go before old nodes with the same amount of disorder. Thus,

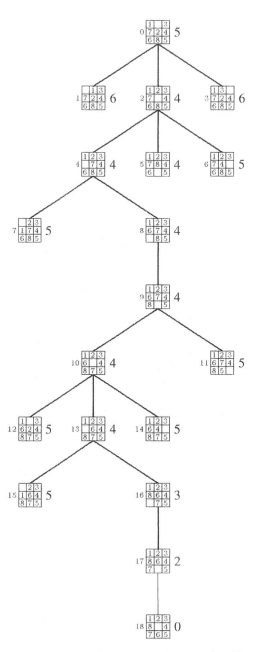

Figure 8.3: A heuristic search finds the goal after generating 18 nodes. The number to the right of each board indicates the "disorder" of the board configuration. When expanding a node, the algorithm always takes the board with the lowest disorder.

the UNUSED list is now:

UNUSED LIST: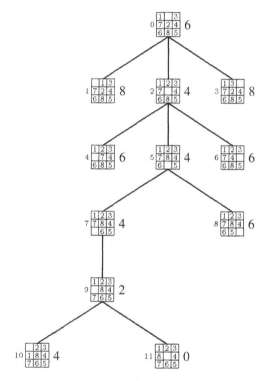

At this point, we select node 4 and generate its children. These are shown in layer 4 of Figure 8.3. Continuing in this manner, we generate a total of 18 nodes before finding the goal node.

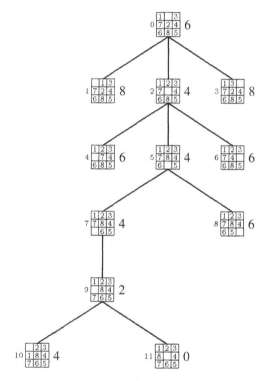

Figure 8.4: The goal is found after generating only 11 nodes using a more powerful heuristic.

The above searching strategy did turn out to be more productive because it does not follow up on matrices that make the board more disorderly. Another possible heuristic to use is to define the degree of disorder to be the sum of the distances that each square is from where it should be. The original matrix is shown below with an additional matrix that shows how many moves each square is from the location it should occupy:

Using this second heuristic, we produce the search shown in Figure 8.4, where the goal is found by generating only 11 nodes. This new heuristic is said to have more heuristic power. Nilsson [143] describes an even better heuristic function for the 8-puzzle.

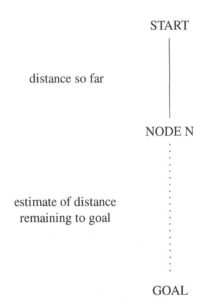

Figure 8.5: The total estimated distance (estimated number of moves) is the distance actually traveled so far plus an estimate of the remaining distance.

Some experiments and mathematical analysis suggest that a better method of choosing a node, N, to follow up on is to estimate the total distance from the start node to the goal node of a path that passes through node N. The estimate consists of two parts, first the actual distance from the start to the node, N, and second, an estimate of the distance from N to the goal node. The disorder can serve as a convenient estimate of this second distance. This strategy is shown in Figure 8.5. The reason for doing this is that heuristic functions are not always very precise in rating the configuration of a node. As we travel through the tree, the function may continue to give a promising rating that remains constant. If we continue following this branch of the tree we may not really be making much progress. "Not making much progress" serves as another valuable heuristic. It is as if we are driving to some town and no matter how many miles we cover, the mileage signs keep saying 10 miles to town. You have to start thinking that something is wrong.

8.1.3   Other Methods

In addition to doing the 8-puzzle with these standard heuristic searching techniques, Lehnert has recently devised a case-based scheme for solving the problem [97, 14, 178]. SOAR has also been used to do the 8-puzzle [91] and with the learning turned on, it improves its performance as it goes along.

8.2   A Geometry Theorem Prover

One of the early, successful, and impressive heuristic search programs was Gelernter's Geometry Theorem Prover [49, 50]. The program takes a problem such as the one shown in

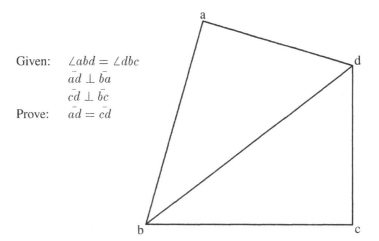

Given: $\angle abd = \angle dbc$
         $\bar{ad} \perp \bar{ba}$
         $\bar{cd} \perp \bar{bc}$
Prove: $\bar{ad} = \bar{cd}$

Figure 8.6: This figure shows one of the simpler problems that the Geometry Theorem Prover program can do.

Figure 8.6 and produces the required proof. This is an easy proof to do. The line segment $\bar{ad}$ is part of $\triangle adb$ and the segment $\bar{cd}$ is part of another $\triangle cdb$. It looks as if these two triangles are congruent. If the two triangles are congruent, then $\bar{ad} = \bar{cd}$ because corresponding sides of congruent triangles are equal. There are several ways to prove triangles congruent. One is to prove that all the corresponding sides of each triangle are equal (the side, side, side method). This does not seem likely. Instead, we have information about two of the three angles in each triangle and we notice that they share a common side. We recall that triangles can be proved congruent if one side and two angles can be proved equal (the side, angle, angle method). Because segment $\bar{bd}$ is equal to itself and it is in both triangles, we can count the "two" corresponding sides as equal. Second, we have $\angle abd = \angle dbc$ because it is given. If we can show that $\angle bad = \angle bcd$, we will have the requirements for proving the triangles congruent. Showing these two angles are equal is easy because we are given that the segments forming these angles are perpendicular, perpendicular segments form right angles, and all right angles are equal. When given this problem, Gelernter's program produced the following proof:

    Angle ABD equals angle DBC
        Premise
    Right angle DAB
        Definition of perpendicular
    Right angle DCB
        Definition of perpendicular
    Angle BAD equals angle BCD
        All right angles are equal
    Segment DB
        Assumption based on diagram
    Segment BD equals segment BD
        Identity

Triangle BCD
    Assumption based on diagram
Triangle BAD
    Assumption based on diagram
Triangle ADB congruent triangle CDB
    Side-angle-angle
Segment AD equals segment CD
    Corresponding elements of congruent triangles are equal

It turns out that the Geometry Theorem Prover uses some simple symbol manipulation techniques and a heuristic tree search. Gelernter's program was done in Fortran, but it is much easier to do the problem in a list processing language like Prolog and we now show how to do so. First, the given facts will be these:

```
tri(a,d,b).            /* adb is a triangle */
tri(c,d,b).            /* cdb is a triangle */
perp(b,a,a,d).         /* segment ba is perpendicular to ad */
perp(b,c,c,d).         /* segment bc is perpendicular to cd */
eqang(a,b,d,c,b,d).    /* angle abd equals angle cbd */
```

The problem to prove is whether or not segment ad equals segment cd:

```
eqseg(a,d,c,d).
```

To do this problem we need some additional facts from plane geometry. First, segment X1Y1 is equal to segment X2Y2 if X1 = X2 and Y1 = Y2:

```
/* A segment is equal to itself.  */

eqseg(X1,Y1,X2,Y2)  :- X1 = X2,
                       Y1 = Y2.
```

Second, two segments, X1Y1 and X2Y2, can be proved equal if each segment is part of a triangle and the two triangles can be proved congruent:

```
/* Corresponding sides of congruent triangles are equal */

eqseg(X1,Y1,X2,Y2)  :- tri(X1,Y1,Z1),
                       tri(X2,Y2,Z2),
                       congruent(X1,Y1,Z1,X2,Y2,Z2).
```

Third, if angle X1Y1Z1 is a right angle and angle X2Y2Z2 is a right angle, then angle X1Y1Z1 equals angle X2Y2Z2:

```
/* All right angles are equal.  */

eqang(X1,Y1,Z1,X2,Y2,Z2)  :- rtang(X1,Y1,Z1),
                             rtang(X2,Y2,Z2).
```

Fourth, the angle X1Y1Z1 is a right angle if segment X1Y1 is perpendicular to segment Y1Z1:

```
/* Perpendicular segments form a right angle.  */

rtang(X1,Y1,Z1) :- perp(X1,Y1,Y1,Z1).
```

Fifth, two triangles, X1Y1Z1 and X2Y2Z2 are congruent if:

> angle X1Y1Z1 equals angle X2Y2Z2,
> angle Z1X1Y1 equals angle Z2X2Y2, and
> segment Y1Z1 equals segment Y2Z2.

In Prolog this is:

```
/* Triangles are congruent by side-angle-angle */

congruent(X1,Y1,Z1,X2,Y2,Z2) :- eqang(X1,Z1,Y1,X2,Z2,Y2),
                                 eqang(Z1,X1,Y1,Z2,X2,Y2),
                                 eqseg(Y1,Z1,Y2,Z2).
```

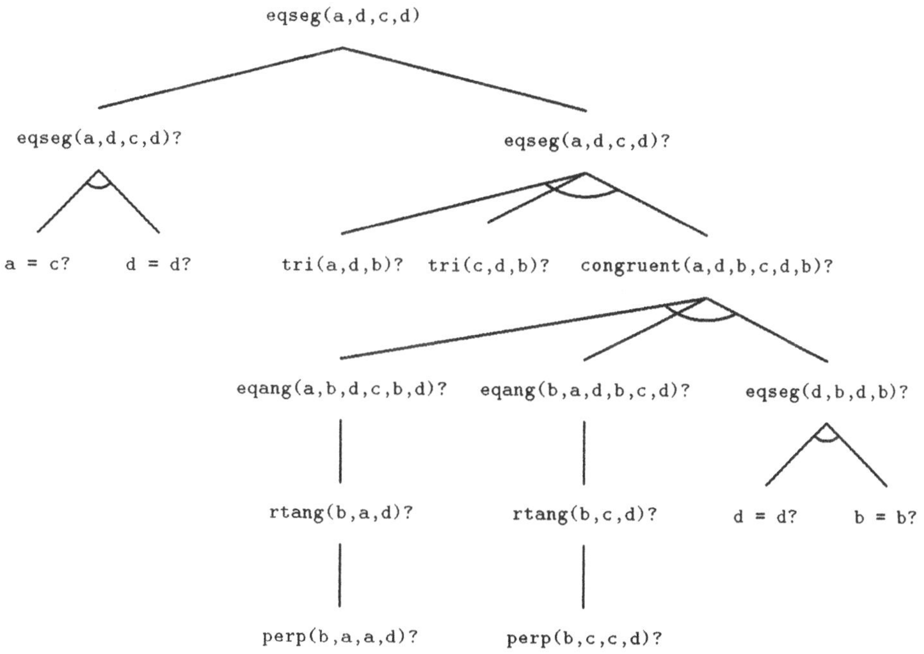

Figure 8.7: A tree showing how the proof will progress using Prolog.

Figure 8.7 shows how the proof progresses. To solve the problem we look through our facts and set of formulas to find a method to prove $ad = cd$. The first one we come across is formula number 1:

```
eqseg(X1,Y1,X2,Y2) :- X1 = X2,
                      Y1 = Y2.
```

We use this formula by setting X1 to a, Y1 to d, X2 to c, and Y2 to d. We now check if X1 (= a) equals X2 (= c). This fails, so we look for another method to prove $ad = cd$. We find the formula that says that corresponding parts of congruent triangles are equal:

```
eqseg(X1,Y1,X2,Y2)  :- tri(X1,Y1,Z1),
                       tri(X2,Y2,Z2),
                       congruent(X1,Y1,Z1,X2,Y2,Z2).
```

We substitute a, d, c, and d into this formula as in the others, giving:

```
eqseg(a,d,c,d)  :- tri(a,d,Z1),
                   tri(c,d,Z2),
                   congruent(a,d,Z1,c,d,Z2).
```

The formulas require us to try to find a triangle that contains the segment ad. Looking in the database we find there is a triangle in the problem, $\triangle adb$, so we can now substitute b for every occurrence of Z1 in the formula giving:

```
eqseg(a,d,c,d)  :- tri(a,d,b),
                   tri(c,d,Z2),
                   congruent(a,d,b,c,d,Z2).
```

The formula now requires looking for a triangle containing the segment cd. We find one, $\triangle cdb$. We substitute b for Z2 giving:

```
eqseg(a,d,c,d)  :- tri(a,d,b),
                   tri(c,d,b),
                   congruent(a,d,b,c,d,b).
```

To finish showing that $\overline{ad} = \overline{cd}$ we have to call the congruent procedure as shown above. Making the substitutions this gives:

```
congruent(a,d,b,c,d,b)  :- eqang(a,b,d,c,b,d),
                           eqang(b,a,d,b,c,d),
                           eqseg(d,b,d,b).
```

It is easy to show that $\angle abd = \angle cbd$ because it is in the list of known facts. Showing $\angle bad = \angle bcd$ is a little harder, but if we search for ways to prove angles equal we find:

```
eqang(X1,Y1,Z1,X2,Y2,Z2)  :- rtang(X1,Y1,Z1),
                             rtang(X2,Y2,Z2).
```

Making the substitutions, the problem becomes:

```
eqang(b,a,d,b,c,d)  :- rtang(b,a,d),
                       rtang(b,c,d).
```

This means we now need to show that both $\angle bad$ and $\angle bcd$ are right angles. The only formula for this is:

```
rtang(X1,Y1,Z1)  :- perp(X1,Y1,Y1,Z1).
```

Substituting for $\angle bad$ we get:

```
rtang(b,a,d) :- perp(b,a,a,d).
```

The right-hand side is true so ∠*bad* is a right angle. In the same way we find that ∠*bcd* is also a right angle so the two angles equal. We are left with the problem of showing that $\bar{db} = \bar{db}$. This is easily shown by substituting into the first of the geometry formulas. Having proved the triangles congruent, we have proved that the segments $\bar{ad}$ and $\bar{cd}$ are equal because they are corresponding parts of congruent triangles.

In the above example we have used only the minimum number of geometry formulas necessary and we have carefully chosen the symbols in the facts and rules so that the proof comes immediately. In general, there are many practical programming problems that occur when the theorem prover is actually programmed.

One problem that complicates the program is illustrated by asking the program to attempt to prove that $\bar{ad} = \bar{dc}$ instead of $\bar{ad} = \bar{cd}$. The system quickly fails to find a triangle that contains $\bar{dc}$, as opposed to $\bar{cd}$. We can remedy this problem by adding the fact:

```
tri(d,c,b).
```

Unfortunately, this minor fix is not enough. Following the formulas rigorously, we need to show:

```
eqang(a,b,d,d,b,c).
```

which again contains just another way of referencing ∠*cbd*. The program will be stopped again. The simple matching and symbol substitution plan that worked so easily is stupid compared to a human being and to make up for this, the Geometry Theorem Prover programmer must add additional procedures to compensate for the lack of common sense in the program.

A second problem is that it is quite easy for the theorem prover to end up going in circles. To illustrate this, we gather together the facts and formulas below and we also add a formula to prove triangles congruent by means of side-side-side:

```
tri(a,d,b).
tri(c,d,b).
perp(b,a,a,d).
perp(b,c,c,d).
eqang(a,b,d,c,b,d).

eqseg(X1,Y1,X2,Y2) :- X1 = X2,
                      Y1 = Y2.

eqseg(X1,Y1,X2,Y2) :- tri(X1,Y1,Z1),
                      tri(X2,Y2,Z2),
                      congruent(X1,Y1,Z1,X2,Y2,Z2).
                      eqang(X1,Y1,Z1,X1,Y1,Z1).

eqang(X1,Y1,Z1,X2,Y2,Z2) :- rtang(X1,Y1,Z1),
                            rtang(X2,Y2,Z2).
```

```
rtang(X1,Y1,Z1) :- perp(X1,Y1,Y1,Z1).

congruent(X1,Y1,Z1,X2,Y2,Z2) :- eqseg(X1,Y1,X2,Y2),
                                 eqseg(Y1,Z1,Y2,Z2),
                                 eqseg(Z1,X1,Z2,X2).

congruent(X1,Y1,Z1,X2,Y2,Z2) :- eqang(X1,Z1,Y1,X2,Z2,Y2),
                                 eqang(Z1,X1,Y1,Z2,X2,Y2),
                                 eqseg(Y1,Z1,Y2,Z2).
```

If we now give the theorem prover the task of proving $\bar{ad} = \bar{cd}$, it will start as before, along the path to prove the triangles congruent. If it finds the side-side-side formula first, it will attempt to prove the two triangles congruent by proving the sides equal. To prove that $\bar{ad} = \bar{cd}$ it will find the formula to prove segments equal by finding triangles that contain the two segments and it will try to prove three sides equal and so on. To circumvent this problem the programmer needs to keep track of all the goals that are active and not repeat them.

Given: $\angle abd = \angle dbc$
         $\bar{ad} \perp \bar{ba}$
         $\bar{cd} \perp \bar{bc}$
Prove: $\bar{ap} = \bar{pc}$

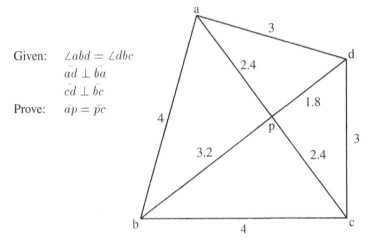

Figure 8.8: A harder problem that illustrates how heuristics can shorten the search time. The numbers near each line segment indicate the length of the segment and a program can use the lengths to eliminate proofs of things that are not possible.

A third serious problem in the Geometry Theorem Prover program is that for every node that occurs in the search of the tree, you can generate a very large number of successors. It turns out that many of these successors represent ridiculous items to prove and human beings would never think to try to do them, but again, the computer program is lacking in common sense. To give an example of a possible ridiculous item that may come up we will use the problem shown in Figure 8.8. It is almost the same as the previous problem except we have added another line from a to c that intersects the line $\bar{bd}$ at point p. The problem will be to prove that $\bar{ap} = \bar{pc}$. The complicating factor is that there are now more triangles in the problem. Given the problem $\bar{ap} = \bar{pc}$?, the program might choose $\triangle apb$ as a triangle containing $\bar{ap}$ and $\triangle cpd$ as a triangle containing $\bar{cp}$. It would then set off on the useless task of trying to prove the two triangles congruent. People would never try to do such a thing, because the two triangles do not look even close to equal. To try to recapture some of the sense that people have Gelernter did the following. When the program is given a problem to be solved it goes and generates its own internal diagram producing lines and angles with actual numerical values, but taking care not to make sides or angles equal unless it is somehow required by the facts. He then had the program perform numerical checks on objectives, such as trying to prove triangles congruent, before actually trying to prove the objective. This feature is easy to add to the formulas that we have presented. We first add facts about the sizes of the triangles involved. Among the facts are these:

```
tri(a,p,b, 2.4, 3.2, 4.0).

tri(c,p,d, 2.4, 1.8, 3.0).
```

In these statements we mean to say that in $\triangle apb$, the sides have lengths 2.4, 3.2, and 4.0 and that in $\triangle cpd$ the lengths are 2.4, 1.8, and 3.0. We can then modify the congruency check procedures such as the side-angle-angle one to be this:

```
congruent(X1,Y1,Z1,X2,Y2,Z2)  :- tri(X1,Y1,Z1,S1,S2,S3),
                                  tri(X2,Y2,Z2,S4,S5,S6),
                                  close(S1,S2,S3,S4,S5,S6),
                                  eqang(X1,Z1,Y1,X2,Z2,Y2),
                                  eqang(Z1,X1,Y1,Z2,X2,Y2),
                                  eqseg(Y1,Z1,Y2,Z2).
```

The idea here is to look up the lengths of the first triangle and let S1, S2, and S3 contain these values. Then we look up the lengths of the sides for the second triangle, S4, S5, and S6. We then pass these sizes to a procedure, close, to determine if the sizes match. If they do, it makes sense to try to prove the two triangles congruent and if not, the program will not attempt to do so. Gelernter reports that heuristic checks like this reduced the number of first-level subgoals generated by his program from on the order of a thousand without these checks to about five with these checks in use.

Another heuristic used in an extended version of the Geometry Theorem Prover was to make an effort to look at the facts in the problem and the list of possible subgoals and to rate the subgoals as to whether or not the goal seemed likely to be provable based on a ratings scheme. Gelernter does not mention a specific example in his paper, however, such an instance is easily seen in the problem shown in Figure 8.6. If the program has the problem of trying to prove two triangles congruent, it could notice that it has data about two angles in each triangle and so, proving the triangles congruent by side-angle-angle seems more likely than proving them congruent by the method of side-side-side. Gelernter reports that with these extended heuristics, "The extended system is able to prove a number of somewhat more difficult theorems that are beyond the capacity of the basic machine." and "The average depth of the problem-solving graph for the refined system, about seven to nine levels, is two-thirds the average depth for the basic system."

8.3   Symbolic Integration and Heuristic Search

In this section we will look at various ways of doing symbolic integration, mainly from the well-known program, SAINT, but also from LEX, a program that learns symbolic integration. We will also show how a small back-propagation network can do some simple steps involved in symbolic integration.

8.3.1   SAINT

SAINT stands for Symbolic Automatic Integrator, a program by Slagle [210] that was also one of the early successful heuristic search programs. SAINT comes with a knowledge of the integrals of the usual elementary functions, such as sine, cosine, exponential, and powers. When given an integral to do, SAINT first looks at the table of elementary forms and tries to match the integral against the patterns. For instance, given the problem, $\int 2^x \, dx$, the program finds that this matches the stored pattern with the solution:

$$\int c^v \, dv = \frac{c^v}{\ln c}$$

SAINT contains 26 such elementary forms. If SAINT finds a match it immediately does the substitution and prints out the solution. If the integral does not match with any of the elementary forms, the program goes on to try some transformations called algorithmic transformations, transformations that are always or almost always appropriate. Some examples are as follows:

a. Factor constant, that is,

$$\int cg(v) \, dv = c \int g(v) \, dv$$

b. Decompose, that is,

$$\int \sum_i g_i(v) \, dv = \sum_i \int g_i(v) \, dv$$

c. Linear substitution, that is, if the integral is of the form

$$\int f(c_1 + c_2 v) \, dv$$

substitute $u = c_1 + c_2 v$, and obtain an integral of the form

$$\int \frac{1}{c_2} f(u) \, du$$

for example, in

$$\int \frac{\cos 3x}{(1 - \sin 3x)^2} \, dx$$

substitute $y = 3x$.

If none of the above kinds of transformations succeed, SAINT starts to make informed guesses about how to proceed. These heuristics appear in two different parts of the program. First, SAINT has certain rules about what kind of substitutions should be tried on an integral. Slagle gives this example of one of the rules:

Below is given only the most successful heuristic, "substitution for a subexpression whose derivative divides the integrand."

Let $g(v)$ be the integrand. For each nonconstant nonlinear subexpression $s(v)$ such that neither its main connection is MINUS nor is it a product with a constant factor, and such that the number of nonconstant factors of the fraction $g(v)/s'(v)$ (after cancellation) is less than the number of factors in $g(v)$, try substituting $u = s(v)$. Thus, in $xe^{x^2}dx$, substitute $u = x^2$.

Rules like the above generate children as in the 8-puzzle, *but only likely children*, not every possible one.

The second major use of heuristics in SAINT is to estimate which node in the tree should be followed up on. This part of the problem is like the problem of deciding which node to follow up on in the 8-puzzle. The heuristic SAINT uses is: *if it is getting simpler, it is getting better*. Simplicity is defined by finding the depth of the expression when it is written in Lisp:

EXPRESSION	LISP	DEPTH
x	x	0
x^2	(* x x)	1
e^{x^2}	(exp (* x x))	2
xe^{x^2}	(* x (exp (* x x)))	3

Note that lists in Lisp use parentheses rather than square brackets as in Prolog and a blank to separate items rather than a comma as in Prolog.

Slagle gives the following example of the program's operation. Note that the timing he mentions is for a very early computer, the IBM 7090, that had very little memory compared to modern computers. Where Slagle uses the term, "cheap," it means the problem will probably use the least amount of computer resources, or in other words, the cheapest problem is the one that will probably be solved fastest:

As a concrete example, we sketch how SAINT solved

$$\int \frac{x^4}{(1-x^2)^{5/2}} \, dy.$$

in 11 minutes. SAINT's only guess at a first step is to try the substitution: $y = arcsin\ x$, which transforms the original problem into:

$$\int \frac{sin^4 y}{cos^4 y} \, dy$$

For the second step SAINT makes three alternative guesses:

A. By trigonometric identities

$$\int \frac{sin^4 y}{cos^4 y} \, dy = \int tan^4 y \, dy$$

B. By trigonometric identities

$$\int \frac{sin^4 y}{cos^4 y} \, dy = \int cot^{-4} y \, dy$$

C. By substituting $z = tan(y/z)$

$$\int \frac{sin^4 y}{cos^4 y} \, dy = \int 32 \frac{z^4}{(1+z^2)(1-z^2)^4} \, dz$$

SAINT immediately brings the 32 outside of the integral.

After estimating that (A) is the cheapest of these three problems, SAINT guesses the substitution $z = tan\ y$, which yields

$$\int tan^4 y \, dy = \int \frac{z^4}{1+z^2} \, dz.$$

SAINT immediately transforms this into

$$\int \left(-1 + z^2 + \frac{1}{1+z^2}\right) dz = -z + \frac{z^3}{3} + \int \frac{dz}{1+z^2}.$$

Estimating incorrectly that (B) is cheaper than

$$\int \frac{dz}{1+z^2},$$

SAINT temporarily abandons the latter and goes off on the following tangent. By substituting $z = cot\ y$, we obtain:

$$\int cot^{-4} y \, dy = \int -\frac{dz}{z^4(1+z^2)}.$$

Now SAINT estimates that

$$\int \frac{dz}{1+z^2}$$

is cheap and guesses the substitution, $w = arctan\ z$ which yields $\int dw$. Immediately, SAINT integrates this, substitutes back, and solves the original problem:

$$\int \frac{x^4}{(1-x^2)^{5/2}} \, dy = arcsin\ x + 1/3\ tan^3 arcsin\ x - tan\ arcsin\ x.$$

As a test of SAINT, Slagle had an assistant select 54 diverse and difficult problems from MIT freshman calculus final examinations. The program solved 52 out of the 54 and failed on the other two.

After SAINT, a more sophisticated program, called SIN [128], outperformed SAINT. SIN does very little searching. A deterministic algorithm developed by Risch [174] can integrate many types of expressions.

8.3.2   A Symbolic Program to Learn Integration

It is also possible to have a program learn how to do integrals quickly and efficiently. The program that has done this is called LEX, produced by Mitchell, Utgoff, Nudel, and Banerji [126]. LEX starts with just a minimal amount of knowledge about integration, but as it does problems it tries to generate rules about how to do problems so it can use the rules in the future. For example, given the problem:

$$\int x \, cos(x) \, dx,$$

LEX will try to solve it every way it can. Eventually, it finds that integration by parts will work. Having found this it can propose this rule for problems with this form:

   IF the problem state contains $\int x \, transc(x) \, dx$,

   THEN use Integration by Parts: $\int u \, dv \Rightarrow uv - \int v \, du$,
         With u bound to x, and dv bound to $(transc(x) \, dx)$,

where $transc(x)$ means a transcendental function of x. After guessing this possible rule, LEX goes on to test the usefulness of the rule by generating cases that fit this pattern, such as these:

$$\int x \, sin(x) \, dx$$

$$\int x \, exp(x) \, dx$$

$$\int x \, tan(x) \, dx$$

One defect that was immediately found in LEX was that the language used to describe heuristics was inadequate. For instance, while integration by parts is useful on the problem $\int x \, tan(x) \, dx$, it would be better to include in this integration by parts rule, the idea of a "twice integrable function." The researchers were hoping to find a way to automate the detection and repair of inadequacies in the generalization language.

8.3.3   A Partial Back-Propagation Solution

It turns out that is it fairly easy to get a back-propagation network to look at simple expressions that need to be integrated and then guess which transformations are most promising, but it appears that it will be fairly hard to improve the network to the point where it could do problems as complex as SAINT or other symbolic programs.

   We will call the back-propagation network AMI, for Associative Memory Integrator and train the network on the cases shown in Figure 8.9. For elementary forms, the function on the right of the figure is the integral of the function on the left. When the problem is the integral of some constant, k, times a function, the transformation to apply is to move the constant outside the integral sign. This is coded on the right of the figure as "mvconst." When the constant, k, occurs within a function, as in $sin(kx)$, the transformation to make is the substitution, $u = kx$. This is coded on the right of the figure as "subst." When the

Problem	Solution
$\int sin(x)$	$-cos(x)$
$\int k\, sin(x)$	mvconst
$\int sin(kx)$	subst
$\int k\, sin(kx)$	mvconst
$\int cos(x)$	$sin(x)$
$\int ln(x)$	$x\, ln(x) - x$
$\int k\, ln(x)$	mvconst
$\int ln(kx)$	subst
$\int cosh(kx)$	subst
$\int cosh(x) + sin(x)$	split
$\int tan(x) + cos(x)$	split
$\int x + tan(x)$	split
$\int ln(x) + x$	split
$\int x\, sin(x)$	parts

Figure 8.9: The patterns to be learned by AMI.

problem is to integrate two functions added together, the solution is to split the integral into two separate integrals, coded on the right of the figure as "split." There is one example of integration by parts.

To produce AMI we first need to code functions as bit patterns. Figure 8.10 shows how we will use 15 bits to encode a function. The network is outlined in Figure 8.11 and it is a 36-20 network. The output units contain a sign bit for the function that is 1 when the function is preceded by a minus sign and 0 otherwise, 15 units to give the function, and four units to give the transformation to apply. The input units are broken into three parts, the first group of 17 units gives the first part of the formula that needs to be integrated, the last 17 units give the second part of the formula to be integrated (if it exists), and the middle two bits indicate whether or not the two parts are added together (coded as 10) or multiplied together (coded as 01). Within each group of the 17 units, 15 units are dedicated to the function name as given in Figure 8.10 plus one unit is used to indicate a leading constant (1 if present, 0 if missing) and one unit describes the argument of the function, either 1 if the argument is of the form, kx, or 0 if the argument is x. Some examples of the coding scheme are as follows:

```
* integral of sin(x) is - cos(x):
0 000011100000000 0 00 0 000000000000000 0 1 0000110100000000000

* to integrate ln(kx), substitute u = kx:
0 000010000010000 1 00 0 000000000000000 0 0 0000000000000000010

* to integrate x * sin(x) use integration by parts:
0 110000000000000 0 01 0 000011100000000 0 0 0000000000000000001
```

After the network was trained with the training data, some new problems that require using integration by parts or splitting the integral into the sum of two integrals were submitted to AMI. The problems and the numerical values that resulted for the units representing

$pow1$	110000000000000	Bit 1: the power of x function
$pow2$	101000000000000	Bit 2: the first power of x
$pow3$	100100000000000	Bit 3: the second power of x
sin	000011100000000	Bit 4: the third power of x
cos	000011010000000	Bit 5: a transcendental function
tan	000011001000000	Bit 6: a trigonometric function
exp	000010000100000	Bit 7: $sin(x)$
ln	000010000010000	Bit 8: $cos(x)$
$sinh$	000010000001000	Bit 9: $tan(x)$
$cosh$	000010000000100	Bit 10: the exponential function
$x\,ln(x) - x$	000010000000010	Bit 11: the natural log function
$ln(|cos(x)|)$	000010000000001	Bit 12: the hyperbolic sine
		Bit 13: the hyperbolic cosine
		Bit 14: $x\,ln(x) - x$, the integral of $ln(x)$
		Bit 15: $ln(|cos(x)|)$ the integral of $tan(x)$

Figure 8.10: The vector codes for the functions used in AMI and the meaning of each bit in the vector.

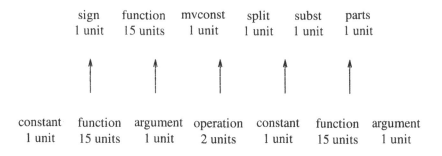

Figure 8.11: The arrangement of the two-layer integration network.

Problem	Split	Parts
$x\,cos(x)$	0.08	0.90
$x^2\,cosh(x)$	0.03	0.91
$x^2\,tan(x)$	0.11	0.71
$x\,cosh(x)$	0.02	0.91
$x\,tan(x)$	0.08	0.71
$x + cos(x)$	1.00	0.03
$x + tan(x)$	1.00	0.00
$x + cosh(x)$	0.96	0.04
$x + tan(x)$	1.00	0.00
$x\,tan(x)$	0.08	0.71

Figure 8.12: After training the network on the data in Figure 8.9, the program was tested on the above problems and got the right answer in every case.

"split" and "parts" are given in Figure 8.12. Notice that while AMI was trained to do the problem, $x\ sin(x)$ by parts, the knowledge gained by the pattern spills over to $x\ cos(x)$ and even to $x^2 cosh(x)$. (Explain why this happens!) From only one example of integration by parts, AMI favors that method in all the cases it should while choosing the split transformation in all those cases where it is appropriate. Also, if AMI is given the choice between using substitution or moving the constant outside the integral in a problem such as:

$$\int k\ sinh(kx),$$

AMI will choose to move the constant first because the training data had an example where moving the constant is preferred over making a substitution. Thus, AMI can learn many of the simplest transformations involved in doing symbolic integration.

It seems quite possible that networks could also learn to guess promising substitutions of the kind that SAINT made (an exercise or research project for the reader), however, the kind of coding scheme we proposed here gets awkward when it comes to encoding functions within functions. It seems likely that there must be a better scheme for representing problems than just using an extremely long vector. Perhaps connectionist state networks will overcome this problem. It should also be clear that at this stage of development it is impossible for networks alone to do this type of problem. Some traditional symbol processing techniques are necessary.

8.4   Other Heuristic Programs

The heuristic search principle was the first important technique studied by AI researchers. All sorts of problems were tackled using this approach. Some of these problems involve playing games and some of these systems are covered in the next chapter. Many early heuristic search programs were involved with proving simple mathematical theorems. Some involved theorem proving in Boolean algebra and some of the proofs found were more elegant than human proofs of the same theorems. One system checked the proofs of theorems in the *Principia Mathematica* and found numerous defects in the proofs. One program solved geometric analogy problems [36]. A few programs, such as MULTIPLE by Slagle [211] and Slagle et al. [212, 213] could do more than one type of problem. GPS, for General Problem Solver [135, 136, 137], is another well-known program that could be adapted to do a number of different types of problems, including simple integrals, the monkey and bananas problem, the tower of Hanoi, the missionaries and cannibals problem, simple natural language parsing problems, and others. For a good review of these early systems, see the book by Slagle [214]. Newell and others have continued to research these methods and the result has been the SOAR program. For more on this research see Laird et al. [91, 141].[1] SOAR has also been used recently to do air combat simulations [228].[2] Another well-known program is the AM program (AM stands for Automated Mathematician) [98] that looks for interesting things to prove in number theory.

---

[1] Also see the Soar home page: http://www.cs.cmu.edu:8001/afs/cs/project/soar/public/www/home-page.html.
[2] For a little more on this see: http://ai.eecs.umich.edu/soar-group.html.

8.5 Exercises

8.1. Exhaustive searches, like the breadth-first search used in the 8-puzzle will always find a solution to a problem, however, sometimes the search takes so long that the algorithm effectively fails. In general, heuristic search algorithms will find an answer much faster but there is no guarantee that a solution will ever be found. Is there any chance that the 8-puzzle heuristic searches given in the text will never find a solution even if they are given an infinite amount of time?

8.2. As noted in the text, the method of estimating the total number of moves you will need to make to solve the 8-puzzle consists of the number of moves you have made so far plus an estimate of how many more moves you need to make. Use this method for the same starting pattern as in the text and for both heuristics given in the text and see if in these cases it cuts down on the search time.

8.3. Solving the 8-puzzle using heuristic searches as was done in Section 8.1 is probably very close to the way people approach it, however, people do not keep careful track of all the configurations that were generated. Try to devise a heuristic search that will solve the 8-puzzle all the time or almost all the time without keeping track of all the configurations that were generated. This should result in a savings of CPU time because fewer board configurations need to be searched. Also, the data structures you use to store the board configurations should be carefully chosen to speed up the programs. Compare the CPU timings for your method with the standard method of saving and searching all the board configurations. (A time consuming project.)

8.4. Consider whether or not you could use some kind of neural network to choose good moves for the 8-puzzle or as an evaluation function to rate the possible moves.

8.5. Program the Geometry Theorem Prover in your choice of languages, either a conventional language such as Fortran, Pascal, or C, or an AI language like Prolog or Lisp. Program it with enough knowledge to do the problem in Figure 8.8 plus one or two others that you make up. You will actually find the problem easier to program if you incorporate heuristic checks on the sizes of triangles, lines, and angles to avoid trying to prove impossible results. Unlike Gelernter, you can input the numerical values yourself. If you are ambitious, have the program neatly print out the steps in the proof the way a person would list the statements and reasons.

8.6. Evaluate Geometry Theorem Proving as a possible candidate for neural networking, memory-based reasoning, and case-based reasoning techniques.

8.7. Explain why AMI can generalize from the patterns it sees.

8.8. Could a memory-based/case-based reasoning scheme be used to do calculus integration problems?

8.9. Try using the data for AMI and a nearest neighbor scheme to guess solutions for integration problems.

8.10. Investigate whether or not a recurrent network can learn to do integration problems like those done by AMI, that is, input the symbols one at a time and identify a likely transformation (not do the transformation!). This may require a lot of computer time. It should be fairly easy to train a network to do at least a few examples, however.

8.11. Using Prolog or Lisp, write a program that will at least do the kinds of integrals and substitutions that AMI did. If you do the problem in Prolog, you may want to represent expressions as lists rather than as structures. Thus, int(3*exp(x)) would become: [int,[*,3,[exp,x]]] when int means $\int$.

Chapter 9

Game Playing

Game playing has traditionally been one of the most active areas of AI research. Researchers usually say that by studying game playing they can get insights into how intelligence works. They also often say that playing a game is a small enough problem that it can be done with ordinary computers and without requiring the large amounts of knowledge that would be needed for something like understanding natural language. Game playing is probably a good way to get insights into intelligence, but another very good reason researchers do it is that it is fun and challenging. There are regular contests for all sorts of game playing programs, especially for chess. The major games we will look at will be checkers and backgammon because so much good research has been done with these games.

9.1   General Game Playing Techniques

In this section we will look at some of the major techniques for game playing. All the techniques involve searching of one sort or another. The most common technique is called minimax and it involves searching the tree of possible moves that can be made.

9.1.1   Minimax

It is best to start the discussion of game playing by considering a very simple game, tic-tac-toe. At the top of Figure 9.1, we show a game of tic-tac-toe in midgame. In the layer below this configuration the boards labeled 1 through 5 show the possible moves X could choose to make. The third layer shows the moves O could make in response to X's move. This looking ahead shows that if X should choose the move to board 1, O has four moves. Of these four moves, the one that produces board 8 has O winning the game. If X moves to the board 1 configuration and O does not make a mistake, O will choose to win. Therefore, X should regard making the move that produces board 1 as very bad. The same goes for moving to boards 2, 3, and 5. The move to board 4 does not produce either a win or loss for X, so based on looking ahead two levels of play, board 4 must have the best move for X. It is very convenient to assign numeric values to each board configuration that represent the goodness of the board. When X has to make a move, we could let the goodness of a board that is a win for X be some large positive number, say a +10. A win for O could be a large negative number, say a −10 and we can assign all other board configurations a value of 0.

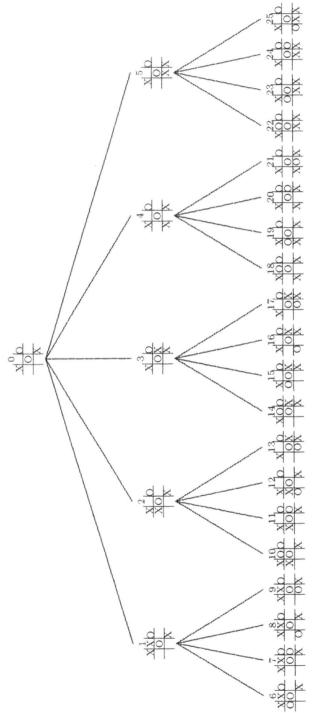

Figure 9.1: A game of tic-tac-toe and how looking ahead helps in choosing a move.

The player X will then be on the lookout for a move that rates a +10. The tree of these values is shown in Figure 9.2. Doing the rating of these configurations begins at the bottom layer, where each of the board configurations gets a value. Ratings for higher levels are calculated in the following way. At board configuration number 1, X will need to look at what kinds of moves O can make. X would like to see a lot of very large numbers among the children of board 1, unfortunately, the largest number is 0 and the smallest is –10, and if O does not make a mistake, he will make the move that gives X a –10, so the rating for board 1 should be the minimum of 0, 0, –10, and 0, which is –10. The same reasoning applies to board configurations 2, 3, and 5, and they each rate –10. By the same reasoning, board configuration 4 rates a 0. To figure out what move to make, X looks for the maximum value in the second layer and thus X should move to board 4.

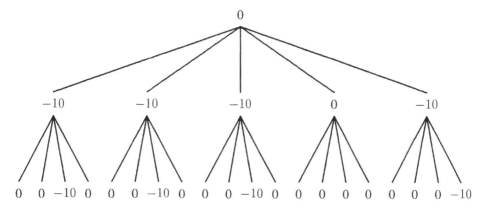

Figure 9.2: Rating the board configurations using minimax.

This general strategy for playing games is called *minimax* and, of course, this is derived from the fact that at alternate layers you are out to maximize or minimize the results from the previous layer. Using it, you search a tree of possible moves to some maximum depth. When you reach the maximum depth you apply some evaluation function to rate the board configurations you get. In this small example we looked ahead only two layers, but normally, it is highly desirable to look ahead as many layers as time allows. This is because the evaluation functions that rate the board configurations are really only doing crude estimates of the value of the board configuration. If we used the simple rating scheme described in the example above and only looked one layer deep, each board configuration would rate a 0, but 0 is in no way a good indication of these configurations. Looking ahead two layers shows four out of the five possible moves to be terrible. Looking ahead four layers shows that board position 4 should really be rated a +10 because it leads to a win for X.

In Figure 9.3 we have a tree for some game with only the values of the nodes shown. The process of moving values back up the tree to the root node is, of course, referred to as "backing up values." Notice that the best move for the maximizing player is to move to the right and then end up with an 8. The largest value on the bottom layer is a 20 and sometimes people feel that the logical thing to do is to move toward the largest value. However, if you try to move toward the board configuration with the 20, the minimizer, if he does not make a mistake, will move to give you a –5. Thus, minimaxing is a very conservative procedure.

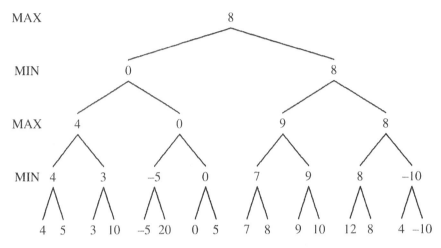

Figure 9.3: A larger tree for minimaxing.

Minimaxing serves as the basis for almost all game playing done with computers. There are several things that can be done to improve upon a program using the minimaxing technique. The simplest of all, in principle, is to find better evaluation functions. For instance, the one given for the tic-tac-toe game above is altogether too simple. It makes good sense to assign a large positive value for wins and a large negative value for losses, but using 0 otherwise is a very poor way to rate board configurations. Exercise 9.3 describes another evaluation function that is better. Some programs learn their evaluation function as they play.

9.1.2 More Sophisticated Searching Methods

Searching deeper usually improves the quality of play,[1] however, search trees can grow very big very quickly. It has been estimated that the search tree for checkers contains 10^{40} nodes and the search tree for chess, 10^{120} nodes. An obvious modification to basic minimaxing is to not search portions of the tree representing moves that seem to be very bad. The alpha-beta pruning technique is one method that cuts down somewhat on the amount of tree searching that needs to be done by ignoring portions of the tree that will turn out to be bad. It is illustrated in Figures 9.4 and 9.5. In Figure 9.4 suppose we have been minimaxing down the left portion of the game tree and have arrived back up at the node labeled B. The 5 at node B can be passed on to node A and we will call the possible move to node B the leading candidate found so far. The player will, of course, be hoping to find a larger value from further searching down the rest of the tree. Next we proceed to node C and then down to E. E is found to have a value of 1. If the player did choose the move to node C, the player's opponent will surely move to node E or to some other node X that could be *even worse*. The player then realizes that a move to C will certainly be worse than a move to B, so there is no point whatsoever in moving to C, and therefore the best

---

[1] There are some strange cases where searching deeper degrades the quality of play. For more on this see [134] or for a summary, see [80].

thing to do is to move on to node D and ignore any other children that C may have. Notice that if we do have a node, X, that is another child of C, it could be +100 and we still know that the move to C should be avoided.

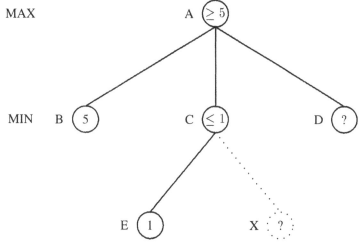

Figure 9.4: Illustrating alpha-beta cutoffs.

The same kind of reasoning applies to Figure 9.5. After searching below node C, we find its value to be a 2. B is at the minimizing level and the player will therefore be looking forward to getting a value less than 2. However, at B the search algorithm descends to D and then to E and it finds that D could choose a move that is rated a 9. Because of this there is no point in evaluating any more descendants of D.

A second possible improvement in searching a game tree is to investigate only the most promising moves and ignore the rest. One approach to doing this is to generate all the possible children of the start node to some given depth, a depth much less than the typical depth of search, then at this shallow depth you can apply an evaluation function. This function may be the same as the one you normally apply at the maximum search depth or it may be a different function that is designed to catch any interesting situation, whether good or bad. Game boards that are rated as interesting should be looked at in more detail and these boards are examined to a greater depth. In Samuel's checker playing program, the search proceeds down to a certain level where the program looks at the game situation, then if, for instance, the next move is a jump, the search continues deeper. In Figure 9.6, only the positions labeled A and B are interesting enough to justify searching deeper.

A new search algorithm for chess called *singular extension* has made it possible to search chess game trees much deeper and has made some recent chess programs much better [10].

9.1.3 Using Experience

Skilled human game players use another way to minimize tree searching: they draw upon all their previous experience with playing the game to pick out the important moves to investigate, it is a case-based/memory-based method. This ability that people have to pick

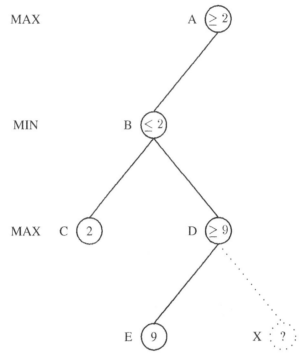

Figure 9.5: Another example of how alpha-beta pruning works.

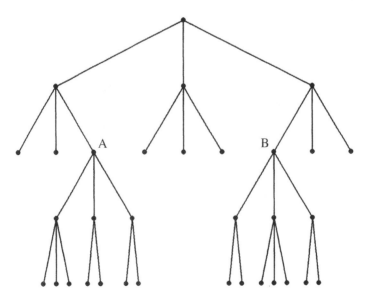

Figure 9.6: The tree is searched to level three and an evaluation function is applied. It finds that it is probably only worthwhile to consider nodes A and B in more detail.

the right patterns is something that has been looked at in more detail for chess playing. In the quotation below, from an article by Simon [209], he describes this pattern matching ability that people have and how they apply it to playing chess.

> Studies made by a number of laboratories, which were first made by the Dutch psychologist A. de Groot, show that a grand master almost never will look at more than 100 possibilities before he selects a move. He does not look at 10 to the 120th; he does not even look at a million. He looks, at most, at 100 possibilities before he selects the move.
>
> It turns out that mediocre players, when they are playing seriously, also look at about a maximum of about 100 possibilities before they make a move. The difference is that the grand master looks at the important possibilities and the tyros look at the irrelevant possibilities, and that's really the only way in which you can distinguish the thinking that they are doing when they are selecting a move. The processes are exactly the same.
>
> How does the grand master recognize those possibilities? Even if he is going to spend ten minutes worrying about a move, he will usually have a hunch in the first five seconds as to the move he's probably going to make. He spends the rest of that ten minutes testing whether the gold is really there, or whether it's fool's gold.
>
> How does he do that? There is a very interesting and simple experiment which you can do at home, if you have a chess board and at least one chess master available. You put before your subject a chess board with the positions arrayed as though they were in an actual game, say twenty moves into the game. Allow your subject to see the board for five to ten seconds. The exact time does not matter. Then take the board away and ask him to reconstruct it. If the experiment is typical, there will be about twenty-five pieces on such a board and the results would be very striking. If the subject that you put to this task were a master, or grand master, the board would be reconstructed almost without error, with about ninety percent accuracy, with only an occasional mistake. If anyone else is your subject he will be lucky to get six pieces back on the board.
>
> What conclusion do we draw from that? One conclusion that has been drawn in earlier days was the idea that somehow you need to have unusual abilities in visual imagery in order to be a chess master.
>
> We can test that hypothesis. Try the same experiment with just one little variation. Instead of arranging the pieces as though they were from a chess game, arrange them at random on the board. Twenty-five pieces arranged at random, five seconds to look at the board, then take it away. Your ordinary player, as before will get six of those pieces right, on average, and your chess master will get six of them right, on average. So it is nothing to do with visual imagery at all.
>
> Well, what is it? The chess master has a long experience of wasted youth, of looking at chess boards, and in the course of looking at hundreds and thousands and tens of thousands of chess boards, that master has learned to recognize all sorts of familiar friends. The master does not see that board as twenty-

five pieces, he sees it as four or five or six clusters, each of which cluster is a familiar friend.

If I wrote on the board here, rapidly, *George Washington* and erased it about as fast as I wrote it then asked you to write down what I have written, you would all write down *George Washington*. You can do that, because I did not write down sixteen separate letters. I wrote down one familiar pattern which all of you have seen many times before.

We have estimated the vocabulary of such patterns that a chess master has, how many of these friendly little collections of three or four pieces that he sees time and again that he can recognize on a chess board, and the number is of the order of magnitude of 50,000. That's probably a minimum estimate of the number with which he is familiar.

If that seems a large number to you, let me point out that probably everybody in this room has an English language vocabulary of at least 50,000 words.

So one of the things that all of us who have any intelligence do is to carry around with us large numbers of familiar patterns, which we recognize instantly when we see them. Instantly in psychology means in a few hundred milliseconds, or a fraction of a second. Instantly on a computer means a nano-second or maybe a pico-second, a millionth or billionth of a second.

Among the large number of patterns which we can recognize instantly are not only English words, but also the faces of our friends.

When we recognize a familiar pattern, we also get access to a lot of information we have about that pattern stored in a long term memory, so this recognition capability acts as an index to the encyclopedia.

Some of us have the good fortune to have stored with the faces of those familiar friends, their names. Others are not so lucky. Politicians are supposed to be very assiduous in storing away the names with the faces, but when we recognize patterns we do get a lot of information which comes with them. If we do not remember the name, we remember vaguely where we knew the person, and in what connection, and so forth.

When the chess master sees a pattern on a chess board, he not only says, "Here's a friendly pattern that I have seen before," but he often gets a lot of information from memory at that point on what to do about such patterns.

This is, in fact, a powerful capability. If you have 50,000 familiar patterns in your mind, and a recipe associated with each one of them, there are quite a few action schemes that you can call up at usually appropriate moments for executing them.[2]

9.2 Checkers

The main work on checkers has been done by A. L. Samuel who started on his checkers playing experiments in 1947 when he was working for IBM. In this section we will

---

[2] Copyright by Herbert A. Simon, reprinted with permission.

look at the methods from his well-known 1959 article [186] and the lesser-known 1967 article [187]. None of these programs is as good as human checkers experts, but a new program called Chinook is and we will mention some of its details and accomplishments.

9.2.1 Rote Learning

Samuel described two different learning methods in this 1959/1963 article. The first one was called *rote learning* in which the program remembered a large number of board configurations and this did not involve learning by modifying the evaluation function. The second series of experiments was called *generalization learning* and it involved modifying the evaluation function. We will look at rote learning first. To do this, Samuel still had to choose some evaluation function to rate the goodness of board configurations. He chose a linear function with four terms:

$$value = c_1 x_1 + c_2 x_2 + c_3 x_3 + c_4 x_4,$$

where:

  x_1 represents the piece advantage where kings count as 1.5 plain men,

  x_2 represents a quantity called "denial of occupancy,"

  x_3 represents the mobility advantage, where mobility represents the number of moves a player can make without being jumped, and

  x_4 is described as "a hybrid term which combined control of the center and piece advancement." The c_is are coefficients that were chosen by Samuel after having the program play through a series of test games.

As was said in the last section, the depth to which Samuel's program searches depends on how interesting the board positions are. Samuel describes the depth of search rules as follows:

> Playing-time considerations make it necessary to limit the look-ahead distance to some fairly small value. This distance is defined as the *ply* (a ply of 2 consisting of one proposed move by the machine and the anticipated reply by the opponent). The ply is not fixed but depends upon the dynamics of the situation, and it varies from move to move and from branch to branch during the move analysis. A great many schemes of adjusting the look-ahead distance have been tried at various times, some of them quite complicated. The most effective one, although quite detailed is simple in concept and is as follows. The program always looks ahead a minimum distance, which for the opening game and without learning is usually set at three moves. At this minimum ply the program will evaluate the board position if none of the following conditions occurs: (1) the next move is a jump, (2) the last move was a jump, or (3) an exchange offer is possible. If any one of these conditions exists, the program continues looking ahead. At a ply of 4 the program will stop and evaluate the resulting board position if conditions (1) and (3) above are not met. At a ply of 5 or greater, the program stops the look-ahead whenever the next ply level does not offer a jump. At a ply of 11 or greater, the program will terminate the look-ahead, even if the next move is to be a jump, should one side

at this time be ahead by more than two kings (to prevent the needless exploration of obviously losing or winning sequences). The program stops at a ply of 20 regardless of all conditions (since the memory space for the look-ahead moves is then exhausted) and an adjustment in score is made to allow for the pending jump. Finally, an adjustment is made in the levels of the break points between the different conditions when time is saved through rote learning (see below) and when the total number of pieces on the board falls below an arbitrary number. All break points are determined by single data words which can be changed at any time by manual intervention.

This tying of the ply with board conditions achieves three desired results. In the first place, it permits board evaluations to be made under conditions of relative stability for so-called dead positions, as defined by Turing.... Secondly, it causes greater surveillance of those paths which offer better opportunities for gaining or losing an advantage. Finally, since branching is usually seriously restricted by a jump situation, the total number of board positions and moves to be considered is still held down to a reasonable number and is more equitably distributed between the various possible initial moves.

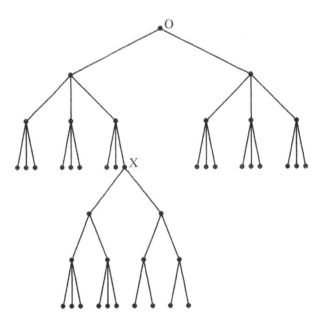

Figure 9.7: An example of rote learning and its benefits. Because the program stored the board, X, and its backed up value when X was encountered in play, using this value now has the effect of extending the search by three ply for a part of the tree.

The essence of rote learning is to save board configurations that were encountered in actual play together with their backed-up values. When the search procedure reaches its maximum depth, the program looks at the board to see if this board configuration has been encountered before. If it has been seen before, the stored value for this board configuration is used, if not, the evaluation function is applied. For instance, suppose that we have searched to a ply of three as shown in Figure 9.7. One of the board configurations on the third ply at the point marked X is a board configuration the program has already encountered before. At the time that X was encountered before, the program obtained a value for it by minimaxing to a depth of three. Now, instead of applying the evaluation function at X, this previously saved backed-up value for X is used. In effect, the program is drawing upon its previous experience with the board configuration X. In some ways this makes a three ply search begin to approach that of a six ply search. The original board that had to be evaluated is labeled as O and now its configuration and the newly obtained backed-up value for the almost six ply search is saved. In this way, the program can continue to gain better and better estimates of each configuration it finds when playing the game. Therefore, its overall play will improve.

Not surprisingly, keeping the large number of saved board configurations that are encountered during play available in a data structure that permits very fast access was a major part of the problem. Samuel reports that the program kept a carefully organized list of 53,000 board configurations. For each configuration, the backed-up value and an age term were kept. The age term is used to eliminate board positions that only rarely occur and age terms were periodically incremented. When existing configurations were accessed or changed, their ages were divided by two. Old entries are removed when they get too old. This then, implements a form of forgetting and "was adopted on the basis of reflections as to the frailty of human memories. It has proven to be very effective."

The results of this learning procedure were that:

> The program learned to play a very good opening game and to recognize most winning and losing end positions many moves in advance, although its midgame play was not greatly improved. This program now qualifies as a rather better-than-average novice, but definitely not as an expert.

> At the present time the memory tape contains something over 53,000 board positions (averaging 3.8 words each) which have been selected from a much larger number of positions by means of the culling technique described. While this is still far from the number which would tax the listing and searching procedures used in the program, rough estimates based on the frequency with which the saved boards are utilized during normal play (these figures are tabulated automatically), indicate that a library tape containing at least 20 times the present number of board positions would be needed to improve the midgame play significantly. At the present rate of acquisition of new positions this would require an inordinate amount of play and, consequently, of machine time.

9.2.2 Generalization Learning

The inability of the IBM 704 to collect and process large enough amounts of data to significantly improve play led to a second series of experiments Samuel called generalization learning. Samuel remarks:

> An obvious way to decrease the amount of storage needed to utilize past experience is to generalize on the basis of experience and to save only the generalizations. This should, of course, be a continuous process if it is to be truly effective, and it should involve several levels of abstraction. A start has been

made in this direction by having the program select a subset of possible terms for use in the evaluation polynomial and by having the program determine the sign and magnitude of the coefficients which multiply these parameters. At the present time this subset consists of 16 terms chosen from a list of 38 parameters. The piece-advantage term needed to define the task is computed separately and, of course, is not altered by the program.

In this series of experiments then, the program has to learn which 16 of 38 parameters are the most important to include in the evaluation polynomial and determine what the values of their coefficients should be. Samuel makes these comments on this approach to choosing parameters:

> It might be argued that this procedure of having the program select terms for the evaluation polynomial from a supplied list is much too simple and that the program should generate the terms for itself. Unfortunately, no satisfactory scheme for doing this has yet been devised. With a man-generated list one might at least ask that the terms be members of an orthogonal set, assuming that this has some meaning as applied to the evaluation of a checker position. Apparently, no one knows enough about checkers to define such a set. The only practical solution seems to be that of including a relatively large number of possible terms in the hope that all of the contributing parameters get covered somehow, even though in an involved and redundant way. This is not an undesirable state of affairs, however, since it simulates the situation which is likely to exist when an attempt is made to apply similar learning techniques to real-life situations.
>
> Many of the terms in the existing list are related in some vague way to the parameters used by checker experts. Some of the concepts which checker experts appear to use have eluded the writer's attempts at definition, and he has been unable to program them. Some of the terms are quite unrelated to the usual checker lore and have been discovered more or less by accident. The second moment about the diagonal axis through the double corners is an example. Twenty-seven different simple terms are now in use, the rest being combinational terms, ...
>
> ...In addition to the simple terms of the type just described, a number of combinational terms have been introduced. Without these terms the scoring polynomial would, of course, be linear. A number of different ways of introducing nonlinear terms have been devised but only one of these has been tested in any detail. This scheme provides terms which have some of the properties of binary logical connectives. Four such terms are formed for each pair of simple terms which are to be related. This is done by making an arbitrary division of the range in values for each of the simple terms and assigning the binary values of 0 and 1 to these ranges. Since most of the simple terms are symmetrical about 0, this is easily done on a sign basis. The new terms are then of the form A & B, A & ~B, ~A & B, and ~A & ~B, yielding values either of 0 or 1. These terms are introduced into the scoring polynomial with adjustable coefficients and signs, and are thereafter indistinguishable from the other terms.

As it would require some 1,404 such combinational terms to interrelate the 27 simple terms originally used, it was found desirable to limit the actual number of combinational terms used at any one time to a small fraction of these and to introduce new terms only as it became possible to retire older ineffectual terms.

Turning to the selecting of the 16 terms used by the program, the method Samuel used was quite intuitively straightforward. At each move the coefficients of the polynomial are changed and the values of the coefficients are examined. The term having the coefficient with the smallest absolute value is the one that is contributing the least to the overall evaluation. It is given a "black mark." If a term accumulates a large enough number of black marks, it gets removed and replaced with the next term in the pool of reserve terms. Eventually this term can circulate back into the polynomial, however, if a term gets rejected too much it is eventually removed altogether.

The basic plan used to modify the coefficients of the polynomial is quite simple and is somewhat similar to how weights are modified in the back-propagation network. While the basic idea is simple, the precise implementation of it while the program is playing the game is quite complicated and will be ignored and just the basic idea will be discussed. To do the learning, one estimate of the goodness of a board configuration can be obtained by just using the evaluation function. A better evaluation can be made by applying minimax to the same board configuration. If minimax reports a better (more positive) value for the configuration than the evaluation function reported, change the coefficients of the polynomial so as to increase the value of the evaluation function. If, for example, we had a four term polynomial such as the following (and where for simplicity we assume all the x_is happen to be 1):

$$value = 8x_1 + 4x_2 - 8x_3 + 2x_4 = 2,$$

then if the backed-up value for this board configuration was 24, we could increase the coefficients a little to give a new polynomial:

$$value = 9x_1 + 5x_2 - 7x_3 + 3x_4 = 10.$$

While the value of this polynomial is still not 24, it is at least closer to the backed-up value of 24. A similar strategy is employed when the evaluation function overestimates the board. The polynomial that Samuel used actually had coefficients that are powers of 2 and the weight changing scheme was more complex. Notice that the polynomial modification procedure Samuel worked out is similar to what happens in back-propagation networks.

The program was designed so that there were two players, Alpha and Beta that used different evaluation functions. Alpha was the only player that learned. If Alpha won a game or was judged to be ahead at the end of a game by a separate part of the program, the Alpha polynomial was given to Beta. However, when Alpha lost a certain number of games (generally three), Alpha was assumed to be on the wrong track and a drastic change was made to the Alpha scoring polynomial by changing its largest coefficient to zero. This becomes the equivalent of giving a back-propagation network a kick when it gets stuck in a local minimum.

There were many experiments done using these methods where many minor aspects were varied. The final results from the 1963 paper were that the program learned to play a better-than-average game of checkers in a relatively short period of time and that the memory requirements were quite modest and remained fixed with time.

Samuel also makes these comments on the value of rote learning versus generalization:

> Some interesting comparisons can be made between the playing style developed by the learning-by-generalization program and that developed by the earlier rote-learning procedure. The program with rote learning soon learned to imitate master play during the opening moves. It was always quite poor during the middle game, but it easily learned how to avoid most of the obvious traps during end-game play and could usually drive on toward a win when left with a piece advantage. The program with the generalization procedure has never learned to play in a conventional manner and its openings are apt to be weak. On the other hand, it soon learned to play a good middle game, and with a piece advantage it usually polishes off its opponent in short order....
>
> Apparently, rote learning is of the greatest help either under conditions when the results of any specific action are long delayed or in those situations where highly specialized techniques are required. Contrasting with this, the generalization procedure is most helpful in situations in which the available permutations of conditions are large in number and when the consequences of any specific action are not long delayed.

An additional indication of the program's level of competence was a game arranged by the editors of the book, *Computers and Thought*, with a human checkers champion. The human expert was Mr. Robert W. Nealey and the computer was an IBM 7090. Mr. Nealey made these comments that are reported in the August 1962 *IBM Research News*:

> Our game ... did have its points. Up to the 31st move, all of our play had been previously published, except where I evaded "the book" several times in a vain effort to throw the computer's timing off. At the 32-27 loser and onwards, all the play is original with us, so far as I have been able to find. It is very interesting to me to note that the computer had to make several star moves in order to get the win, and that I had several opportunities to draw otherwise. That is why I kept the game going. The machine, therefore, played a perfect ending without one misstep. In the matter of the end game, I have not had such competition from any human being since 1954, when I lost my last game.

9.2.3 Samuel's Later Work

After the above research, Samuel continued to research learning methods for checkers [187]. This time the emphasis was on adjusting the program's static evaluation function so that it would give the same moves an expert would make (taken from a book of collected moves), rather than learning through actual play. The main portion of this research centered on using a technique with a structure called a *signature table*. For a formal definition of a signature table and its use in a general pattern recognition problem see [145]. The signature table approach works as follows: first, suppose there are only two parameters to measure on a board, x, that takes on four possible values and y, that takes on three possible values. In a game, you can simply measure the values of x and y and use these as indices

0.3	0.7	1.0	1.0
0.2	0.3	0.4	0.5
0.1	0.1	0.2	0.2

Figure 9.8: A two-dimensional signature table is quite easy to work with and makes it possible to capture nonlinear relationships within the measured board parameters, however, checkers demands much more than two parameters so an approximate solution had to be devised for checkers.

into a table like the one shown in Figure 9.8 to give you a value for the goodness of the current board. An important feature of this approach is that the evaluation function can be a nonlinear function of the board conditions. Samuel actually used a hierarchy of such tables and used a learning algorithm to train the structure to produce the same moves as experts recommended.

Samuel reports that this version of checker playing was superior to the 1959 versions, however, Dreyfus and Dreyfus [28] also report this from an interview (1983) with Samuel:

> ... Samuel said in a recent interview at Stanford University, where he is a retired professor, the program did once defeat a state champion, but the champion "turned around and defeated the program in six mail games." According to Samuel, after thirty-five years of effort, "the program is quite capable of beating any amateur player and can give better players a good contest." It is clearly no champion. Samuel is still bringing in expert players for help, but he fears he "may be reaching the point of diminishing returns." ... Samuel thinks the experts are poor at recollecting their compiled heuristics: "The experts do not know enough about the mental processes involved in playing the game."

9.2.4 Chinook

A new checker playing program by Schaeffer is called Chinook.[3] One of the key innovations in the Chinook program is that it has a database of endgames that contains all the possible moves that can be made and the results of each move. In December 1990 the world champion human checkers player, Marion Tinsley played a match against a version of the program that had data for all endgames with six or fewer pieces on the board and in 14 games Tinsley won one game and the other 13 ended in draws. A new version checks the 35 billion positions with seven or fewer pieces on the board. In a 40 game match played in August 1992, Tinsley won four games and Chinook won two. One loss by the program was caused by an error in a book of checker moves. In the second loss, Chinook stalled and ran out of time resulting in it forfeiting the game. The cause of this problem was not immediately apparent. Solving some of Chinook's mistakes under tournament conditions would require a 1000-fold faster computer. Schaeffer and his team think that if they have the 400

---

[3] See *Science News*, Volume 140, No. 3, July 20, 1991, page 40; *Computerworld*, February 17, 1992, page 28; and *Science News*, Volume 142, No. 14, October 3, 1992, page 217. See also [189, 190, 191, 192].

billion positions that go with eight pieces on the board, the program would be virtually unbeatable. They even hope to develop a program that knows all the possible outcomes of any given position—a program that would play perfect checkers.

9.3 Backgammon

Backgammon is another game that has received a great deal of attention. Two major series of experiments have been done with this game. The first series, by Berliner [8, 9], resulted in a program that by good fortune happened to beat the world champion. The second series of experiments by Tesauro and Sejnowski [229] and then by Tesauro alone [230] was done using back-propagation networks. In another experiment Tesauro used a new technique called temporal difference learning [231, 232, 233]. Some of the results are of general importance to anyone interested in using back-propagation to solve problems. Backgammon game playing is a little different from the checkers and chess style of game playing in that at every possible move the branching factor is so great (around 400) that no one has attempted to do minimax searching. Instead, the emphasis has been on producing evaluation functions that simply try to choose the best move at each step.

9.3.1 Berliner's BKG Program

Berliner's experiments consisted of three major attempts at producing an effective evaluation function. His first attempt was to use a single linear evaluation function for all types of situations throughout the whole game. This method did not work very well because, while the coefficients of the function gave good average results, there were many special situations where it simply was not very good. In his first attempt to remedy this problem he divided up board situations into a number of different classes and used a different evaluation function for each class, for instance, one function for a "running game" and another function for a "blockading game." This did improve the quality of the play, but the program still made serious blunders when the board configurations were near the boundaries between classes. It is at the boundaries that the program should be thinking about shifting its approach from one strategy to another and this shift should be done gradually, but instead, it turned out that it could be quite abrupt. In one class near the boundary, a small value for the function may result, while just across the boundary in a neighboring class, a large value will result. For an example of this phenomenon Berliner calls the "blemish effect," suppose the program has board configuration A, a configuration in class X, as shown on the left in Figure 9.9. Suppose two possible moves are available, move B keeps the game in class X with a value for the evaluation function of 1, while move C is in class Y and its evaluation function gives a value of 9. Naturally the program will be "in a hurry" to make the move that takes it to class Y because of its larger value. Or, as shown on the right in Figure 9.9, the program will try to delay making the transition to class Y.

Berliner fixed the problem by eliminating the classes and instead he used variable values for the coefficients of just one polynomial. In different circumstances the coefficients would change depending on the board situation. The change in coefficients would be gradual so that there were no sharp boundaries between one type of board situation and another. This version of the program produced a dramatic improvement in the quality of play, but it was

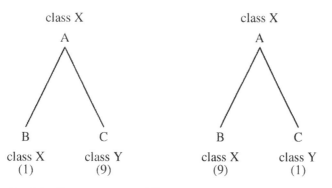

Figure 9.9: The blemish effect occurs when different classes use different evaluation functions and the values do not vary smoothly. On the left, the program will be "in a hurry" to cross into class Y. On the the right, it will delay moving into class Y.

still not as good as the best human players. Despite this, the program (called BKG 9.8) was taken to play against the winner of the world championship in Monte Carlo in July 1979. The program got lucky and won by a score of 7 games to 1.

9.3.2 Backgammon using Back-Propagation

	dark pieces before					light pieces before					light pieces present after move					
pieces:	$\geq$5	4	3	$\geq$2	1	1	$\geq$2	3	4	$\geq$5	0	1	$\geq$2	3	4	$\geq$5
data:	0	0	0	0	0	0	1	0	0	0	0	0	1	1	0	0
bit:	1	2	3	4	5	6	7	8	9	10	11	12	13	14	15	16

Figure 9.10: One of the data coding schemes used to train the network.

In the first set of experiments by Tesauro and Sejnowski, the most important problem was representing the state of the board to the back-propagation networks. Tesauro and Sejnowski experimented with a number of different ways to represent the input patterns. The simplest representation was a unary coding of the number of light- and dark-colored pieces at each location before the move was made plus the number of light-colored pieces at each location after the move was made plus the roll of the dice. All the data is encoded as if each move is made by the light-colored player. An example of the coding scheme used to represent the state of one of the 24 points on the board is shown in Figure 9.10. Bits 1 through 5 code the number of dark pieces present before the move, bits 6 through 10 are for the number of light pieces present before the move, and bits 11 through 16 are the number of light pieces present after the move. In this example, two light pieces were present before the move and three were present after the move.

This simple representation turned out to have some limitations and other more complicated ones were also devised. These involved coding the type of move that was made and also including some precomputed features[4] before giving the new set of inputs to the net. For training, 3202 board positions were collected. For each board position and for every possible roll of the dice, an input pattern was created. Some of these patterns, patterns that were exceptionally good or bad, were rated on a scale from +100 to −100. Other patterns that were not exceptionally good or bad were given a random value each time the pattern was presented to the network. The random values were in the range from −65 to +35. These patterns that were not humanly rated were actually skipped by the training program 75 percent of the time for an ordinary roll of the dice and 92 percent of the time for double rolls and this randomness constitutes noise in the training process. Two-, three-, and four-layer networks were tested and the number of input units in the experiments ranged from 393 to 651. The various networks were trained for from 10 to 24 cycles. The network was never trained as to what a legal move could be. When the program was playing, it took the current board position and the roll of the dice and generated every possible legal move, then each move was input to the network and the highest rated one was taken as the answer.

The program was compared with both a human expert (Tesauro) and a backgammon playing program, gammontool, of Sun Microsystems, Inc. that used conventional techniques. Tesauro says that gammontool is a good benchmark for measuring the playing strength of a program. He says that human beginners can beat gammontool about 40 percent of the time, intermediate level players win about 60 percent of the time, and human experts win about 75 percent of the time [231]. One test involved playing the best network against gammontool. The result was that the network won almost 60 percent of the time. This best network had 459 input units and included input units that had precomputed features. In principle, multilayer networks can tune hidden units to combine input features, but the use of precomputed features gave better results. Without the use of the precomputed features, the network only beat gammontool 41 percent of the time. The authors say that the use of precomputed features is the single most important factor in improving the quality of play and it was more significant than using hidden units.

In a second important experiment, Tesauro and Sejnowski used the same 410 features that the gammontool program used, but the network was allowed to learn its own weights. The result was that the back-propagation network beat the conventional program 54 percent of the time. They take this as an indication that a machine-tuned polynomial is likely to be better than a handcrafted one.

In one series of 20 games against the network, the network beat Tesauro in 11 of them. He reports that the program was basically lucky to do that well and in general it would not. In analyzing the play, they found the program made a number of serious and possibly fatal blunders but it was lucky and managed to win anyway. They estimate that the current program could be expected to beat an expert about 35–40 percent of the time. They think that by adding some specific handcrafted examples, many of the obvious blunders can be eliminated and they then expect the performance to be 65 percent against gammontool and 45 percent against human experts. Some more subtle mistakes are due to problems that the researchers currently think will be difficult to correct. They say that correcting them may require "either an intractably large number of examples, or a major overhaul in either the precomputed features or the training program." The researchers note that the program

---

[4] For a simple example of a precomputed feature, consider the XOR problem where there are normally two inputs, x_1 and x_2. A good precomputed feature to add is $x_3 = x_1 x_2$. Adding this precomputed feature typically speeds up training a lot.

is working without much of the information that is available to a human player. All the network received was an indication that a move was good or bad. The program had no idea what the object of the game was or that its actions have consequences that lead to the achievement of the objective. After this research, they came to believe that adding the knowledge that people have about the game will be necessary to achieve a large increase in performance.

One general principle emerged from their experiments. Tesauro and Sejnowski came to the conclusion that simply giving a network raw data from random examples is not sufficient to produce a good system. Producing a good system requires a substantial amount of human effort "both in the design of the coding scheme and in the design of the training set."

9.3.3   A Second Back-Propagation Approach

The above method of producing a backgammon network was called the relative score paradigm. The second method of doing the problem used another method called the comparison paradigm. In this method, a network is presented with two possible final board positions and it is told which one is better. Also, instead of using one large network, the problem was partitioned into five specialized smaller networks for what are described as "engaged positions." "Running positions" were not considered in these experiments. When the game player had to produce a move for the running positions, the gammontool program was used instead. This comparison paradigm method produced better results.

Each of the five networks for the engaged positions had a special configuration like that shown in Figure 9.11. The network is not fully interconnected between layers as is the usual case. Instead, it is divided into two halves. Given that you are at a certain board configuration, one possible configuration for after the move is placed on the left input units and another one goes on the right input units. The network is trained to produce a 1 on the single output unit when the board configuration on the left is better than the one on the right, otherwise it should produce a 0. Now, this means that if a given pair of board configurations produces a 1, then exchanging this pair should produce a 0. To insure that this is the case, the set of weights, W_1, on the lower left will have the same values as the set of weights, W_2, on the lower right and the set of weights, W_3, going from the hidden units on the left to the output unit will be the negative of the set of weights, W_4, going from the hidden units on the right to the output unit. For example, suppose the pattern on the left input units produces a value of 0.67 on the left hidden unit, while the pattern on the right input units produces a value of 0.33 on the right hidden unit. To get a value of 0.9 (close enough to 1) on the output unit requires that the weight from the left hidden unit to the output unit must be 6.9 and the weight from the right hidden unit to the output unit must be –6.9. This situation is shown in Figure 9.11. Now, if the two input patterns are reversed, the hidden unit on the left have to register 0.33, the hidden unit on the right will have to register 0.67, and then the output unit will be just under 0.1.

The engaged positions were divided into five categories: bearoff, bearin, opponent bearin, opponent bearoff, plus another category that covered all the other cases. The first four categories had about 1,000 training positions each and the fifth category had about 4,000. Tesauro found that the addition of hidden units did not improve performance. He also compared a 651-12-1 relative score network and the five specialized 289-1 networks

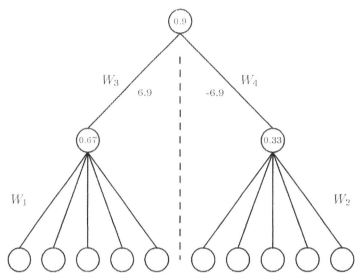

Figure 9.11: A network like that used in the comparison paradigm for backgammon. The network is divided down the middle into two halves. One input position is placed on the left input units and the other one goes on the right. The weights leading into the hidden units are the same on both sides. The weights going from the hidden unit on the right (here, only one weight) are the negative of the weights on the left-hand side.

with the human choice of the best move. Figure 9.12 shows the percentage of times the networks got the same answer as the human expert. These results show a general noticeable improvement for the comparison paradigm networks. Tesauro observed that it is very remarkable that these smaller networks with less than 300 weights each were able to outperform the network with almost 8,000 weights.

Type of test set	Relative Score Network	Comparison Network
bearoff	82	83
bearin	54	60
opponent bearoff	56	54
opponent bearin	60	66
other	58	65

Figure 9.12: A comparison of the results of a 651-12-1 relative score network with a 289-1 comparison network.

In another set of tests, complete games were played and in these experiments the five specialized networks were used to compute the best move for the engaged positions and the gammontool program was used to choose moves for the running positions. In this series of games, the percentage of wins by the networks against gammontool and Tesauro are given in Figure 9.13. Tesauro also reports that the comparison networks made fewer severe blunders than the relative score network.

Opponent	Relative Score Network	Comparison Network
gammontool	59	64
Tesauro	35	42

Figure 9.13: A comparison of the results of the two types of networks.

9.3.4 Temporal Difference Learning

In all the above training methods, experts told the programs how good a move was and the networks learned the expert's preferences. Another way to train a program would be to let it play out a game, let it win or lose, and then have the program go back and assess which move or moves were the key to winning or losing. Moves that led to wins get positive reinforcement and moves that led to losses get negative reinforcement. This *credit assignment problem*, assigning credit to individual moves has been a barrier to machine learning, however, Sutton [227] proposed a solution to this problem called temporal difference learning. Tesauro [231, 232, 233] has applied this algorithm to learning backgammon and it has worked surprisingly well.

The temporal difference formula used was:

$$w_{t+1} = w_t + \alpha (Y_{t+1} - Y_t) \sum_{k=1}^{t} \lambda^{(t-k)} \nabla_w Y_k.$$

In the algorithm there is the network of weights at time t, w_t. At time t using w_t, the network outputs Y_t, a value that estimates its chance of winning the game. Now w_t is used to evaluate all the possible moves that could be made. Of course, the highest rated move is the move to make and its rating is Y_{t+1}. The $\nabla_w Y_k$ term is the gradient of Y_k with respect to the weights, w_k. All these terms from the first move at $k = 1$ to the current term $k = t$ must be kept. The λ parameter controls how much of each of these terms will count toward the current weight change. With $\lambda = 1$, all the terms count equally, with $\lambda = 0$, only the current term counts, and with a value of λ between 0 and 1, the ith term will count more than the $i - 1$st term. Tesauro used $\lambda = 0.7$. The factor α is simply the learning rate parameter. The network starts with random initial weights.

Tesauro did a series of tests with the algorithm. One of the series involved giving TD networks raw board data, that is, data without any precomputed features, and letting it play itself in a large number of games. With 40 hidden units and after playing 200,000 games,[5] the program won 66.2 percent of its games against gammontool and a later version that included precomputed features reached 71 percent.

The newest reported version [232, 233] is reportedly the best backgammon program ever produced. The estimates are that Berliner's BKG program would lose an average of about 0.3 to 0.4 points per game against the very best human players. The TD-Gammon 2.1 program trained using 1.5 million games lost only 0.02 points per game in a 40 game match against one of the best human players. The program discovered new strategies that human players now agree are better than the old human strategies.

---

[5] This required two weeks of training on an IBM RS/6000 workstation.

9.4   Exercises

9.1. Apply minimax to the following tree:

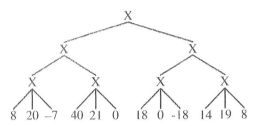

9.2. Apply alpha-beta minimax to the tree of Exercise 9.1, but take care to only visit those nodes that must be evaluated, not those that are cut off. How many fewer nodes were visited with the alpha-beta technique as compared with plain minimaxing?

9.3. Write a program to do tic-tac-toe using minimaxing without alpha-beta searching. One evaluation function to experiment with is the one in the text. Another evaluation function is the following one. Look at the board and count the number of ways X can win and subtract the number of ways O can win. This difference should be the rating of the board. Thus, given the board:

X can win by completing the vertical set on the left, the vertical set on the right, and the horizontal sets on the top and bottom, for a total of four. O on the other hand, can potentially win in six possible ways. Thus, the board should be scored as $4 - 6 = -2$ for X. Tie games should be scored as 0. Wins and losses should be given very large positive and very large negative values.

Have your program play against itself using the two different evaluation functions and analyze whether or not one is better than the other. If you can think of any other evaluation functions, try them as well. The depth to which your program should search depends on how fast your computer is. For slower computers searching three or four deep may be all that is practical. The program can be speeded up by using clever data structures. In Pascal, representing the board positions as sets can be efficient. In other languages, bit strings are the equivalent of Pascal's sets. If your computer is very slow, you may have to give the program games to do that already have several moves on the board. Needless to say, use recursion and *do not* create the search tree in memory as a linked list.

9.4. Is there any way to apply back-propagation to tic-tac-toe?

9.5. Suppose you want to produce a back-propagation-based checker playing program. Consider some possible ways of going about doing this and evaluate their merit. If you are ambitious and familiar with checkers, get the network to learn a few simple moves. Naturally, if you are really interested you can make the project as large as you have time for.

9.6. Consider whether or not case-based or memory-based methods would be useful for tic-tac-toe and backgammon.

9.7. In game playing, human players often use specific strategies to guide their moves, but programs do not do this. Can you think of any way for conventional game playing programs or conventional programs augmented with neural, case-based, or memory-based methods to use specific strategies? How do people devise new strategies?

Chapter 10

Natural Language Processing

Natural language understanding by computers has always been one of the most important problems for AI researchers. One of the reasons is that the use of language has often been viewed as *the* defining characteristic of intelligence, so producing an intelligent computer would require producing a computer that could understand natural language. Another reason is the more practical one that communicating with computers through their limited artificial languages has always been difficult and time consuming and so it is highly desirable to shift this burden from people to the computers. Another practical reason for interest in natural language is that it would be highly advantageous to have computers translate from one natural language to another.

Many of the earliest attempts to process natural language involved trying to translate text from one natural language to another. The first of these attempts involved trying to take a word at a time from the text and replace it with the equivalent word in the other language. This did not work very well and the next attempt was to try to do phrase by phrase translations. This did not work very well either, so researchers fell back to the position that the computer should try to do the best it can translating as much of the text as it can and leave the harder translation problems for human beings who would then clean up these hard parts. This approach was never very useful either. Machine translation became one of the greatest early failures of the natural language processing efforts. One of the most famous mistranslations from an early system was when the system took the following statement in English:

> The spirit is willing but the flesh is weak,

translated it to Russian and then back to English where it became:

> The vodka is strong but the meat is rotten.

Linguists then proposed that in order to do natural language processing, you needed to have the computer reach some deep understanding of the sentences that it was processing. They proposed that the natural road to understanding sentences was to look at the words in the sentence, identify what parts of speech they were, for example, nouns, verbs, adjectives, and so forth, and from there, you needed to in effect "diagram" the sentence in a manner similar to the one students are taught in grade school. Only then, after analyzing the grammatical structure, could you go ahead and analyze what the sentence means. Then, if you were doing language translation, you could then go and produce the equivalent sentence in the other language.

These efforts to understand natural language assumed that syntax was the key to producing computers that understand language, however, research along these lines has shown that considering syntax alone is inadequate and the research has now shifted to trying to take the syntax and semantics of a sentence into account more or less in parallel. Other methods assume that semantics is the key and syntax only provides some extra clues to the problem.

Currently, the ability of programs to understand natural language is still severely limited. Early on, people working on language understanding thought that a relatively simple theory of language was possible, but it now seems unlikely that there is any neat simple theory of language that will enable a relatively small computer to understand language. Indeed, it might turn out that to get a computer to understand language it will be necessary to give the computer a body and raise it much as human children are raised. In the meantime, approximate solutions to understanding natural language are all that will be available. In this chapter we will look at the major symbolic proposals for processing natural language plus some of the new neural networking approaches. The simple neural models of language that have been developed so far are much less useful than traditional symbolic methods, but they can exhibit aspects of human thinking that are difficult to produce in symbolic models, so pursuing improved neural models should prove to be useful.

10.1   Formal Languages

$$
\begin{aligned}
S \rightarrow \quad & a \\
& b \\
& aa \\
& bb \\
& aXa \\
& bXb \\
\\
X \rightarrow \quad & a \\
& b \\
& aa \\
& bb \\
& aXa \\
& bXb
\end{aligned}
$$

Figure 10.1: A simple grammar that produces strings of palindromes consisting of only the letters a and b.

As people began experimenting with programs to use natural language, a theory of formal languages was being developed. This theory proposed that there were four basic types of language, classified according to the kinds of rules used to generate sentences in the language. The theory was initially developed around abstract and often meaningless "languages" that just consisted of strings of characters. As an example of how rules are typically

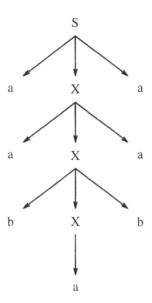

Figure 10.2: The series of steps involved in generating the string "aababaa" produces a parse tree.

written and used, Figure 10.1 gives a small grammar that can generate strings of letter a's and letter b's where the strings are also palindromes. In these rules the capital letters are called *nonterminals* while the lower-case letters are *terminals*. S is a special nonterminal known as the *starting symbol*. The rules say, that starting with an S "written down on a piece of paper," you can cross out the S and replace it with the string of letters, "a," "b," "aa," "bb," "aXa," or "bXb." Suppose we choose "aXa." The string on paper is now:

aXa.

The nonterminal, X, can be replaced by other letters and the second rule describes what they are. If we choose "bb" we get the final string of terminals: "abba." If we choose "aa" we get the final string of terminals: "aaaa." In either of these cases, replacements can no longer be made because there are no nonterminals left in the string. If we chose to replace X with "bXb" or "aXa," we would have either:

abXba or aaXaa

and the replacement game could continue. Using these rules you can generate every string in the language and never generate a string such as "aabb" that is not in the language. Figure 10.2 is a *parse tree* that shows how the string "aababaa" is generated.

The four types of formal languages are:

Type 0: the *unrestricted* language, in which any kind of replacement rule can be used,

Type 1: the *context-sensitive* language, in which there are rules in which replacements can only be made within a certain context,

Type 2: the *context-free* language, in which context does not affect the choice of replacements within a string, and

Type 3: the *finite state* language, in which the rules are restricted even more than in type 2. The only rules allowed must be of the form: X → aY, where a nonterminal can be rewritten as a single terminal followed by a single nonterminal, or X → a, where a nonterminal can be rewritten as a single terminal. Finite state languages are also called *regular* languages.

Type 2, the context-free language, is by far the most important one for computer scientists because most programming languages can be defined using this type of language. A notable exception to this is the unusual, little used, but theoretically important language, Algol 68, which is defined using a context-sensitive grammar. Also, natural languages (with a few possible minor exceptions) can be defined using context-free grammars, although it is often more convenient to define them using context-sensitive rules.

```
<variable> ::= a | b
<mult-op> ::= * | /
<adding-op> ::= + | -
<factor> ::= <variable> | ( <simple-exp> )
<term> ::= <factor> | <term> <mult-op> <factor>
<simple-exp> ::= <term> | <simple-exp> <adding-op> <term>
```

Figure 10.3: Grammar rules for creating a simple-expression like the one defined in Pascal.

To illustrate the use of a context-free language for defining arithmetic expressions, we will look at an example of a *simple-expression* derived from Pascal where the simple-expression will be even simpler in that the only variables allowed will be the letters a and b. In these rules we will use another notation for giving grammar rules, called BNF, or Backus Normal Form (and sometimes called Backus-Naur Form). In these rules, a non-terminal consists of a descriptive name enclosed in angle brackets (<>) and where the ::= symbol is used instead of →, a vertical bar (|) means "or," and terminal symbols are themselves. The rules are given in Figure 10.3. For simplicity, the variables are the letters a and b. Starting with <simple-exp> we show how to generate the expression, b * (a + b) in Figure 10.4.

Besides using formal language theory to define correct statements in a language, another important use of formal language theory is to construct compilers that are used to translate computer languages into machine language instructions. In this use, language statements are analyzed as to how they were generated from the rules. Knowing this enables the compiler to correctly understand the statement so it can generate the correct code. There are several ways to do this, the simplest one being the recursive descent technique. Another method is to take in an arithmetic expression and produce a *parse tree* as in Figure 10.4. From a parse tree a compiler can generate code. Creating a parse tree is pretty much the computer language analog to diagramming a sentence.

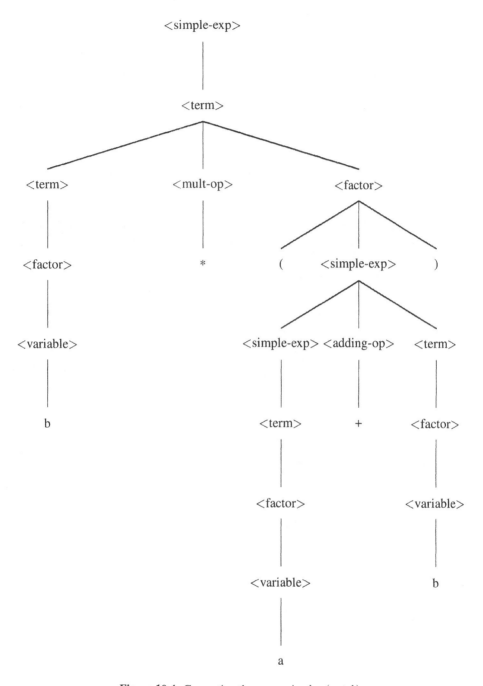

Figure 10.4: Generating the expression $b * (a + b)$.

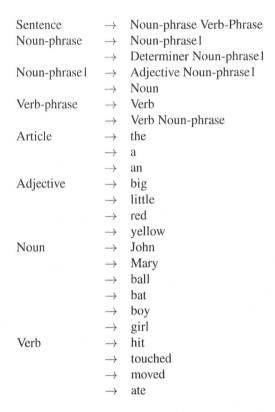

Sentence	$\rightarrow$	Noun-phrase Verb-Phrase
Noun-phrase	$\rightarrow$	Noun-phrase1
	$\rightarrow$	Determiner Noun-phrase1
Noun-phrase1	$\rightarrow$	Adjective Noun-phrase1
	$\rightarrow$	Noun
Verb-phrase	$\rightarrow$	Verb
	$\rightarrow$	Verb Noun-phrase
Article	$\rightarrow$	the
	$\rightarrow$	a
	$\rightarrow$	an
Adjective	$\rightarrow$	big
	$\rightarrow$	little
	$\rightarrow$	red
	$\rightarrow$	yellow
Noun	$\rightarrow$	John
	$\rightarrow$	Mary
	$\rightarrow$	ball
	$\rightarrow$	bat
	$\rightarrow$	boy
	$\rightarrow$	girl
Verb	$\rightarrow$	hit
	$\rightarrow$	touched
	$\rightarrow$	moved
	$\rightarrow$	ate

Figure 10.5: Some rules for generating a subset of English.

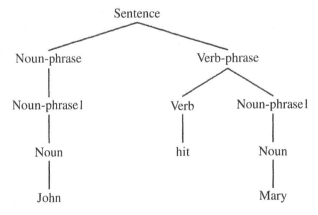

Figure 10.6: The parse tree of the sentence: "John hit Mary."

Some rules to generate some simple English sentences are shown in Figure 10.5 and a parse tree of the sentence "John hit Mary" is shown in Figure 10.6.

It is also worth mentioning that formal grammars can be used to generate totally meaningless and irrational statements. One of the most famous meaningless sentences is:

Colorless green ideas sleep furiously.

This shows that much is lost when a language is condensed down to *just* a grammar.

The fact that natural language can be generated by explicit rules using recursive calls has always been a major factor in the traditional assumption that people do form and carry around explicit rules in their heads, not just for language, but for all types of thinking. Moreover, the fact that recursion is involved in generating and in parsing sentences has also been a major factor in the traditional assumption that people do recursive Lisp-like/Prolog-like processing in their heads. Now, connectionists and others are beginning to question these long-held assumptions about how people process language.

10.2   The Transition Network Grammar

The fact that linguists had always analyzed sentence structures by diagramming them and the fact that generating parse trees can be useful in translating computer languages led to the idea that the first phase in the analysis of a sentence *must* be to find the parse tree of the sentence. Having found that, you could move on and try to figure out what the sentence meant. The most typical means of trying to parse a sentence is to define a graph, or network, called a *transition network grammar*. This network provides an organized means of parsing a sentence, extracting the subject, verb, direct object, and so forth, and also creating a parse tree in the process. In the description given here, we will not try to form the parse tree.

The transition network grammar approach may be the most common means for analyzing sentences that is available at the moment, and, like other techniques, it is generally adequate for very small problem domains. Some commercial natural language understanding systems use this approach. The systems are usually designed to serve as front ends to a database system so that users can ask questions about facts in the database in natural English without having to learn the database system's language. The sales materials supplied by companies selling natural language front ends typically state that if you want to ask the system a question, it will probably understand you, but if it does not, you can always rephrase the question in such a way as to make it understand. One of the early successful systems based on the transition network grammar approach was the Lunar Sciences Natural Language Information System (LSNLIS or LUNAR). This system was commissioned by NASA to answer questions about rocks brought back from the moon. When the system was taken to a conference of lunar scientists, the system was able to understand 90 percent of the questions that these scientists put to it [266].

10.2.1   A Simple Transition Network

The idea behind the transition network grammar is very simple and it is much like recursive descent compiling. The process is also very much like a rat trying to find its way through a maze. In Figure 10.7 we show the network necessary for just processing a noun-phrase

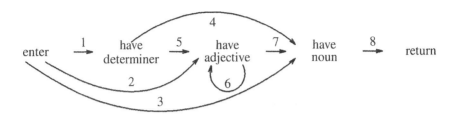

Figure 10.7: A simplified transition network grammar to process a noun-phrase.

as defined by the grammar in the previous section. The object of going through the noun-phrase network is to find the determiner (if any), adjectives (if any), and the noun. When this network is finished it returns this information to the state that called it. When entering this network, you apply a test for each arc until you find the first test that succeeds. Arc 1 has a test for a determiner. If you find that the first word is a determiner, the network makes the transition to the state labeled 'have determiner' and you also save the particular determiner in some convenient variable. The convenient variable is referred to as a register. If the first word is not a determiner, you try arc 2 which tests for an adjective. If the test succeeds, go to the state labeled 'have adjective' and save the adjective you found in a list of adjectives. If this test fails, you try arc 3 that tests for a noun. If this test succeeds, go to the state labeled 'have noun' and save the particular noun in a register. If none of these tests work, report failure. To handle a noun-phrase with more than one adjective, as in "the big red ball," you end up at the 'have adjective' state when you analyze "big." Looking at the next word, "red" will return you via arc 6 to the 'have adjective' state while adding "red" to a list of adjectives.

Notice what may happen if you give the network the phrase, "the baseball bat." First, it moves to the state labeled 'have determiner.' At 'have determiner' it tests for an upcoming noun or an upcoming adjective. Taking baseball as a noun, it moves on to the state labeled 'have noun.' Upon checking the next word, the program finds it is a noun so it needs to back up in the network to look for another path. This time it will find baseball is an adjective and then move forward again to 'have noun.'

Before continuing with a network for the rest of a sentence, notice what you would do if you had to make this noun-phrase network handle prepositional phrases like "in the hat" as in the sentence, "The cat in the hat strikes back." After finding the noun cat, you need to check ahead to see if the next word is a preposition. If it is, any sort of noun-phrase can follow, including even:

The cat in the hat in the backyard under the tree . . .

so it is quite convenient to recursively call up the noun-phrase network all over again to work on "the hat in . . . ."

We can now look at the quite simple pieces for the rest of the transition network for processing a sentence in Figure 10.8. In the sentence network, arc 1 calls up the noun-phrase network. Since there are no other alternatives in this network, it had better succeed,

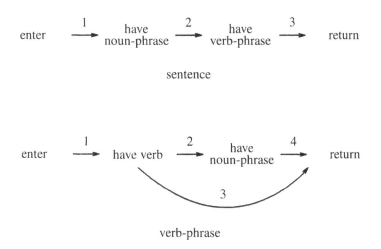

Figure 10.8: The rest of the transition network grammar.

if we are to continue. If it succeeds, you move on to the 'have noun-phrase' state where arc 2 will test for a verb-phrase by calling up the verb-phrase network. Again, it had better succeed or the whole process returns failure. In the verb-phrase network, arc 1 tests for a verb. If you have one, you move to the 'have verb' state. There, arc 2 calls the noun-phrase network to try to find a noun-phrase. If it fails, you move on to try arc 3. Arc 3 is a test for "no more words to process." If it succeeds, return success along with the verb stored away in a convenient variable. Arc 4 also tests for no more words to process.

10.2.2   A Prolog Implementation

	the	baseball	broke	the	window	
1	2	3	4	5	6	

Figure 10.9: We think of the words in the sentence as coming between the numbered points.

Transition network grammars are normally implemented in Lisp, but it is quite easy to implement the above network in Prolog. To begin with, we assume the data for the sentence to parse will be coded using the method illustrated in Figure 10.9. In this figure, the words in the sentence are positioned between numbers, so we could say that "baseball" runs from positions 2 to 3, the subject noun-phrase runs from 1 to 3, the verb-phrase runs from 3 to 6, and so on. We can encode the word positions in Prolog as follows:

```
word(the,1,2).
word(baseball,2,3).
word(broke,3,4).
word(the,4,5).
word(window,5,6).
```

We will also need to know the parts of speech for each word in the language. To do this, we have a dictionary predicate that stores some words and contains a part of speech entry for each word like so:

```
dictionary(the,det).
dictionary(big,adj).
dictionary(brown,adj).
dictionary(baseball,adj).
dictionary(baseball,noun).
dictionary(bat,noun).
dictionary(window,noun).
dictionary(broke,verb).
dictionary(picture,noun).
dictionary(picture,adj).
```

Our goal will be to figure out the correct part of speech for each word. Given the sentence "The bat broke the window," we will produce the following listing of facts in the Prolog database:

```
isa(the,1,2,det).
isa(baseball,2,3,noun).
isa(broke,3,4,verb).
isa(the,4,5,det).
isa(window,5,6,noun).
```

The first of these is taken to mean that "'the' is the word that runs from 1 to 2 and it is a determiner." We will add these facts to the database by using the store predicate, which is defined as follows:

```
store(W,X,Y,Z) :- retract(isa(W,X,Y,P)),
                  assert(isa(W,X,Y,Z)).


store(W,X,Y,Z) :- assert(isa(W,X,Y,Z)).
```

Store(W,X,Y,Z) means that the word W, that runs from X to Y, has been determined to be the part-of-speech, Z, and this fact should be stored away. Remember, however, that in going through the network on a sentence like: "The baseball broke the window," the network might first assume that "baseball" is an adjective. If the network does this we will have store called like so:

```
store(baseball,2,3,adj)
```

When the network later determines that this is wrong, we need to retract this fact. The first store predicate does this by first attempting to retract any existing fact about the word, W, that runs from X to Y with part-of-speech, P. If the retract succeeds, we continue on to put in the new fact. If the retract fails, Prolog moves on to try the second store which only stores the new fact.

Next, we need to define a set of rules to recognize a sentence. This is easily done by following the grammar. To recognize a sentence that runs from X to Y, we must have a noun-phrase (np) that runs from X to Z and a verb-phrase (vp) that runs from Z to Y:

```
sent(X,Y) :- np(X,Z), vp(Z,Y).
```

To parse a sentence that starts at X and runs to Y, we just call up the "sent" predicate. In the case of the sample sentence in Figure 10.9 we therefore use: "sent(1,6)." As in the original grammar, an np that runs from X to Y could be either a noun-phrase, np1, without a determiner at the beginning:

```
np(X,Y) :- np1(X,Y).
```

or a noun-phrase that has a determiner at the beginning followed by the words that make up the np1 structure:

```
np(X,Y) :- word(W,X,Z), dictionary(W,det),
           store(W,X,Z,det), np1(Z,Y).
```

This definition says that, "there is a noun-phrase from X to Y if you look up the word, W, that runs from X to Z, and that word, W, is a determiner, and we store away this fact and there is an np1 structure that runs from Z to Y." The following rule looks for an adjective followed by an np1 type structure:

```
np1(X,Y) :- word(W,X,Z), dictionary(W,adj),
            store(W,X,Z,adj), np1(Z,Y).
```

This np1 rule simply looks for a noun:

```
np1(X,Y) :- word(W,X,Y), dictionary(W,noun),
            store(W,X,Y,noun).
```

A verb-phrase, vp, consists of a verb followed by a noun-phrase:

```
vp(X,Y) :- word(W,X,Z), dictionary(W,verb),
           store(W,X,Z,verb), np(Z,Y).
```

or just a verb without a following noun-phrase:

```
vp(X,Y) :- word(W,X,Y), dictionary(W,verb),
           store(W,X,Y,verb).
```

The whole program is shown in Figure 10.10. Tracing through the program for any given sentence is very straightforward.

10.2.3   A Neural Analog

We will now look at a neural network analog to the symbolic transition network grammar. Parse trees, such as the ones in Figures 10.2 and 10.4, obviously seem to have some resemblance to multilayer neural networks. In Figure 10.4, the parse tree looks very much like a neural network designed to detect the presence of a sentence, with nodes such as NP and VP functioning as hidden units. Also, networks like those shown in Figures 10.7 and 10.8 look a lot like conventional neural networks.

To mimic the transition network grammar approach in a neural network, the basic idea is, as in the transition network grammar, that certain kinds of words come in some pre-

```
sent(X,Y)  :- np(X,Z), vp(Z,Y).

np(X,Y)    :- np1(X,Y).

np(X,Y)    :- word(W,X,Z), dictionary(W,det),
              store(W,X,Z,det), np1(Z,Y).

np1(X,Y)   :- word(W,X,Z), dictionary(W,adj),
              store(W,X,Z,adj), np1(Z,Y).

np1(X,Y)   :- word(W,X,Y), dictionary(W,noun),
              store(W,X,Y,noun).

vp(X,Y)    :- word(W,X,Z), dictionary(W,verb),
              store(W,X,Z,verb), np(Z,Y).

vp(X,Y)    :- word(W,X,Y), dictionary(W,verb),
              store(W,X,Y,verb).

store(W,X,Y,Z)  :- retract(isa(W,X,Y,_)), assert(isa(W,X,Y,Z)).
store(W,X,Y,Z)  :- assert(isa(W,X,Y,Z)).

dictionary(the,det).
dictionary(big,adj).
dictionary(brown,adj).
dictionary(baseball,adj).
dictionary(baseball,noun).
dictionary(bat,noun).
dictionary(window,noun).
dictionary(broke,verb).
dictionary(picture,noun).
dictionary(picture,adj).
```

Figure 10.10: A simple Prolog transition network grammar program.

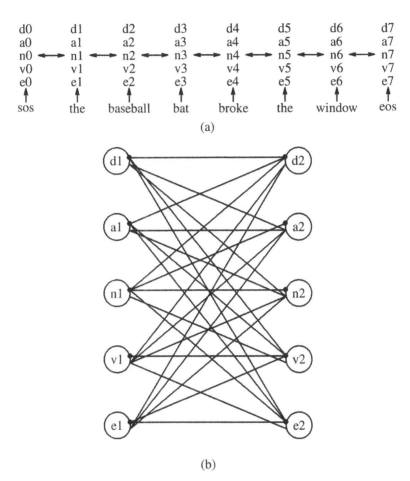

$$
\begin{array}{cccccccc}
d0 & d1 & d2 & d3 & d4 & d5 & d6 & d7 \\
a0 & a1 & a2 & a3 & a4 & a5 & a6 & a7 \\
n0 \leftrightarrow & n1 \leftrightarrow & n2 \leftrightarrow & n3 \leftrightarrow & n4 \leftrightarrow & n5 \leftrightarrow & n6 \leftrightarrow & n7 \\
v0 & v1 & v2 & v3 & v4 & v5 & v6 & v7 \\
e0 & e1 & e2 & e3 & e4 & e5 & e6 & e7
\end{array}
$$

sos   the   baseball   bat   broke   the   window   eos

(a)

(b)

Figure 10.11: In (a) is an outline of a neural network to determine the parts of speech in a sentence. Each column, n, has a node that will turn on if the word should be a determiner (dn), an adjective (an), a noun (nn), a verb (vn), or an endpoint (en) of the sentence. Within each column, every node inhibits all the other nodes in the column. Words activate nodes in each group, for instance, baseball, which could be either an adjective or a noun, will activate a2 and n2 in column 2 and inhibit d2, v2, and e2. Each node in a column will also activate or inhibit nodes in a neighboring column to enforce the rules of grammar. A more detailed view of some of these connections (without the intracolumn inhibition links) is shown in (b) where the lines that end in a • are bidirectional inhibition links and the other lines are bidirectional activation links.

scribed order. For example, adjectives or a noun can be expected to follow determiners, nouns should precede verbs, and so on. With these grammar rules built into a network in the form of activation and inhibition links, a network can then settle into a state that satisfies these constraints. Figure 10.11 illustrates the network we will use. The way this network will function is that we will look up the possible parts of speech for each word in a dictionary and then activate the corresponding parts-of-speech units in each group. We also add a start-of-sentence marker, sos, at the beginning of the sentence and an end-of-sentence marker, eos, after the last word of the sentence. Sos and eos activate the endpoint units (e). Working with the sentence in Figure 10.11, we look up "the" and find it is a determiner, so we have it activate the determiner node, d1. When we look up "baseball," we find it could be an adjective or a noun. We therefore use the "baseball" node to activate the adjective node, a2, and the noun node, n2. For "bat," we have it activating only the noun node, n3, for "broke" we have it activating only the v4 node, and so on. Between each group of nodes we have activation and inhibition links to enforce the rules of English grammar. We arrange to have the determiner node in column 1 (d1) activate the possibilities that a noun or adjective follows in column 2 by having activation links to n2 and a2. Similarly, a determiner in column 1 should inhibit a determiner in column 2 (d2), a verb in column 2 (v2), and the end-of-sentence marker for column 2 (e2). Every node in each column will activate or inhibit every node in the following column. Also, every node in each column will be activating or inhibiting every node in the preceding column. Thus, if we have a verb in column n, we must activate the noun node in column $n - 1$ and inhibit the verb, adjective, determiner, and endpoint nodes in column $n - 1$. Finally, within each column of nodes we arrange to have each node inhibit all the others, so that only one node wins.

The network in Figure 10.11 can indeed take sentences like, "The baseball broke the window" and "The baseball bat broke the window" and settle down to assign the correct parts of speech to each word. Using the sentence, "The baseball broke the window," the presence of the verb, broke, in the third position inhibits the possibility that baseball is an adjective and tries to make it a noun. In the transition network grammar version, if the program first took baseball to be an adjective, the presence of a verb following it causes the network to backtrack, the transition network equivalent of strong inhibition.

There is a nice feature that comes from the analog nature of the network. The analog neural network can, like Rumelhart and McClelland's word recognition program that handled missing features, handle missing parts of sentences. For example, if we give the network in Figure 10.10 the sentence:

The baseball ___ broke the window,

where the noun has been removed, the network can still conclude that "baseball" is an adjective and the missing word must be a noun. Such a string of words will still be recognized as close to being a sentence. Classical transition network grammars do not naturally have this ability to deal with problem sentences, although some research has been done on trying to get transition networks to degrade gracefully rather that just fail completely.

The analog neural network parser we have shown here is only a toy network designed to clearly illustrate that the symbolic and neural network approaches are closely related. Fanty [38, 39] has demonstrated a method for constructing parse trees dynamically using a different connectionist approach and his approach works for any context-free grammar.

Santos [188] discusses another connectionist natural language parser with some learning ability. In a similar vein, Hanson and Kegl [60] used back-propagation on a three-layer network to do pattern completion experiments using sentences that range from 2 to 15 words. At this time, the neural networking methods have not been thoroughly researched. Symbol processing methods, such as the transition network grammar, are currently the only ones that can be considered useful methods.

10.2.4   Syntax is not Enough

There are still very many problems associated with syntax-directed interpretation of natural language. For instance, in Chapter 2 we had the examples:

> Time flies like an arrow.
> John ate up the street.

There are several possible ways through the network for each of these sentences corresponding to several different structures and meanings, and only an understanding of the world can produce the correct parse. Also, there are sentences such as the following pair where the parts of speech are identical but where they both have different parses:

> The boy ate the pasta with the fork.
> The boy ate the pasta with the sauce.

Only knowledge of the world enables people to realize that "with the fork" modifies "ate" and "with the sauce" modifies "pasta." In the following sentence you need to have specific knowledge about the circumstances involved in order to parse it correctly:

> John saw the girl with the telescope.

This could mean either that John used an optical instrument to see the girl or it could mean that John saw a girl and that girl was using a telescope. From examples like these, it has become clear that the approach that linguists proposed of first looking at syntax and then looking at semantics simply cannot work very well.

To deal with the problems involved in syntax-based methods, one approach has been to try to consider syntax and semantics more or less in parallel. In neural networking terms, you could graft on another network that looked at the possible meanings that the words in the sentence are trying to convey and let knowledge of the real world influence the parsing of the sentence. Of course, this is the approach of Waltz and Pollack discussed in Section 2.4. Their approach is an interesting development, but it remains to be seen how successful it can be. Conventional symbol processing AI has already tried something like this approach in the sequential symbol processing format. In this approach, you start at the beginning of the sentence and first you do some syntax, then you try to analyze what it may mean. Use this knowledge to do more syntax, then do more semantics. The most well-known version of this approach was the SHRDLU program of Winograd [261] in which there was a robot in a simple environment consisting of blocks of different shapes, sizes, and colors. The robot could understand commands to move blocks around and could answer questions as to why it did what it did. It created a lot of excitement at the time, however, as with all AI programs, the principles failed to "scale-up" and at least this particular method of considering syntax and semantics in parallel, has been abandoned.

10.3   Semantics-Based Methods

All hope for understanding natural language, however, does not rest on considering syntax alone or on considering syntax and semantics in parallel. Other innovative symbolic methods have been proposed. In this section we will look at two of them that take semantics as the most important ingredient, semantic grammar and conceptual dependency theory.

10.3.1   Semantic Grammar

\<sentence\>	→	\<person\> ate up \<food\>
		\<person\> ate \<place\>
		\<person\> broke \<object phrase\>
		\<animal\> broke \<object phrase\>
\<person\>	→	\<article\> \<indefinite person\>
		\<definite person\>
\<indefinite person\>	→	boy | girl | man | woman
\<definite person\>	→	John | Mary
\<article\>	→	a | the | some
\<animal\>	→	\<article\> \<indefinite animal\>
	→	\<specific animal\>
\<indefinite animal\>	→	dog | cat | hamster | parrot
\<specific animal\>	→	Heathcliffe | Fido | Spike
\<object phrase\>	→	\<article\> \<definite object\>
\<object\>	→	rock | bat | hammer | window | vase | glass
\<food\>	→	the cookies
		the candy
\<place\>	→	up the street
		down the block
		across town
		at McDonald's
		at Burger King

Figure 10.12: A simple semantic grammar uses semantically meaningful nonterminals rather than syntactic ones.

In a semantic grammar the nonterminals are semantically meaningful categories like \<person\> or \<place\> or \<food\> and so on, rather than abstract syntactic categories. An example of a small semantic grammar is shown in Figure 10.12. One of the problems with semantic grammars is that the grammar is not as highly condensed as a traditional syntactic grammar. On the other hand, because some knowledge about the world is included, many parsing problems like those represented by:

John ate up the street, and

John ate up the cookies

are neatly avoided. Semantic-grammar-based natural language systems are often used as front ends for database systems. In questions submitted to database systems, the basic need is to "pick off the parameters" that are needed in the search to answer the question. Systems using the semantic grammar concept use the transition network grammar approach to process sentences. A major example of the semantic grammar approach is the SOPHIE program by Brown and Burton [15] that converses with students about electronic circuits.

10.3.2 Conceptual Dependency Notation

Of the many other approaches proposed to get computers to understand natural language, one of the most prominent, carefully described, and thoroughly researched methods is by Schank and various students and associates of his. Schank proposes that the purpose of language is communication and any theory that fails to put communication first, such as those theories emphasizing syntax, is not very realistic. His approach is quite different from the approach proposed by linguists in the previous section where the assumption was that before you could try to understand the meaning of the sentence you must first diagram or parse the sentence. We will be examining some of Schank's methods for understanding sentences and then in the next section go on to look at his method for understanding simple stories using scripts. We cannot cover every part of this research in very great detail. The book, *The Cognitive Computer* [195], is a popular-level book that discusses Schank's theories and the book, *Inside Computer Understanding* [193] discusses the methods in much more detail. This latter book includes Lisp programs to illustrate the methods.

The first assumption of Schank's method is that when a sentence is understood by the human mind, it is mapped into some kind of language-independent notation. As evidence of this, consider what happens if Fred tells you that, "John is in love with Mary." It is highly unlikely, after a few minutes, or even after a few seconds, that you will have those *exact* words still stored in your mind. Instead, you remember the content and the meaning of the words so that if someone later asks you what Fred said, you may say that he said: "John loves Mary." This rewording of sentences without changing the underlying thoughts is quite common. Many comedies are based on a chain of such rewordings that ultimately *do* change the content of the message as it is passed from person to person. In Schank's method, both of the above sentences would be mapped into the same internal representation. The representation is independent of syntax and independent of the exact words that were originally used. Schank and his associates call their language-independent notation, "conceptual dependency" theory, or CD for short.

The first example of CD notation we look at is for the sentence:

Gina gave Dennis a cookie.

This sentence translates to an internal representation that in Prolog could be:

```
[atrans, [actor, [person, [name, gina]],
         [object, [cookie]],
         [from, [person, [name, gina]],
         [to, [person, [name, dennis]]]
```

Here, "atrans" is a "primitive action" meaning "a transfer of possession." The entries, actor, object, from, and to are *slots* that need to be filled in and the goal of processing a sentence is to fill in slots like these. The level of detail you find for the content of each slot will vary according to what might be needed in a specific application. It might suffice to let the actor slot be:

```
[person, [name, gina]]
```

or you may want to add the detail that Gina is female, like so:

```
[person, [name, gina],
        [gender, female]]
```

Schank and his associates maintain that there are in fact only a very few primitive actions that are necessary for the CD representation. One set consists of only these 11 primitive actions:

atrans	mtrans	speak	ingest
ptrans	mbuild	grasp	expel
propel	attend	move	

They also assert that many verbs, like "amaze," "hurt," "comfort," "please," or "annoy" only refer to "state changes" and so these verbs do not qualify as actions.

The meaning of most of the primitive actions listed above are obvious. One that is not obvious is ptrans. Ptrans stands for a physical transfer of location. An example of its use comes from the sentence:

Dennis went to Honolulu,

which in CD notation becomes:

```
[ptrans, [person, [name, dennis]],
        [object, [person, [name, dennis]]],
        [to, [city, [name, honolulu]]],
        [from, []]]
```

This sentence is a little strange because Dennis is both the object and the actor in this sentence. It could be interpreted as saying that "Dennis took it upon himself to move himself to Honolulu." Since from this sentence we do not know where Dennis started out from, the from slot is empty.

Mtrans stands for a transfer of information and is used for verbs like "read," "tell," "see," "forget," and so on. Attend stands for making use of something. The mbuild primitive is concerned with mental actions that concern the state of the mind. There is no particular need to use these primitives in what follows so we will not discuss them further.

Sometimes one sentence will produce more than one CD form, as for example, the sentence:

Margaret bought lemonade from Dennis

gives the two CD forms:

```
[atrans, [actor, [person, [name, margaret]]],
        [object, [money]],
        [from, [person, [name, margaret]]],
        [to, [person, [name, dennis]]]]
```

```
[atrans, [actor, [person, [name, dennis]]],
         [object, [lemonade]],
         [from, [person, [name, dennis]]],
         [to, [person, [name, margaret]]]]
```

Notice that whereas money is never explicitly mentioned in the sentence, it must neverthe-less be supplied to complete the details of the exchange.

Another aspect of CD theory is that a simple verb like "take" can be used in many different ways. These different ways can map into different primitive actions as shown by the following sentences:

Joey took a train.	(ptrans)
Joey took an aspirin.	(ingest)
Joey took a cookie.	(atrans)
Joey took my advice.	(mtrans)

We now want to look at a description of how the translation of sentences is done. This method is based on the idea that as a sentence is read, certain expectations come up in the human mind. As an example of this process we use the sentence:

Mary ate a cookie.

When we read the word, "Mary," the fact that Mary is a person comes to mind. We will also now have the expectation that a verb will be following. When we find the word, "ate," we recognize it as a verb. At this point we would have a CD form where the primitive action is ingest, the actor is Mary, but the object is, as yet, unknown:

```
[ingest, [actor, [person, [name, mary]]],
         [object, []]],
```

We can now expect that the object she ate will be some kind of food. The next word is "a," meaning that whatever it is she ate, it is something indefinite and not some very particular or definite item that we might have some reason to know about. We find the word, "cookie" and our expectation of food has been confirmed. Ultimately the CD form becomes:

```
[ingest, [actor, [person, [name, mary]]],
         [object, [cookie, [ref, indef]]]]
```

This says that Mary ingested a cookie and the reference to the cookie was indefinite.

The name of the program that transforms English to CD form is called ELI, for English Language Interpreter. Again, we must say that it is impossible to cover the whole program here, however, the Schank group has produced a small version of ELI, called McELI (for Micro ELI). The original McELI was written in Lisp but our version of McELI will be done in Prolog.

The following are some simple global variables used in McELI:

stack,	a stack of instructions to perform
word,	the current word being processed
triggered,	a list that temporarily stores instructions
sentence,	the unprocessed part of the sentence
cd_form,	a temporary variable (and not the final answer)
predicates,	a list of adjectives for each noun
part_of_speech,	the most recent type of word or structure encountered
concept,	the final CD form for the whole sentence

These variables will receive values using a predicate called assign. For example, the following call of assign will set cd_form to the symbol, [person]:

```
assign(cd_form,[person]).
```

The way assign works it will store the value of cd_form like so:

```
cd_form([person]).
```

We can then look up the value of a variable like cd_form, by:

```
cd_form(X)
```

and then X will be instantiated to the current value of cd_form. Assign looks at its second argument and if it is already defined to be some value, assign uses this value. So for instance, if the statement is:

```
assign(concept,cd_form),
```

assign will assign concept the value of cd_form instead of the symbol, cd_form. Thus, the two assignments:

```
assign(cd_form,[person])
assign(concept,cd_form),
```

will give concept the value, [person].

The next important component of McELI is a dictionary that holds definitions of words. The definitions include what associations (assignments) are made when the word is encountered and what expectations are generated. Each entry will be in the following form:

```
d( <word>, packet([ [ <option 1> ],
                    [ <option 2> ],

                           .
                           .
                           .

                    [ <option n> ] ] ).
```

where <word> will be the particular word. The packet predicate contains a list of some actions that might be done. Each <option i> entry has the following form:

```
[ <test>, <action 1>, <action 2>, ...  , <action m> ]
```

In operation, the Prolog program will search down the list of options until it finds a test that evaluates to true. Then the actions in the remainder of the list following the test are executed. The actions can be arbitrary Prolog predicates. For an example of a dictionary entry, here is the dictionary definition for "Mary":

```
d(mary,packet([[true,
                assign(part_of_speech,noun_phrase),
                assign(cd_form,[person, [name, mary]])]])).
```

In this case, there is only one option in the list. This option has the test, "true," which is a special Prolog predicate that always evaluates to true. Therefore, this definition says that when the word "mary" is encountered, you should assign cd_form the value, [person, [name, mary]], and part_of_speech, the symbol, noun_phrase.

```
d(went,packet([[true,
                assign(part_of_speech,verb),
                assign(cd_form,[ptrans, [actor, go_var1],
                                        [object, go_var1],
                                        [to, go_var2],
                                        [from, go_var3]]),
                assign(go_var1,subject),
                assign(go_var2,[]),
                assign(go_var3,[]),
                packet([[word(to),
                         packet([[part_of_speech(noun_phrase),
                                  assign(go_var2,cd_form)]])],
                        [word(home),
                         assign(go_var2,[house])]])]])).
```

Figure 10.13: The definition of "went."

The definition for the word "went" is more complicated and it is shown in Figure 10.13. This definition says that when you find the word, "went," you assign the global variable, part_of_speech to be the symbol, verb, and the variable, cd_form to be the list:

```
[ptrans, [actor, go_var1],
         [object, go_var1],
         [to, go_var2],
         [from, go_var3]],
```

The variables, go_var1, go_var2, and go_var3 are variables that will take on values as the information becomes available. For instance, if we have the sentence, "Mary went to the store," then go_var1 will be assigned the value, [person, [name, mary]], and go_var2 will be assigned the value, [store]. Go_var3 will remain undefined. After the variable, cd_form is assigned the structure above, still more assignments will be made for the "went" dictionary entry. Go_var1 will be assigned the value of the variable, subject. Go_var2 and go_var3 will be assigned the value, []. The way the program is arranged, the final analysis of the sentence will be in the variable, concept, and it will be exactly this:

```
[ptrans,  [actor,  go_var1],
          [object,  go_var1],
          [to,   go_var2],
          [from,  go_var3]],
```

where the symbols, go_var1, go_var2, and go_var3 will still be present. Other processes assign values to these symbols and an additional set of steps can be used to produce a CD form with these symbols replaced by their values.

The final action in the went entry is another packet:

```
packet([[word(to),
         packet([[part_of_speech(noun_phrase),
                  assign(go_var2,cd_form)]])],
        [word(home),
         assign(go_var2,[house])]]).
```

When a packet instruction is encountered in processing, the packet is placed at the end of the list, triggered. The whole list of triggered items will be placed on the top of the stack before the next word is processed. When this particular packet is found on the top of the stack, the program first tests for the word, to. If the current value of word is "to," then the action following the test is performed. If the current word is not "to," then the program tests to see if it is the word, "home." If it is, the one action following this test is performed.

Figures 10.14 and 10.15 show the dictionary entries for various words. The first word is "sos," for start-of-sentence, a dummy word that is placed at the beginning of each sentence to initialize the process. The final entry is designed to return the empty list if the word is not found. A few of these definitions use the append predicate. Append takes its first two arguments, which should be lists, and runs them together to form its third argument, as in:

```
append([big],[red],[big,red]).
```

The definition of append is:

```
append([],L,L).
append([A|L1],L2,[A|L3])  :- append(L1,L2,L3).
```

In addition, Figure 10.16 gives most of the McELI Prolog program (with the remaining details left as an exercise).

The program basically does the following. The predicate, start, is given a sentence, like so:

```
start([mary,  went,  home]).
```

The start predicate adds the dummy word, sos, to the sentence and calls process to process the whole sentence. Process first assigns the variable, word, the value of W, the first word in the list. Process then looks up the word, W, in the dictionary and finds the instructions associated with the word. The instructions are placed on the top of the stack by the statements:

```
stack(S),
```

```
d(sos,packet([[true,
                assign(predicates,[]),
                packet([[part_of_speech(noun_phrase),
                        assign(subject,cd_form),
                        packet([[part_of_speech(verb),
                                assign(concept,cd_form)]])]])]])])).
d(mary,packet([[true,
                assign(part_of_speech,noun_phrase),
                assign(cd_form,[person, [name, mary]])]])).
d(she,packet([[true,
                assign(part_of_speech,noun_phrase),
                assign(cd_form,[person])]])).
d(went,packet([[true,
                assign(part_of_speech,verb),
                assign(cd_form,[ptrans, [actor, go_var1],
                                        [object, go_var1],
                                        [to, go_var2],
                                        [from, go_var3]]),
                assign(go_var1,subject),
                assign(go_var2,[]),
                assign(go_var3,[]),
                packet([[word(to),
                        packet([[part_of_speech(noun_phrase),
                                assign(go_var2,cd_form)]])],
                        [word(home),
                        assign(go_var2,[house])]])]])).
d(got,packet([[true,
                assign(part_of_speech,verb),
                assign(cd_form,[atrans, [actor, get_var1],
                                        [object, get_var2],
                                        [to, get_var1],
                                        [from, get_var3]]),
                assign(get_var1,subject),
                assign(get_var2,[]),
                assign(get_var3,[]),
                packet([[part_of_speech(noun_phrase),
                        assign(get_var2,cd_form)]])]])).
d(a,packet([[part_of_speech(noun),
                assign(part_of_speech,noun_phrase),
                cd_form(F),
                predicates(P),
                append(F,P,New_cd_form),
                assign(cd_form,New_cd_form),
                assign(predicates,[])]])).
```

Figure 10.14: Some definitions for words used in the Prolog version of McELI.

```
d(the,packet([[part_of_speech(noun),
               assign(part_of_speech,noun_phrase),
               cd_form(F),
               predicates(P),
               append(F,P,New_cd_form),
               assign(cd_form,New_cd_form),
               assign(predicates,[])]]])).
d(doll,packet([[true,
                assign(part_of_speech,noun),
                assign(cd_form,[doll])]]])).
d(ball,packet([[true,
                assign(part_of_speech,noun),
                assign(cd_form,[ball])]]])).
d(store,packet([[true,
                 assign(part_of_speech,noun),
                 assign(cd_form,[store])]]])).
d(_,[]).
```

Figure 10.15: More definitions for words used in the Prolog McELI.

```
start(Sentence) :- assign(stack,[]),
                   process([sos|Sentence]).

process([]) :- write('all finished'), nl.

process([W|Rest]) :- assign(word,W),
                     d(W,Packet),
                     stack(S),
                     gluon(Packet,S,Newstack),
                     assign(stack,Newstack),
                     checkstack,
                     process(Rest).

checkstack :- stack(S),
              assign(triggered,[]),
              check(S,Rest_of_stack),
              triggered(T),
              append(T,Rest_of_stack,Newstack),
              assign(stack,Newstack).
```

Figure 10.16: The main components of the Prolog version of McELI.

```
                    gluon(Packet,S,Newstack),
                    assign(stack,Newstack),
```

The stack predicate instantiates S to be the value of the stack. Gluon takes the new instruction packet, Packet, for the word and the existing stack, S, and forms a new value for the stack, called Newstack. Newstack is then stored away as the new value for the stack.

Now a call to checkstack will have the program start checking the instruction packets on the stack. Every one that succeeds will be removed from the stack. Some packets may cause smaller packets inside the main packet to be added to the list, triggered. Packets on the stack are examined until one packet fails to have any of its tests succeed. At this point, the variable Rest_of_stack has the failed packet and all the rest of the packets on the stack. Append and assign produce a new stack with the triggered list on top. Control is returned to the last statement in process: process(Rest), where the rest of the words in the sentence are processed.

We will now go through the process of interpreting the sentence:

Mary went home.

The first word encountered will be the dummy word, sos. Its packet is transferred to the top of the stack, giving:

```
packet([[true,
         assign(predicates,[]),
         packet([[part_of_speech(noun_phrase),
                  assign(subject,cd_form),
                  packet([[part_of_speech(verb),
                           assign(concept,cd_form)]])]])]])
```

Now, in examining the stack, we find this packet has an option with a test that evaluates to true, so the list of actions following the test are performed. First, predicates is assigned the value, [], then the following packet is placed on the list, triggered:

```
packet([[part_of_speech(noun_phrase),
         assign(subject,cd_form),
         packet([[part_of_speech(verb),
                  assign(concept,cd_form)]])]])
```

The program now goes on to try to execute more packets on the stack but none are found because the stack is empty. Next the contents of the list, triggered, are placed on the top of the stack and then the next word is processed.

The next word is mary. The definition of mary is found in the dictionary and placed on the top of the stack. The stack is now examined and the following actions take place. The variable, part_of_speech is assigned the symbol, noun_phrase. The variable, cd_form is assigned the list, [person, [name, mary]]. The program examines the next packet on the top of the stack. The test is: part_of_speech(noun_phrase), which is true, so subject is assigned the value of cd_form, or [person, [name, mary]]. The following packet goes on the list, triggered:

```
packet([[part_of_speech(verb),
         assign(concept,cd_form)]])]])
```

Now, looking at the top of the stack, it is empty so this one packet is placed on the stack. Process works on the next word.

The next word is went. Its dictionary entry is placed on the stack and then these instructions are executed:

```
assign(part_of_speech,verb),
assign(cd_form,[ptrans, [actor, go_var1],
                        [object, go_var1],
                        [to, go_var2],
                        [from, go_var3]]),
assign(go_var1,subject),
assign(go_var2,[]),
assign(go_var3,[]),
```

The following packet is placed on triggered:

```
packet([[word(to),
         packet([[part_of_speech(noun_phrase),
                  assign(go_var2,cd_form)]])],
        [word(home),
         assign(go_var2,[house])]])
```

The next packet on the stack is:

```
packet([[part_of_speech(verb),
         assign(concept,cd_form)]])]])
```

Since part_of_speech(verb) is true, concept is assigned the value, cd_form. The packet on triggered is moved to the stack and the next word is processed.

The next word is home. It is not found in the dictionary and the empty list is returned, but it is not added to the stack. The one packet on the stack is examined. The first test in it is for the word, to, but this fails. The second test is for the word, home, and this succeeds so go_var2 is assigned the value, [house]. Now, with no more words to process, the program ends.

If we now take the value of concept and substitute in the values of its variables we have:

```
[ptrans, [actor, [person, [name, mary]]],
         [object, [person, [name, mary]]],
         [to, [house]].
```

The larger version of the program, ELI, is capable of processing many more sophisticated sentences. Among the differences between ELI and Micro ELI is that McELI uses a stack to store expectations but ELI uses a list. This arrangement leaves ELI with the problem of removing expectations that fail so that they do not get in the way of later processing.

10.4   Scripts and Short Stories

In the preceding sections we have covered some symbolic methods for processing sentences. We now turn to the processing of short stories. The program we will look at that understands short stories is called SAM for Script Applier Mechanism and it comes from Schank's research group. First we will look at a simpler version based on a small version of SAM called McSAM (for Micro SAM). In this version, there will be only one event available, going to the store. Here is an example of this type of story:

> Mary went to the store after school.
> She got cookies.
> She went home.

In the last section, when the single sentence, "Margaret bought some lemonade from Dennis," was translated to CD form, recall how it was necessary to add the fact that there was also a transfer of money taking place. In this shopping story, some other missing details need to be brought in by the computer program. Among them are the facts that the store had cookies to sell, Mary wanted some, she picked them up, she gave some money to the store and the store gave the cookies to Mary. Other items of the story that would be brought to a person's mind would be that Mary went to the kind of a store that sold items like cookies, the money came out of her pocket and went into a cash register, she may have had to wait in a line at the cashier's stand, and so on. The little script we will use will not add this many details, it will have only the following understanding of the story:

> Mary went to the store from school.
> Mary picked up the cookies.
> The store gave the cookies to Mary.
> Mary gave some money to the store.
> Mary went home from the store.

The script in Figure 10.17 can add these details and the notation is pretty straightforward. First, the props entry lists variables used in the script that need to be filled in. In this script they will be the object bought and the money used to buy the object. The entry for locations lists the various scene locations in the script. The actor symbols are `shopper` and `store_name`.

Following the variables, we have a list of actions that may be specified in the script. When the program gets the sentence, "Mary went to the store from school," suppose the resulting CD form is:

```
[ptrans, [actor, [person, [name, mary],
                          [gender, female]],
         [from, [school]],
         [to, [store]]]
```

The script applier program will take this form and try to match it against actions in the script. The first action that can be matched is the first CD form where the person entry will match with shopper and so all instances of shopper in the script can be replaced with the entry:

```
script(shopping,
       props(object_bought, money),
       locations(start_loc, store_name, finish_loc),
       actors(shopper, store_name),
       actions([[ptrans, [actor, shopper],
                         [object, shopper],
                         [from, start_loc],
                         [to, store_name]],

                [ptrans, [actor, shopper],
                         [object, object_bought],
                         [from, store_name],
                         [to, shopper]],

                [atrans, [actor, store_name],
                         [object, object_bought],
                         [from, store_name],
                         [to, shopper]],

                [atrans, [actor, shopper],
                         [object, money],
                         [from, shopper],
                         [to, store_name]],

                [ptrans, [actor, shopper],
                         [object, shopper],
                         [from, store_name],
                         [to, finish_loc]]]])).
```

Figure 10.17: A script in Prolog to handle shopping.

```
[person, [name, mary],
         [gender, female]]
```

Start_loc will be [school] and store_name will be [store].

Next, when the program reads "She got cookies," the CD form for this will be:

```
[atrans, [actor, [person, [gender, female]],
         [object, [cookies]],
         [from, []],
         [to, []]]
```

The program searches down the list of actions and finds that this matches:

```
[atrans, [actor, [person, [name, mary],
                          [gender, female]],
         [object, object_bought],
         [from, [store]],
         [to, [person]]
```

All instances of object_bought found in the script can be replaced by [cookies]. Finally, when the program reads "She went home," it finds that this closely matches the last CD form in the script.

It is interesting to consider what would have happened if, instead of the second sentence being, "She got cookies," it was instead:

John got cookies.

What happens is that the person in the script has already been identified as Mary, so trying to match "John got cookies" against the particular version of the script being filled in would cause the program to fail. Also, it is important to note what would happen if the second sentence of the story turned out to be:

She worked behind the counter.

This event is not in the shopping script, so again, the system would fail. In this particular case, a more sophisticated program with more scripts available could go looking through them to find another script that matched more closely. SAM is capable of doing this. It keeps a rating as to how strongly the events in the story match the contents of each likely script. SAM is also more complicated in that for each script there are optional tracks that contain possible variations on the activities that can happen within a script. So, in a "take a plane" script there are the variations about buying a ticket at the airport, checking or not checking baggage, waiting at the end of the flight for baggage, or not waiting for baggage when all the actor in the script had was a carry-on piece of luggage, and so on.

We will now take a look at an example of some of what SAM did with a specific story. The story (from [194]) is:

> Sunday morning Enver Hoxha, the Premier of Albania, and Mrs. Hoxha arrived in Peking at the invitation of Communist China. The Albanian party was welcomed at Peking Airport by Foreign Minister Huang. Chairman Hua and

Mr. Hoxha discussed economic relations between China and Albania for three hours.

The first sentence translates to quite a complicated CD form. Below we show only the two larger parts:

```
((ACTOR TMP32 <=> (*PTRANS*) OBJECT TMP32
                TO (*INSIDE* PART (#POLITY POLTYPE (*MUNIC*)
                                            POLNAME (PEKING)))
                FROM (NIL) INST (NIL))

TMP32 = (#GROUP MEMBER (#PERSON GENDER (*FEM*)
                                LASTNAME (HOXHA))
                MEMBER (#PERSON GENDER (*MASC*)
                                FIRSTNAME (ENVER)
                                LASTNAME (HOXHA)
                                TITLE (PREMIER)
                                POLITY (#POLITY POLNAME (ALBANIA)
                                                POLTYPE
(*NATION*))
                                REF(DEF)))
```

The first of these lists translates to the fact that some person or group identified with the variable TMP32 PTRANSed themselves to the municipality of Peking. The exact composition of the group TMP32 is described in the second list, it is a group whose first member is a female with the last name Hoxha and whose second member is a male with the first name, last name, title, and political affiliation shown. Note that in the first list the place where the group came from is NIL because the English language analyzer does not make inferences about such things and all it can report is that there is nothing in the sentence that explicitly identifies where they came from. Notice too, that the item, INST, is also NIL because the instrument that brought them to Peking is also unknown at this point.

The CD form of the sentence now gets passed on to SAM. This version of SAM has been programmed with knowledge of car accidents, train wrecks, and VIP visits. SAM looks at the given sentence and tries to match it against patterns in each script. Naturally, it finds the VIP visit is the closest match. Using the VIP script, SAM continues on to look for typical aspects of a VIP visit in the CD forms. One item to look for is the identity of the instrument that moved the group to Peking, but SAM does not find any instrument. If, by the time the story is finished, there is no indication of how the group was moved, the script contains directions to assume it was by plane. In this case, however, SAM will go on to find the sentence that says that the group was welcomed at the airport in Peking. When SAM sees this, it goes on to reason that a group that is welcomed at an airport must have flown in by plane so there is no need to guess that the group was moved by plane and so the INST entry can be filled in. Other minor inferences are also made by SAM at this point.

From the final sentence of the story more inferences are made. For instance, it is typical for VIP talks to happen in the city in which the group arrived, so SAM concludes that the site of the talks is Peking. Now that the story is over, the program goes on to fill in other blanks in the story with values that are typical for VIP visits. If SAM had not already found that the visitors arrived by plane it would now make that assumption.

The system is also capable of answering questions. For instance the question:

```
Who went to China?
```

translated to CD form is:

```
((ACTOR TMP8 <=> (*PTRANS*) OBJECT TMP8
      TO (*PROX* PART (#POLITY POLTYPE (*NATION*)
                                    POLNAME (CHINA)))
      FROM (NIL) INST (NIL))
  MODE(MOD0) TIME(TIM2))

TMP8 = (*?*)
```

This says that the actor, TMP8, is the unknown, (*?*), and that TMP8 was PTRANSed to China. From here it is a relatively simple matter to do pattern matching, much as Prolog would do, against the facts in the filled-in script. The program finds TMP8 to be:

```
PREMIER ENVER HOXHA AND MRS. HOXHA
```

As another example, the question:

```
How did they get to China?
```

becomes in CD form:

```
((ACTOR TMP103 <=> (*PTRANS*) OBJECT TMP103
              TO (*PROX* PART (#POLITY POLTYPE (*NATION*)
                                            POLNAME (CHINA)))
              FROM (NIL) INST (*?*))
  MODE(MOD1) TIME(TIM5))

TMP103 = (#GROUP MEMBER (#PERSON) REF(DEF))
```

The item to report back is again flagged as (*?*). This CD form is again matched against details of the story. The system reports in English:

```
MRS. HOXHA AND PREMIER ENVER HOXHA FLEW TO COMMUNIST CHINA.
```

Schank and his associates went on to do other related kinds of AI programs. PAM (Plan Applier Mechanism) is a story understander that tries to understand stories by estimating what the goals of the characters in the story may be. Some other projects were POLITICS, FRUMP, IPP, and BORIS. Some of these later programs also integrated into one program both the natural language understanding capability of ELI and the script applying ability of SAM and this resulted in more efficient programs. Another program, called TALE-SPIN generated short stories. For more details of these projects, see [194] and/or [195].

Schank believes that the current bottleneck in trying to improve upon these kinds of programs is the problem of making the system learn facts about the world just by reading a story. It is this shortcoming in these programs that led Schank to the idea of case-based reasoning. Indeed, Schank also regards scripts as implemented in SAM as unrealistic and he has moved on to research ways that scriptlike processing can be done by organizing memories (see [194]).

10.5   A Neural-Network-Based Approach

In this section we will look at one back-propagation-based approach to understanding natural language. Following sections will look at more recent and more sophisticated systems. In effect, there is a neural analog available for all the traditional symbolic methods we have already looked at.

The program we will look at is a small experimental one produced by McClelland and Kawamoto [108]. The network they used was a back-propagation-like network, but for our purposes, their most important results can be demonstrated just as easily using an ordinary two-layer back-propagation network. The experiment shows that some of the kinds of sentence processing that have traditionally been done by recursive symbol processing methods can also be accomplished by neural networks.

McClelland and Kawamoto's program worked with "sentences" such as the following ones that were stripped of the words "the" and "with":

woman ate cheese	(The woman ate the cheese)
woman ate chicken pasta	(The woman ate the chicken with the pasta)
girl ate pasta spoon	(The girl ate the pasta with the spoon)
woman broke window	(The woman broke the window)
boy broke window hammer	(The boy broke the window with the hammer)
girl hit boy	(The girl hit the boy)
man moved	(The man moved)
woman moved plate	(The woman moved the plate)

The major goal of their experiments was to have a network look at the stripped down sentences and determine:

> 1) the 'agent' in the sentence,
> 2) the 'patient' in the sentence,
> 3) the 'instrument' in the sentence, and
> 4) the 'modifier' in the sentence.

Some of these fields may be empty. It is easy to describe these four fields with a few examples. In the sentence:

> boy ate pasta fork (The boy ate the pasta with the fork),

boy is the agent, pasta is the patient, fork is the instrument used in the action of eating, and there is no modifier. In the sentence:

> boy ate pasta cheese (The boy ate the pasta with the cheese),

there is no instrument, but the modifier of pasta is cheese. In:

> vase broke (The vase broke),

the patient is the vase and the agent responsible for the breaking is unspecified. In:

> girl moved (The girl moved),

the agent is girl and the patient is girl. The terms agent, patient, instrument, and modifier come from a method proposed for representing sentences, called a *case grammar*. With a case grammar the goal of parsing is to produce a notation describing the sentence similar to the conceptual dependency notation used by Schank.

bb-bat	01010	01100	01000	10101	00000
boy	10101	00010	00010	10100	00010
paperwt	01010	01100	10001	00100	00100
cheese	01100	01100	00100	11010	00000
chicken	01100	01100	10000	11010	00000
food	01100	01100	10000	11010	00000
fork	01010	01100	01001	00100	01000
girl	10100	10010	00010	10100	00010
hammer	01010	01100	01000	10100	10000
man	10101	00001	00010	10100	00010
woman	10100	10001	00010	10100	00010
plate	01010	01100	00100	11000	00100
rock	01010	01100	00011	00100	00001
potato	01100	01100	10000	11010	00000
pasta	01100	01100	01000	11010	00000
spoon	01010	01100	01000	10100	01000
carrot	01010	01100	01001	01010	00000
vase	01010	01100	01000	11000	00100
window	01010	01010	00101	01000	00100
dog	01101	00010	00010	10100	00010
sheep	01100	10010	00010	11000	00010

Figure 10.18: The coding for the words used in the network. Bb-bat is a baseball bat.

The nouns and verbs used in McClelland and Kawamoto's experiment were coded as sets of vectors with 25 microfeatures each. The vectors for some of the nouns used are shown in Figure 10.18 and the meaning of each bit is listed in Figure 10.19. The codings for the five verbs used by McClelland and Kawamoto included 25 microfeatures, but here, we will use only a three-bit pattern for each of three verbs:

ate	100
broke	010
hit	001

We will have 78 input units in four groups that will have the following meanings:

units 1–25:	a bit pattern for the subject noun,
units 26–28:	a bit pattern for the verb,
units 29–53:	a bit pattern for the direct object,
units 54–78:	a bit pattern for the "with" noun-phrase.

The output units will be divided into four sets of 25 units each. Set 1 represents the agent (if any) that is doing something. Set 2 represents the patient, the person or thing to which

1)   object is human
2)   object is nonhuman
3)   object is soft
4)   object is hard
5)   object is male
6)   object is female
7)   object is neuter
8)   volume is small
9)   volume is medium
10)   volume is large
11)   form is compact
12)   form is roughly one-dimensional (baseball bat, carrot, and so on)
13)   form is roughly two-dimensional (curtain, window, and so on)
14)   form is roughly three-dimensional (people, animals, and so on)
15)   object is pointed
16)   object is rounded
17)   object is fragile
18)   object is unbreakable
19)   object is food
20)   object is toy
21)   object is tool
22)   object is utensil
23)   object is furniture
24)   object is animate
25)   object is inanimate

Figure 10.19: The meanings of the bits used in Figure 10.18.

something was done. Set 3 represents the instrument (if any) that was involved in the event. Set 4 represents a modifier field, a field that modifies the patient:

units 1–25:   the agent
units 26–50:   the patient
units 51–75:   the instrument
units 76–100: the modifier

The patterns in Figure 10.20 represent the patterns that were put on the input units. The patterns on the right represent the answer that the input pattern should produce on the output units. A '-' means the field is empty and in the network the field is a pattern of all zeros.

We now look a set of test patterns that was used to see how effectively the system had learned to generalize. In the results described here, the two-layer network was trained until all the output units were within 0.1 of the desired value. The first set of values is the correct response and the second set is the network's response. A value of '1' in the network's response indicates the output value was greater than 0.9. A value of '˄' indicates the response was between 0.5 and 0.9. A value of 'v' indicates the response was between 0.1 and 0.5. Responses below 0.1 are indicated by a '0.' In the first test pattern, the sentence is: "boy ate." Implied in this is the fact that the boy ate some kind of food. The network has been trained to recognize this and it supplies the default pattern for food over the patient output units:

```
boy                  food                        −                        −
1010100010000101010000010 0110001100100001101000000 0000000000000000000000000 0000000000000000000000000
1010100010000101010000010 0110001100100001101000000 0000000000000000000000000 0000000000000000000000000
```

In the test sentence: "boy ate carrot pasta," some errors in the answer occur when the network fails to flag pasta as being one-dimensional, fragile, and food and it flags carrot as rounded. (McClelland and Kawamoto's network gives these kinds of errors as well. The purpose of their program was not to produce a perfect solution, but only to investigate possibilities and potential.) However, the network does have pasta modifying carrot:

```
boy                  carrot                      −                        pasta
1010100010000101010000010 0101001100010010101000000 0000000000000000000000000 0110001100010001101000000
1010100010000101010000010 0101001100v100˄˄101000000 0000000000000000000000000 011000˄100000001v0v000000
```

In: "man ate chicken fork," the network fails to flag chicken as compact, but it does have the fork as an instrument:

```
man                  chicken                     fork                     −
1010100001000101010000010 0110001100100001101000000 0101001100010010010001000 0000000000000000000000000
1010100001000101010000010 01100v1100v00v01101000000 0101001100010010010001000 0000000000000000000000000
```

In: "dog broke plate," the network is correct, except it fails to flag dog as nonhuman. Notice that in the training data, the two examples it had were: "woman broke window" and

man ate - -	man food - -
girl ate - -	girl food - -
sheep ate - -	sheep food - -
dog ate - -	dog food - -
woman ate pasta -	woman pasta - -
woman ate cheese -	woman cheese - -
man ate pasta chicken	man pasta - chicken
girl ate pasta cheese	girl pasta - cheese
woman ate chicken pasta	woman chicken - pasta
boy ate chicken fork	boy chicken fork -
man ate pasta fork	man pasta fork -
girl ate pasta spoon	girl pasta spoon -
vase broke - -	- vase - -
window broke - -	- window - -
woman broke window -	woman window - -
boy broke plate -	boy plate - -
paperwt broke vase -	- vase paperwt -
bb-bat broke vase -	- vase bb-bat -
rock broke window -	- window rock -
bb-bat broke plate -	- plate bb-bat -
girl broke window rock	girl window rock -
woman broke plate hammer	woman plate hammer -
man broke window bb-bat	man window bb-bat -
boy broke plate hammer	boy plate hammer -
boy moved - -	boy boy - -
dog moved - -	dog dog - -
man moved - -	man man - -
woman moved - -	woman woman - -
woman moved plate -	woman plate - -
man moved hammer -	man hammer - -
girl moved paperwt -	girl paperwt - -
woman moved vase -	woman vase - -
girl moved pasta -	girl pasta - -
girl moved carrot -	girl carrot - -

Figure 10.20: The input sentences on the left and the output patterns on the right that will be trained into a 78-100 back-propagation network.

"boy broke plate" and it did not have an example of a nonhuman animate object breaking anything:

```
dog                     plate                    -                    -
0110100010000101010000010 0101001100001001100000100 0000000000000000000000000 0000000000000000000000000
0010100010000101010000010 0101001100001001100000100 0000000000000000000000000 0000000000000000000000000
```

In: "girl broke vase hammer," the network is correct:

```
girl                    vase                    hammer                    -
1010010010000101010000010 0101001100010001100000100 0101001100010001010010000 0000000000000000000000000
1010010010000101010000010 0101001100010001100000100 0101001100010001010010000 0000000000000000000000000
```

In: "hammer broke vase," the network correctly makes hammer the instrument and leaves the agent blank but makes the instrument a toy rather than a tool:

```
-                       vase                    hammer                    -
0000000000000000000000000 0101001100010001100000100 0101001100010001010010000 0000000000000000000000000
0000000000000000000000000 0101001100010001100000100 01010011000`000`010`00000 0000000000000000000000000
```

In: "plate broke," there is no agent or instrument, but it makes four mistakes in the patient bits:

```
-                       plate                    -                    -
0000000000000000000000000 0101001100001001100000100 0000000000000000000000000 0000000000000000000000000
0000000000000000000000000 0101001vv00`10`v100000100 0000000000000000000000000 0000000000000000000000000
```

In: "girl moved," the network makes girl the agent and the patient:

```
girl                    girl                     -                    -
1010010010000101010000010 1010010010000101010000010 0000000000000000000000000 0000000000000000000000000
1010010010000101010000010 10100`001000101010000010 0000000000000000000000000 0000000000000000000000000
```

In: "man moved window," the patient that moved has all the correct characteristics of window:

```
man                     window                   -                    -
1010100001000101010000010 0101001010001010100000100 0000000000000000000000000 0000000000000000000000000
1010100001000101010000010 0101001010001010100000100 0000000000000000000000000 0000000000000000000000000
```

Another capability that we would expect to see is, that given an incomplete pattern, the network could fill it in sensibly. For instance, take the sentence, "The xxx ate the chicken with the spoon," where xxx is either badly garbled or some completely unknown word. When people encounter such a situation, they can still predict certain properties of xxx. In the example below, we let xxx be a string of zeros and the network still produces some responses on the agent units. These units indicate that xxx is human, soft, female, medium, 3D, rounded, unbreakable, and animate:

```
-                       chicken                  spoon                    -
0000000000000000000000000 0110001100010001101000000 0101001100010001010001000 0000000000000000000000000
1010v`00`v000101010000010 0110011100vv0v0110`000000 0101001100100`1010001000 0vv0000v000`000v000000000
```

Another such garbled sentence might be: "The girl broke the xxx with the hammer." In this case, the network flags xxx as being nonhuman, soft, neuter, medium, rounded, fragile, but unexpectedly, it makes the patient animate:

```
girl                            --                      hammer                          --
1010010010000101010000010 00000000000000000000000000 0101001100010001010010000 00000000000000000000000000
1010010010000101010000010 0`100v`v10000v011000000`0  0101001100010001010010000 00000000000000000000000000
```

The McClelland and Kawamoto system has other interesting properties as well but we will neglect them here.

10.6   Defining Words by the Way they are Used

These has been an interesting follow-up experiment to the McClelland and Kawamoto system done by Miikkulainen and Dyer [112, 113, 114]. Their system uses the same sentences, but it lets an extended back-propagation network develop its own internal representation for all of the words used as input. That is, the network is not given a fixed vector of micro-features for a concept such as boy. The system starts with random initial patterns for the words and, as the system is learning to produce the correct output, these patterns are allowed to change. The patterns for each word converge to a vector of values and similar concepts have similar vectors of values. We might say the network develops its own neural code for each concept. The system is called FGREP (Forming Global Representations with Extended back Propagation).

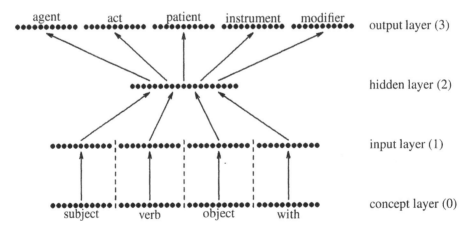

Figure 10.21: The general layout of the extended back-propagation network. Layers 0 and 1 are divided into four parts. The input layer has to develop 'neural codes' for each word represented on layer 0. The subject part of layer 0 only connects to the subject part of the input layer. The verb part of layer 0 only connects to the verb part of the input layer, and so on.

One FGREP network is shown in Figure 10.21. It consists of a standard three-layer back-propagation network with a fourth layer, layer 0, that comes before the input layer. The modifications of the weights between layer 0 and layer 1 are performed by a slightly different formula. The output layer is divided into five groups of units, for the agent, act, patient, instrument, and modifier. Layers 0 and 1 are divided into four parts, for the subject, verb, direct object, and the "with" noun-phrase. The group of layer 0 subject nodes only connects to the layer 1 subject nodes and *not* to all the other layer 1 nodes. The same is true for the other three groups of nodes.

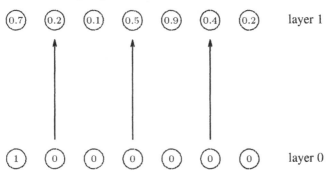

Figure 10.22: The bit pattern for a concept such as "boy" is on layer 0. Each of the concept patterns contains only one 1 bit and the rest are 0. The numbers of layer 1 nodes can vary, but there must be one layer 0 node for each word the system has to use. Every layer 0 node connects to every layer 1 node, but as is so often the case with larger networks, it is difficult to show every weight. The arrows represent this collection of weights.

Figure 10.22 shows more details of the part of the network from layer 0 to layer 1. In the system, each word (concept) is a bit pattern consisting of one 1 and the rest of the bits are 0. For example, bit 1 may mean "boy," bit 2 may mean "girl," bit 10 might mean "vase," and so forth. There needs to be one position for each concept that the system has to learn about. During training, the pattern on the concept layer activates the input layer. Whatever pattern appears on the subject units of layer 1 is now copied to the agent units of the output layer and it becomes the current target pattern for the network to produce. Likewise, whatever pattern appears on the "with" units of input layer is copied to the instrument or modifier units as appropriate, and so on.

The activation of an input unit, o_j, is given by:

$$o_j = \sum_i w_{ij} o_i,$$

but note that there will be only one nonzero term in this sum because there is only one nonzero unit in the concept layer.

The errors in the network are passed back in the usual way, except when the error is passed back to the input layer. Here, the following formula is used instead:

$$\delta_i = \sum_j \delta_j w_{ij},$$

where δ_i is the error for the ith unit of layer 1, δ_j is the error for the jth unit of layer 2, and w_{ij} is the weight between the units. This formula is similar to what back-propagation uses, but it is simpler because the activation function of input layer units will be equal to the

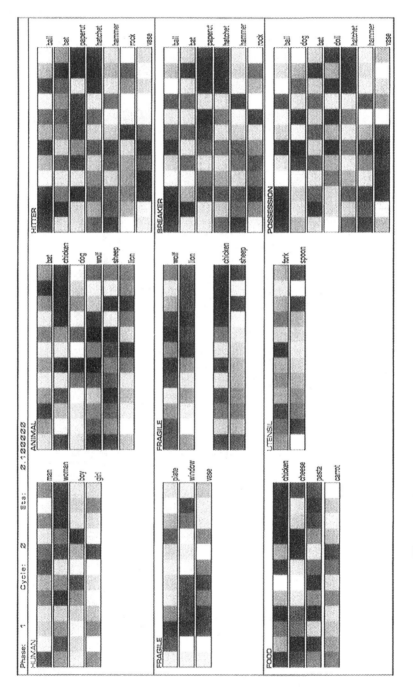

Figure 10.23: The values that words took on. The system starts out with random values.

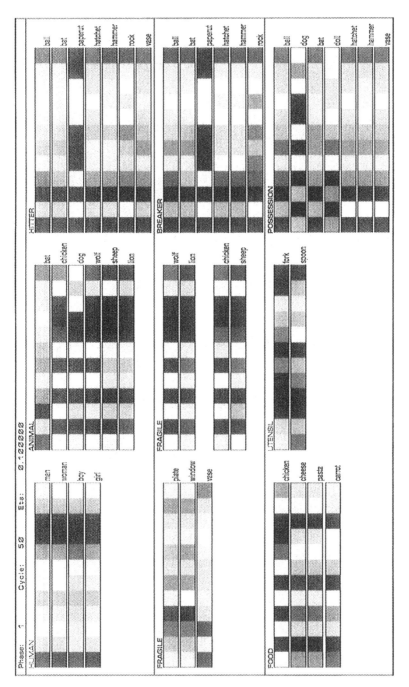

Figure 10.24: The values that words took on. The network has settled on a consistent representation.

unit's total input and then the derivative of this function is 1. The new weight between layer 0 and layer 1 for concept c is computed by:

$$w_{ci}(t+1) = max[0, min[1, w_{ci}(t) + \delta_i w_{ci}(t)]].$$

Notice that because there is only one 1 in layer 0, (bit c), the weights, w_{ci}, end up becoming the input representation of the concept.

Figures 10.23 and 10.24 show the values of weights in the range 0 to 1 that the words took on for one particular run. In Figure 10.23, the initial random weights are shown. In Figure 10.24, the weights have converged to stable values after training. Notice that similar concepts take on similar values. Keep in mind that within each list there is no particular meaning to an individual entry. This is in sharp contrast to the microfeature coding of McClelland and Kawamoto where each bit does stand for something. The values in each vector here only take on meaning when they are used in combination with other patterns of vectors. Miikkulainen and Dyer report that the networks trained with this method seem to do a better job of handling unknown patterns than the Rumelhart and Kawamoto network. Also, the network is less sensitive to damage to its connections.

This experiment of Miikkulainen and Dyer is interesting because words end up taking on meanings that are defined by the way the words are used in the sentences. Every sentence in which a word is used contributes a part to the word's meaning. Also, a word's meaning can shift as new, different uses of the word occur. This approach of deriving the meaning of a word from specific instances in which it is used is radically different from the traditional AI approach of trying to put a fixed, condensed knowledge of words into a system. It is also interesting to notice that in this system, a symbol (word) is a vector of real values rather than just an integer as it is in standard symbolic theories.

10.7 A Recurrent Network for Sentences

It must be pretty obvious that inputting words to a network all at once, as in the methods of the last two sections, is not going to be a very good solution to understanding natural language. It also must be pretty obvious that connectionist-state networks that take words in sequentially should be more useful. Of course, such networks do work and we will now look at an example of one by St. John and McClelland [183, 184]. The sentences used in this system are a little more complicated than the ones used by McClelland and Kawamoto.

In their experiments to understand sentences, St. John and McClelland used a five-layer back-propagation network. The network is outlined in Figure 10.25. In this network, the object is to train the 100 units in the third layer to produce a CD-like or case grammarlike representation of an entire sentence. This representation is called the "sentence gestalt." As with all hidden layers, it is fairly difficult to figure out what any of the hidden units actually stand for. In order to find out if the network is extracting the right things about a sentence and in order to actually train it to extract the right things, this sentence gestalt is used as input together with the layer 3 probe units to produce answers on the output units. When a pattern goes on the probe units, in effect, the network is being asked a question about the sentence it has read. For example, suppose the network has sequentially read the sentence, "The pitcher threw the ball." The sentence gestalt units now have some kind of code that represents the important features of the sentence. If we want to know the agent

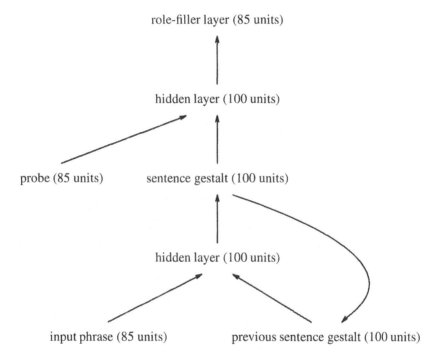

role-filler layer (85 units)

hidden layer (100 units)

probe (85 units)   sentence gestalt (100 units)

hidden layer (100 units)

input phrase (85 units)   previous sentence gestalt (100 units)

Figure 10.25: The general layout of the sentence understanding network. Phrases are input in layer 1 and probes to identify the roles and their values are input in the third layer. The value for the role appears on the output layer. The sentence gestalt units in the third layer gather the role-filler information and the values of these units are repeatedly copied to the 100 sentence gestalt units in layer 1.

in the sentence, then we place the code for agent on the probe units, then on the output units we get a code for agent/pitcher and the pitcher is a person, and so forth. If we want to know the patient, we put the code for patient on the probe units and the pattern, patient/ball appears on the output units. Likewise for the action, where the network responds with a code for "threw."

Training the network is done as follows. Suppose the network has received all the parts of the sentence one-at-a-time. Suppose the sentence again is: "The pitcher threw the ball." The goal is to have the sentence gestalt units store some kind of a code that represents an understanding of the whole sentence. To do this, the code for agent is placed on the probe units and the answer, agent/pitcher and person is placed on the output units. As in all back-propagation networks, the errors are now passed back.[1] Next, the code for patient is placed on the probe units and the code for patient/ball is placed on the output units. Errors are passed back. The same process occurs for the action/threw pair. This training scheme is a little more complicated than training standard back-propagation networks and there are some other unusual twists in the rest of the training scheme as well.

The input patterns are not individual words, but instead, each input is a pattern for a phrase. We will look at these patterns for phrases shortly. One of the unusual twists in the training scheme is that after the first phrase, the second phrase, the third phrase, and so on, the network goes through the entire training sequence we indicated above. So, if the network is working on learning "The pitcher threw the ball" and it has only seen the first phrase for "the pitcher," it is trained that the agent is the pitcher *and* the action is "threw" *and* the patient is "ball," despite the fact it has not even seen all these inputs yet. This may seem strange but it does make sense. One of the goals of this experiment was to get the network to anticipate, as people do, the possible meaning of the whole sentence given only the beginning of the sentence. With this training scheme, whenever the network sees "the pitcher," it will somewhat anticipate that the action will be "threw" and the patient will be "ball."

Next, we will look at a couple of the complexities involved in the representation of the input. The first of these is that the phrases were not just simply submitted to the network in order, they were submitted in order with an extra clue for each phrase to indicate the position of the phrase in the sentence. These positions are called preverbal, verbal, first-postverbal, and n-postverbal. In the sentence, "The ball was hit by someone with the bat in the park," "the park" is flagged as preverbal, "was hit" is flagged as verbal, "by someone" is first-postverbal, and "with the bat" and "in the park" are both flagged as n-postverbal. This strategy was used because it made it possible for the network to learn faster. St. John and McClelland report that the network could learn the task without these additional cues, but that the training time was much longer.

The input units do not code microfeatures of words as in the McClelland and Kawamoto experiment. For the input units there is one unit for each different word in the system's vocabulary. For example, again, given the sentence, "The ball was hit by someone with the

---

[1] In this network, however, a different measure for the error is used. This different measure is designed so that the output units register probabilities. This detail is not important for the rest of the description.

bat in the park," here are the phrases and the bits that will be set for each of the phrases:

the ball	preverbal, ball
was hit	verbal, was, hit
by someone	first-postverbal, by, someone
with the bat	n-postverbal, with, bat
in the park	n-postverbal, in, park

There are also ambiguous words in the input that need to be understood by the network. Among the input units there is a code for bat, but in the output units there are codes for bat (animal) and bat (baseball). When "bat" occurs as part of the input, the network needs to figure out what kind of bat is being talked about in the sentence. Some of the output units correspond to microfeatures like those in the McClelland and Kawamoto network.

After the usual large amount of training, the network learned the training data and also performed well with test sentences. It is interesting to look at some of the examples handled by the network. First, we look at the processing of the sentence, "The adult ate the steak with daintiness." The network's "state of mind" after each of the four phrases in this sentence are input is shown in Figure 10.26. At $t = 0$, all the activations are 0. After giving the network, "the adult," the number of predictions the network made ($t = 1$) is limited and no substantial predictions are made by the network. Among adults that the network knows about there is a bus driver and a teacher and both these units light up to a medium amount. Next, given the phrase, "ate," a few small predictions about the type of food come up. Steak, soup, and crackers all become slightly lit. Next, given "the steak," the network increases the activation of bus driver and decreases the activation of teacher because in the network's training experience, bus drivers are more apt to eat steak than teachers. Finally, when "with daintiness" is input, the network changes its mind and suspects that the adult was a teacher, since again, in its training experience, bus drivers eat with gusto while teachers eat with daintiness.

In some other sentences, the network makes sensible predictions as well. Given: "The pitcher hit the bat with the bat," the network makes the agent a person, the patient an animal, and the instrument a baseball bat. Given: "The schoolgirl spread something with a knife," the network suspects the something was jelly. Given: "The teacher kissed someone," the network suspects the patient is a male and possibly a child because in the network's experience, teachers are female and females only kiss males.

There are a couple of other systems using similar recurrent architectures. One simpler experiment was done by Elman [35]. In this experiment, Elman used short sentences that included pronouns and the network was able to do a good job of learning to whom the pronoun referred. Understanding pronoun references has been a difficult task for symbolic systems. In other related experiments, Allen [2] designed a network to answer short questions about a microworld that consists of just a few objects and in another experiment [3] used a similar network to understand pronoun references.

10.8 Neural-Based Scripts

The traditional AI idea of scripts has also been implemented with neural networks by Miikkulainen and Dyer. In their first project, DISPAR (from DIStributed PARa-

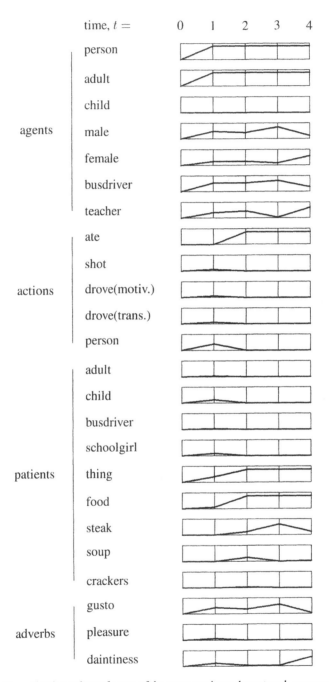

Figure 10.26: The activation values of some of the output units as the network processes the sentence, "The adult ate the steak with daintiness" as four phrases.

phraser), a set of networks reads in a single short story and adds obvious details to the story [115, 116, 118]. This project used only back-propagation networks and will be described below. In a more recent project, the program, DISCERN (from DIstributed SCript processing and Episodic memoRy Network), can deal with three types of short stories, fill in obvious details, and answer simple questions about a story. This project used back-propagation and another type of neural network and will not be described in this section. For details of this project see [117, 119, 120, 121].[2] Both projects are not as capable as the symbolic counterparts.

The DISPAR program could deal with a single story such as the following:

> John went to Leone's. John asked the waiter for lobster. John gave a large tip.

The system produced this paraphrase of the story:

> John went to Leone's. The waiter seated John. John asked the waiter for lobster. The waiter brought John the lobster. John ate the lobster. The lobster tasted good. John paid the waiter. John gave a large tip. John left Leone's.

In the training phase, the networks were given the three prototypical tracks of a restaurant script shown in Figure 10.27. There is not as much variety in these examples as you would find in a symbol processing implementation of a script. The types of food, the name of the customer, and the name of the restaurant can vary from story to story. Also, in the coffee shop track, if the food tasted good, then the customer will leave a large tip, but if it tasted bad, the customer will leave a small tip. So, if the system recognizes a coffee shop script and the story says the food is good, the system will infer that the tip was large, even if the story does not say so, and if the food is bad it will infer that the tip was small.

The program uses four networks as outlined in Figure 10.28 to take in a story and then paraphrase it. The program uses the FGREP method of representing words with an extra twist. Each word has a content part developed with the FGREP method and an ID part used for different actors, restaurant names, and foods. For the ID part, whatever the pattern in this portion of the vector is, it is simply copied over from input to output. If we use three units for the ID portion, then we could code the ID part of the person vector for John as (1,0,0), Mary as (0,1,0), Fred as (0.5,0.5,0), and so on. A sample is shown in Figure 10.29. The number of IDs that can be used is limited only by the resolution ability of the network.

The first network takes in words one-at-a-time and uses them to produce a *case-role* representation of the sentence, a representation where different segments of a vector store the "parameters" of the sentence. The network is a connectionist-state network that cycles the values of the hidden units back to the input layer. A sample of this network is shown in Figure 10.30. Suppose in the training sequence that the first sentence is: "John went to Leone's." When "John" is submitted to the input units, the target on the output units will have the agent portion coded as "John," but the act will be "went" and the location will be "Leone's," despite the fact that the network has not yet seen the entire sentence. This is the same approach as in the St. John and McClelland network, and again, the idea is that networks should be able to make humanlike predictions about what ideas will come next.

---

[2]More information, including an X11-based demo is available at: http://net.cs.utexas.edu/users/nn/-discern.html.

Fancy-restaurant track:

Customer went to fancy-restaurant.
Waiter seated customer.
Customer asked the waiter for fancy-food.
Waiter brought customer the fancy-food.
Customer ate the fancy-food.
The fancy-food tasted good.
Customer paid the waiter.
Customer gave a large tip.
Customer left fancy-restaurant.

Coffee-shop track:

Customer went to coffee-shop.
Customer seated customer.
Customer asked the waiter for coffee-shop-food.
Waiter brought the customer coffee-shop-food.
Customer ate the coffee-shop-food.
The coffee-shop-food tasted good/bad.
Customer paid the cashier.
Customer gave a large/small tip.
Customer left coffee-shop.

Fast-food track:

Customer went to fast-food-restaurant.
Customer asked the cashier for fast-food.
Cashier brought customer the fast-food.
Customer paid the cashier.
Customer seated customer.
Customer ate the fast-food.
The fast-food tasted bad.
Customer gave no tip.
Customer left fast-food-restaurant.

Figure 10.27: Three different variations on restaurant scripts.

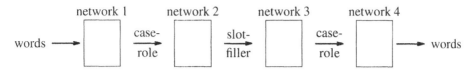

Figure 10.28: The series of four networks used in DISPAR. The words of each sentence go into network 1. When a sentence is complete, the network outputs a case-role representation of the sentence that goes into network 2. Network 2 collects all the case-role vectors, one for each sentence. When the last case-role representation is input, the result is the slot-filler representation, a complete coding of the entire story. Now, network 3 is used to output one case-role representation at a time. This goes into network 4 which outputs one word at a time.

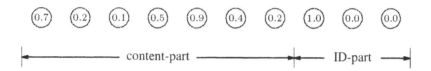

Figure 10.29: Each word is coded with a content-part derived from training and an ID-part that is assigned a unique value for each unique person or object in the story.

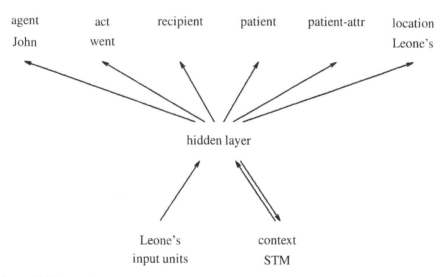

Figure 10.30: The first of four networks takes in words of a sentence one-at-a-time and forms a case-role representation. The output layer shows the representation for "John went to Leone's" after "Leone's" is input.

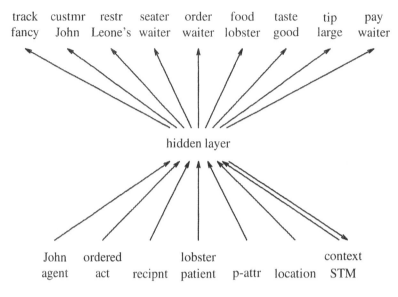

Figure 10.31: The case-role representations of each sentence are input to this network to form a slot-filler representation of the story. Here, the coding for "John ordered lobster," derived from the first network is input.

The second network sequentially takes in the case-role representation of each sentence to form a *slot-filler* representation, a vector where different segments store the "parameters" of the whole story. A sample of this is shown in Figure 10.31.

Once an entire story is in the second network, the process can be reversed in the next two networks. The third network is trained to take the slot-filler representation as input and output a series of case-role representations. Each case-role representation is used as input for a fourth network that sequentially outputs single words. A sample of the third network is shown in Figure 10.32 and a sample of the fourth network is shown in Figure 10.33.

10.9   Learning the Past Tense of Verbs

In this section we look at a very important and controversial experimental network that learns the past tenses of verbs and we also look briefly at a traditional rule-based model of the same process. The network was done by Rumelhart and McClelland [181] and it is designed to try to produce some of the same mistakes children make in learning the past tense of verbs. This is an important experiment because Rumelhart and McClelland go on to argue that their model indicates that language may be processed realistically by connectionist pattern recognition networks and that traditional symbol processing rule-based theories are not necessary to explain language behavior. Proponents of the traditional viewpoint of language, notably Prince and Pinker [152] disagree strongly with their work and conclusions and the result is that this work has become a big issue in the debate between Connectionism and Classical Symbol Processing.

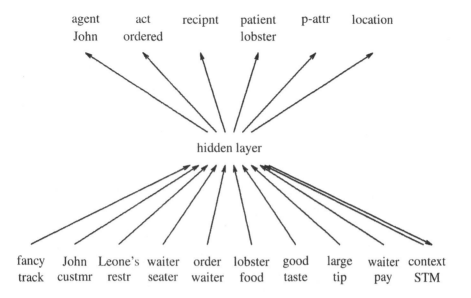

Figure 10.32: Given the slot-filler representation of the story, this network sequentially outputs case-role representations of a sentence. Each case-role representation is passed to a fourth network in Figure 10.33 that sequentially outputs words.

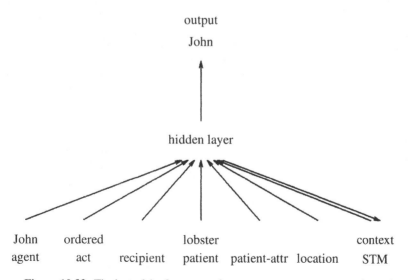

Figure 10.33: The last of the four networks outputs a sentence one word at a time.

10.9.1   Over-Regularization

First we want to describe the phenomenon that the Rumelhart and McClelland model is most interested in explaining. Most English past tense verbs are regular and are created by taking the present tense of the verb and adding "ed" to the end. Some examples are *walk/walked*, *open/opened*, and *mail/mailed*. Among irregular verbs, there are a number of categories within which there is some regularity. First, there are verbs such as beat, cut, and hit that are not changed at all. Second, there are verbs where the final "d" is changed to "t," as in *send/sent* and *build/built*. There are a number of categories where there are vowel changes made, as in *sing/sang*, *drink/drank*, *sting/stung*. Then there are completely arbitrary mappings like *go/went*. If there is one aspect involved in the learning of the past tense of verbs by children that everyone agrees with it, is that they "over-regularize." That is, they occasionally apply the regular rule of adding "ed" to verbs instead of using the correct irregular past tense. Thus, for example, children might say, "I goed to school today" or "I builded a sand castle" or "I hitted the ball." They do this even after having learned and used the correct form of the past tense of the verb, so their mistakes actually represent a step backwards in their use of language. Another type of error children make is to create a past tense using the correct past tense + "ed," as in "wented." Both these types of errors occur randomly in children's speech so that the correct and incorrect forms may be used in the space of one short conversation. After a while, they stop this over-regularization and go back to using the correct forms. During this whole process, their vocabulary is expanding from a relatively few verbs to hundreds of different verbs.

The traditional explanation of this phenomenon is that initially children memorize the irregular endings, then at some point and for some reason a rule-finding mechanism kicks in and it is this mechanism that will produce the over-regularization errors. Finally, as the correct rules are discovered, the errors disappear. In this final stage the child's mind will have a set of rules that handles regular verbs and "regular" irregular verbs plus a database to handle irregular verbs that do not follow any rules. In contrast, neural network models can give much the same behavior within a single mechanism, a single network. To a considerable extent, the difference between the neural and classical approaches is one of condensed versus uncondensed knowledge. In the classical approach, the goal is to produce condensed knowledge in the form of explicit rules as much as possible, whereas in the neural approach, the network gives rulelike behavior as well as storing cases.

10.9.2   The Rumelhart and McClelland Network

Rumelhart and McClelland produced a network that will make much the same kinds of mistakes that children make. This is fairly easily accomplished in their model by simply training the network on the right set of data, data that they believe is representative of the kinds of verbs that children encounter as they grow up. In particular, they used a two-layer back-propagation-like network to learn verb tenses.[3] Representing the data in a reasonable way was a problem. People hear the sounds of words in order over a short period of time, yet a single network of the sort they had to work with could not see this temporal order

---

[3] This was just before back-propagation came into general use.

of sounds.[4] The solution Rumelhart and McClelland came up with was to use a little-known method of representing words using *wickelfeatures*. Wickelfeatures allow words to be stated with the time dimension included in the data. Some researchers suggest this is not an ideal representation. We will neglect most of the details of the network and the details of the wickelfeature representation of the data. Another problem they had to deal with has to do with processing the output coming out of the network. The features coming out of the network are not clear-cut and unambiguous. The network can light up many features to varying degrees and it can become hard to figure out what answer the features are closest to. To solve this problem, another network was hard-wired to receive the past tense representation as input and then output a phonological string. We will neglect this problem as well.

The verb-tense network was trained in two stages as follows:

1) A set of 10 verbs, 2 regular and 8 irregular, was trained for 10 iterations. This represents a selection of verbs that were 20 percent regular and 80 percent irregular. These verbs are described as high-frequency verbs on the basis of their common use.

2) A set of 410 verbs, 334 regular and 76 irregular, was added to the original set of 10 and the network was trained for 190 iterations. This larger training set represents a selection of verbs that are 80 percent regular and 20 percent irregular. These verbs are described as middle-frequency verbs because they are used less often than the high-frequency verbs.

The Rumelhart and McClelland approach is based on the observations of some researchers that early on when children know only a small number of verbs, most of these verbs are irregular. There are no rules to be learned and children and the network end up storing 10 different answers for these 10 different verbs. At some point, children start acquiring large numbers of new verbs. Most of these are regular verbs and so to represent this stage, the training set is increased by 410 verbs and now it is a mixture of 80 percent regular and 20 percent irregular verbs. The early irregular verbs were well-learned after 10 iterations, but the influx of large numbers of regular verbs upsets the weights in the network against the irregulars although it does not upset the learning of the regular verbs. Figure 10.34 shows the percentage of features correct in regular and irregular verbs as the number of trials increases. At the eleventh iteration the new verbs are added and the network's performance on irregular verbs decreases. At this point the network will over-regularize for a while before finally learning the correct forms.

Since the Rumelhart and McClelland network, Marchman and Plunkett [156] have shown that over-regularization occurs in a back-propagation network when the patterns are added gradually. The Marchman and Plunkett network used "artificial verbs," a set of patterns that closely resemble the real verb mapping problem but where the patterns are easier to handle.

This disrupting of the weights in the network by the influx of a new set of patterns needs to be looked at more closely. Prince and Pinker quote a researcher who studied three children and found that these children did not ever go through a change from a 20 percent/80 percent regular/irregular mixture to an 80 percent/20 percent mixture during the time when the children started to over-regularize. Instead, the mixture of regular and

---

[4]There were no recurrent networks at this time either. For a simpler version of verb-tense learning using a recurrent network, there is now [131].

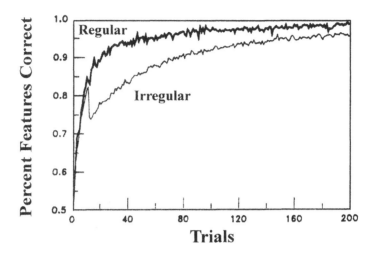

Figure 10.34: The percentage of correct features for regular and irregular high-frequency verbs as a function of trials.

irregular verbs that the children used stayed constant at about 50 percent/50 percent during this time. This becomes a major point of their criticism and they suggest that the over-regularization is achieved in the network by an unrealistic change in the percentage of regular input patterns. It is interesting, though, that even in a general pattern learning network where there is *no change* in the percentages of regular and irregular patterns, an influx of new patterns can still disrupt a network in such a way as to decrease performance on the irregulars and sometimes somewhat enhance the chances of over-regularization. To see how this can happen, suppose some regular and some irregular verbs are trained into a network. Next, add a new set of regular and irregular patterns. If the new irregular patterns have quite a lot in common with the old irregular patterns (as is likely with real verbs), the weight changes you have to make for the new irregular verbs will be quite small. Now, if the new regular verbs have little in common with the old regular verbs (again this is also likely with real verbs), then the weight changes for these new patterns will be quite large, especially at first. The result is that old irregular verbs can be regularized. The degree of disruption and over-regularization will vary according to all the various back-propagation learning parameters, the values of the initial random weights the network starts out with, and the patterns that are used (see Exercise 10.15). So actually, this criticism of Prince and Pinker may be unimportant. Prince and Pinker offer many other criticisms as well and interested readers should see [152].

In addition to the over-regularization phenomenon, the final Rumelhart and McClelland network was tested on a set of low-frequency regular and irregular verbs that were not in the training set. Regular verbs had an average of 92 percent of their wickelfeatures correct and irregular verbs had an average of 84 percent of their features correct. Some of the mistakes the network made were the following:

type	$\rightarrow$	typeded
step	$\rightarrow$	steppeded
sip	$\rightarrow$	sept
slip	$\rightarrow$	slept
smoke	$\rightarrow$	smoke
mail	$\rightarrow$	membled

and critics have pointed to these bad results as showing that the network model is not correct.

10.9.3 The Classical Rule-Based Model

We now turn to a rule-based explanation of the past tense learning phenomena. This description comes from the article by Prince and Pinker [152] where the description is based on [151]. The description we will give here is even more sketchy. Readers interested in more details should consult [152] and [151]. This model has apparently never been programmed. It is also fairly close to the back-propagation approach.

The model assumes that, in the beginning, a child's knowledge of the past tense of verbs is simply a memory phenomenon, so that the correct past tenses are just memorized. At some point, either because of physical developments or because the number of verbs the child knows about has reached a critical threshold, a rule-finding mechanism starts to operate. The rule-finding mechanism consists of several modules. One module proposes possible rules and does some elementary reorganization and then after a large number of rules have been proposed a more sophisticated process organizes the rules in a systematic way. This organizing process takes into account other aspects of the language and the end result is to produce a typical rule-based system. The system will store a set of special cases for verbs that display no regularity whatsoever plus it will store a collection of rules that can be applied to the irregular verbs that display some regularity. Verbs that do not fit either of these categories get the default regularized ending.

The proposed rule proposal module works in the following way. Suppose the word, "speak," and its past tense, "spoke," are encountered. A rule pattern is formed that includes the facts of this transformation. In the following rule, the important features of the word are the vowel that changes and its neighbors on each side:

IF a verb contains "pik" THEN change the "i" to "o" giving "pok"   (1)

The features of the words are written out in their phonetic form. The "i" in "pik" is the pronunciation of "ea" in "speak." Now, when the pattern "pik" is encountered in a verb, the system will guess that it becomes "pok" in the past tense form of the verb. Rules like this one will also have a "strength" factor associated with them. The strength factor is proportional to the number of times the rule has been found to work. For simplicity, each time a rule is found to work, its strength will be increased by 1. For a second example, take the pair, *tip/tipped*. This pair will give rise to the rule:

IF a word ends in "p" THEN add the ending "t"   (2)

where "t" is the phonological code for the sound of "ped." Also, given the pair, *walk/walked* the following rule is derived:

IF a word ends in "k" THEN add the ending "t"   (3)

The rule module could now notice that rules (2) and (3) can be regarded as special cases of a word ending in a consonant. Thus, it could collapse rules (2) and (3) into rule (4):

IF a word ends in a consonant THEN add the ending "t"   (4)

This collapsed rule would have a strength factor that is the sum of the strength factors of (2) and (3) giving in this case a strength of 2.

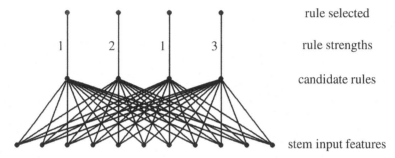

Figure 10.35: A networklike illustration of the competition between candidate rules. The candidate rule nodes sum the input features that are present in the stem of a verb, then this signal is passed through the rule strengths (weights) to give output values for each rule. Among the output nodes, the one with the largest value gives the exact rule to apply.

When rule (4) was added, a problem developed. The word, "speak," ending in a consonant, also matches rule (4) so the system now has the problem of deciding which rule should be used. To resolve this conflict, some other details of the system need to be explained. We have been assuming that rules either match or do not match, but Pinker's scheme allows for partial matches and competition among candidate rules. The arrangement is illustrated in a networklike fashion in Figure 10.35. The nodes labeled "candidate rules" look at the stem input features and light up in proportion to the number of stem features that are matched. So, if a given rule matches three out of four input features present, then that rule's node lights up to 0.75 and if a given rule finds one out of three features present it lights up to 0.33, and so on. These activations are then multiplied by the strength factor on the links between the candidate rules and the output nodes. The highest-rated candidate output node indicates the rule to apply. Over-regularization can occur in the following way. For a while the original rule will provide the correct answer, however, after a number of inputs, a rule that does not match as closely as the original may have had enough inputs to increase its strength beyond that of the original, correct rule, hence, the rule for "speak" giving "spoke" can be overwhelmed by another rule that does not match as closely but whose strength now makes its final activation greater than that of the original rule. "Speak" can become over-regularized to "speaked." Notice this happens here because of the input history of the system and it is similar to why over-regularization occurs in the Rumelhart and McClelland model. In Pinker's scheme, it is a set of specific instances that alter weights in a pattern matching process that uses explicit rules, whereas in the Rumelhart and McClelland scheme, it is a flood of regular verbs that alters weights in a pattern matching process where the rulelike structure is less clear-cut.

In this model, producing the final set of rules that represent the adult stage is fairly complicated and we will neglect trying to describe it.

10.9.4   A Hybrid Model

In a later article, Pinker [153] gives evidence for a model of verb past tense formation that uses a rule to generate the endings of regular verbs and an associative network to generate the past tense value of irregular verbs. This proposal comes after an experiment in which people were shown the present tense of a verb and then had to give the past tense as quickly as possible. When subjects were given low-frequency irregular verbs and high-frequency irregular verbs, the subjects gave the answers to the high-frequency verbs 16 to 29 milliseconds faster than for the low-frequency verbs. On the other hand, when the same thing was done with high- and low-frequency regular verbs, the difference was about 2 milliseconds. These results suggest to Pinker that a rule is applied for the regular verbs but for the irregular verbs an associative memory is consulted and this memory produces answers more quickly for those verbs that are more deeply worn into memory.[5]

10.10   Other Positions on Language

Still more radical innovations may be needed to process language. In Chapter 6 we noted that people apparently ground their understanding of the world in sensory inputs, and image processing in particular is an important component of understanding natural language. As another example of this, suppose you read this phrase from Touretzky [235, 236]:

the dog on the hood in the car

This phrase makes no sense because you cannot have the dog on the hood of a car and in the car at the same time. An impossible situation like this is evidently detected when a person tries to construct a picture of the situation. This example then gives some evidence that people are always constructing images when they process language. (Grammar school reading teachers have always claimed that to understand what you read, you have to form an image of what is going on in the story in your mind using your imagination.) Touretzky (and others) have made this point about the need for a language processing system that envisions what it reads about. Touretzky also makes the point that so far, all neural networks do is to recognize some basic patterns that are trained into them, as, for instance, in the McClelland and Kawamoto network. Training almost every possible combination of sentences and phrases into a network is just not going to be practical or be good enough.

---

[5] For instance, if you go and train patterns into a Hopfield network via back-propagation and there are many occurrences of some patterns and only a few occurrences of others, the patterns that are seen quite a lot dig deeper energy minima than the patterns that occur much less often. If you apply the Boltzman machine relaxation technique to produce the past tense of a verb given its present tense, the network will settle into the deeper minima faster than the shallow ones.

Language understanding programs are going to have to be able to build up mental images from component parts.

A radical new viewpoint comes from Lakoff and Johnson ([92, 93, 75]) and other linguists who challenge the traditional approach to understanding language. Lakoff and Johnson argue that most of the world is understood through metaphor and that language reflects this process of understanding. They believe that people use some fundamental schemas to understand the world. For example, one of the most fundamental schemas that people use to describe the world revolves around starting at a destination, going in some direction along a path, and arriving at a goal. They call this the *source-path-goal* schema. Here are some examples of the sorts of statements that fit this schema:

> An idea came to me.
> The system is coming up.
> The system is going down.
> The system crashed.
> Time flies when you are having fun.
> I was railroaded into that decision.
> I got sidetracked by a minor problem.
> Unix is coming on fast.
> We have come a long way since the days of Fortran II.
> You are going the wrong way by writing it in assembler.

A second fundamental schema is the *container* schema. Things are either in or out of the container. The container schema is the basis for set theory and, therefore, also for logic. Here are some examples of this schema in use in language:

> I am trapped in a boring job.
> My life is stuck in a rut.
> They left me in the dark about their plans.
> The computer can do that in five minutes.
> John came out of retirement.
> He was filled with pride.
> Working overtime for so long on this project has drained me.
> As we rounded the bend the glacier came into sight.

Another basic human experience is war, but arguments can be understood as a kind of war. Lakoff and Johnson [92] give these specific instances of arguments as war:

> Your claims are *indefensible*.
> He *attacked every weak point* in my argument.
> His criticisms were *right on target*.
> He *shot down* all of my arguments.

Thus, to understand language, you have to understand war.

Lakoff and Johnson list many other such basic schemas that people use to understand their experience. Among the important conclusions of their thinking is that to really understand language and to really understand the world, you need to have the same outlook on the world that you get from being a human being with a body and senses. Lakoff goes on to argue for "cognitive semantics" based on fundamental schemas and metaphor to replace the traditional ideas about language and thought. In [94], Lakoff discusses how connectionist methods might be applicable to his ideas. Reiger [169] discusses a simple implementation of these ideas.

10.11   Exercises

10.1. Design a grammar to generate legal prefix expressions, assuming that the only operations permitted are +, −, *, and /, and the only operands are simple integers.

10.2. Can you generate palindromes with a finite state grammar?

10.3. In Ada, integers are of two types, the ordinary integer and the based integer. Based integers are integers given in a base other than 10. Here are some examples of based integers:

> 8#177
> 2#101010
> 2#101_010
> 16#923
> 16#ae342d
> 16#ae_34_2d

The number before the '#' indicates the base of the integer and this number can range from 2 to 16. Underscores are allowed to improve readability. Write grammar rules that can be used to generate all the correct and only the correct integer expressions.

10.4. Design a slightly more complex grammar for English than the one given in Figure 10.5 that will include the use of prepositional phrases so it can produce sentences like:

> The winter in Chicago is very cold.
> The lion from the zoo chased the cat with the hat.
> The cat in the hat in the backyard under the tree ran home.

10.5. Trace through the Prolog transition network grammar in the text by hand with the sentence: "The baseball bat broke the picture window."

10.6. If you know enough Lisp or Prolog, program a simple version of a transition network grammar that handles prepositional phrases within noun-phrases by calling up the noun-phrase rule recursively.

10.7. Trace through the McELI program by hand for the sentence, "John got a ball."

10.8. With the sentences:

> Joey took the aspirin.
> Joey took the book.
> Joey took the plane.
> Joey took the advice.

how do you think you would program McELI to determine the correct primitive action, ingest, ptrans, atrans, or mtrans?

10.9. If you know enough Prolog, finish programming Micro ELI in Prolog. Or, if you know enough Lisp, program McELI in Lisp. Or program McELI in a conventional language. At a bare minimum, make your program translate the sentences:

> Mary went to the store.
> Mary got a doll.
> Mary went home.

You could also have your program do more, including the four sentences in the previous problem.

10.10. If we had the sentence, "John got a big red kite," the CD form for this sentence would be:

```
[atrans, [actor, [person, [name, john]]],
         [object, [kite, [size, big], [color, red]]]]
```

Add instruction packets to the Prolog version of McELI in the text that would do this translation.

10.11. One of the failings of the McClelland and Kawamoto network approach to sentence understanding was evident when the network attempted to process the sentence, "dog broke plate." It failed to flag "dog" as "nonhuman." This was because it never had an example of a nonhuman breaking something. For another example of this, suppose you trained a network with data like this:

> girl ate pasta spoon
> woman ate pasta spoon
> girl ate chicken spoon
> woman ate chicken pasta
> woman ate pasta chicken
> girl ate pasta cheese

If you give this network, "boy ate pasta fork," the boy will be turned into a girl on the output units and the fork will be turned into a spoon. If the networks were given comprehensive training sets that included these other possibilities, this problem would be eliminated. The problem with this solution is that the training set will be huge and the training time will become outrageous. This leads to a series of questions. Can you think of any other solutions to this problem that are not outrageous? Would people be able to generalize correctly if they are given the same limited amount of data (bit strings) that the networks received? If they would, how would they do it, and how could this be programmed? If people could not generalize correctly given only the bit strings, then how *do* people manage to generalize correctly for this situation?

10.12. McClelland and Kawamoto's neural network converted sentencelike sequences such as:

> boy broke window hammer

into a case grammar notation. In this exercise, we want to look at other methods to do the same task.

a) Write a Prolog (or Pascal or some other language) program that will take in the training sentences, their correct answers, and the bit string codes as data. Next, have the program read in test sentences (as words) and have the program find the correct agent, patient, instrument, and modifier. Thus, "boy hammer window broke" should become:

agent:	boy
action:	broke
patient:	window
instrument:	hammer
modifier:	

Needless to say, do not use back-propagation or any of its variations, but other pattern recognition techniques are acceptable. Try to find as many distinctly different methods for doing this as you can. How well do your methods work on sentences with missing parts and on sentences that are nonsense, such as perhaps "window broke hammer boy."

b) Typically, symbol processing programs do not store bit strings of microfeatures for each symbol. Typically, each symbol has a list of properties associated with it where the property is basically the analog to a microfeature in the PDP approach. For example, the microfeatures for 'boy' are: human, soft, male, medium, three-dimensional, rounded, unbreakable, and animate. In a traditional symbolic program the properties of a boy might normally be coded in a hierarchical fashion over several property lists such as:

```
property(boy,[object(human), sex(male), size(medium)]).

property(human,[texture(soft), shape(three_dimensional),
        features(rounded), breakable(no), animate(yes)]).
```

When the program looks up the properties of boy, it finds three properties. One of these properties is 'human.' The program can now look up the property 'human' and collect more features. Handling the sentence to case representation is now harder. Design a program to do so. Obviously, a program could convert these properties to a list of microfeatures and then proceed as in a), however, that is not allowed. No numerical methods are allowed. The method in part c) below is not allowed. Find as many ways as possible to do the transformation to case representation using property lists. Program any or all of the methods you discover. How well will your solutions work on sentences with missing parts and on nonsense sentences?

c) This problem can be done by defining an appropriate semantic grammar. Program the problem this way. How well will this solution work on sentences with missing parts and on nonsense sentences?

10.13. Given CD forms such as these, stated in Prolog:

```
[ptrans, [actor, [person, [name, [mary]]]]
         [object, [person, [name, [mary]]]]
         [to, [store]]]

[atrans, [actor, [store]]
         [object, [cookies]]
         [from, [store]]
         [to, [mary]]

[ptrans, [actor, [person, [name, [mary]]]]
         [object, [person, [name, [mary]]]]
         [from, [store]]
         [to, [home]]]
```

and the necessary script, have a Prolog program (or if you are a Lisp programmer do this in Lisp) answer questions like, "Who went to the store?" "What did Mary buy?" and "What did Mary give to the store?" Assume the questions are given in CD format. (Of course, if you want to program the translation to CD form, feel free to do so.) Output the answers in CD form, or, if you are more ambitious, devise a way to turn the CD form into plain English sentences.

10.14. In the Chapter 1 exercises there was the problem of giving a program some simple third and fourth grade arithmetic word problems and having the program guess the solution. The methods suggested were simple pattern recognition schemes. These schemes did not use any of the more sophisticated means of processing natural language described in this chapter. Now, suppose you want to do a more sophisticated job of handling those sentences and then have a program solve the problems. Consider how well each of the natural language understanding methods described in this chapter can be adapted to do problems of this sort. Feel free to program any or all of the methods you come up with.

10.15. Actually programming a network to do the Rumelhart and McClelland verb tense model is very hard and time consuming, however, it is easy to show that over-regularization effects occur in very small networks using a small number of short, simple patterns. The effects can even occur when a network learns an initial set of patterns with a 50 percent/50 percent mixture of regulars and irregulars and is then given an additional group with a 50 percent/50 percent mixture if the errors being back-propagated from the new regular patterns are larger than the errors coming from the irregulars. There will be some over-regularization effects in most networks and for most parameters and variations on the back-propagation algorithm, but perhaps the greatest effects occur when the program uses the derivative term from the differential step size algorithm. Recall that that algorithm uses a derivative term of 1 rather than $o_j(1 - o_j)$ for the output layer units. Design a set of initial training patterns together with a second set of patterns that when added to the network produce over-regularization effects. To keep it simple you could just have three output units, one for a regular sequence and two for irregular responses. Note that, whereas over-regularization effects can be observed with general patterns, this is no guarantee that the effects will be significant with actual verbs.

10.16. A big point of this book is to show that AI is mostly pattern recognition and that there are three important ways to do pattern recognition: first, devise rules, second, train networks, and third, store lots of cases. The first two of these possibilities applied to natural language have been discussed in this chapter because they have been extensively researched. Now, consider whether or not a case-based approach would work.

Here are a couple of arguments for a case-based approach to understanding language. First, suppose you had to learn a foreign language, like perhaps, Spanish. One technique used to teach foreign languages has been to make students memorize sentences or even short stories in the language. Students also memorize the names of various objects as well. For instance, in Spanish, you can ask "Where is the library?" by saying, "¿Dónde está la biblioteca?" where "la biblioteca" is "the library." Now, if you learn that a church is "la iglesia" and a school is "la escuela," you can perform a "cut and paste" operation on the question about where is the library and ask "¿Dónde está la iglesia?" and "¿Dónde está la escuela?" This kind of cut and paste operation is much like what CHEF does, except CHEF works with recipes.

A second small argument for this position could come indirectly from William James. Anyone that has spent a small amount of time reading William James will readily perceive that the writer's style contains patterns of speech that are no longer common. If people only operate using abstract rules of grammar it is hard to understand why James' writing should seem peculiar. Any statement that is grammatically correct should seem just as normal as any other. On the other hand, if people understand language by storing large numbers of language patterns, then when archaic patterns come up they will have to seem slightly odd. Thinking in terms of computational energy, the archaic patterns represent shallow minima.

Give more arguments for the idea that language can be understood using case-based methods. Give arguments against. If you think a case-based solution seems like a good idea, consider how you could program it.

Afterword

Artificial intelligence has a dual nature. One goal of AI is to produce small special purpose programs to do small tasks. Here you can produce quite a few useful programs, but in many cases the performance suffers because to do it right the program must know all about the very big real world. So, for example, to understand speech, correct grammar, or translate from one language to another, the program must understand the world. If you skimp here the results are going to be poor, but of course, in some cases relatively poor can still be fairly useful.

There is a point of view among some AI researchers that simply automating the small tasks that in people require intelligence produces programs with artificial intelligence. This viewpoint says that the game playing program that simply minimaxes and manages to beat some people should be thought of as having artificial intelligence. First, there is the problem that the performance can be poor, but even beyond that, the programs do not end up behaving the way intelligent people behave, in almost every case the programs do not learn and when a bug shows up they cannot cope. To the general public the program looks stupid, not intelligent. Likewise, under this definition, people do arithmetic and if someone writes a program to do arithmetic then the program must be considered an artificial intelligence program. But once a bug shows up and the system cannot correct its mistake, it will look stupid, not intelligent. Then people do accounting, should a program that does accounting be called an artificial intelligence program? Much of what goes on in AI should probably be called "advanced programming techniques" rather than artificial *intelligence*. Programs that have only some characteristics of intelligence ought to get a label like "semi-intelligent" or "partially intelligent" to keep the public from becoming skeptical of the whole field.

The other goal of AI is to produce an artificial system with all the flexibility to deal with the whole real world you find in a human being. This grand goal has proven to be elusive. Early on, AI researchers thought that they had found the necessary components and that they would scale up. But now many people believe new and fresh ideas are called for and they have turned to researching these ideas. Worse still, in my opinion, a system that comes pretty close to that of a human being will have to have a body and senses much like those of a human being and learn about the world the way a human being learns, including starting out knowing as little as a human baby. Producing a human level of intelligence in a system is going to be very hard and it is likely to be some time before much progress is made.

Appendix A

A.1 A Derivation of Back-Propagation

In this derivation of back-propagation (the generalized delta rule) we will start with a derivation of the original delta rule, where the activation function for the units is linear. One consequence of the linear activation function is that networks with more than two layers can be collapsed down to a two-layer network, so there is nothing to be gained by having more than two layers. Note that the delta rule is not quite the same thing as the linear pattern classifier, however, it can also be used to separate linearly separable patterns.

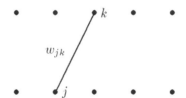

Figure A.1: A two-layer network that uses the delta rule where, for simplicity, only the weight, w_{jk} is drawn in. In deriving the delta rule, the main problem is to find out how o_k varies as w_{jk} changes.

A.1.1 The Delta Rule

The delta rule operates like the generalized delta rule, but with a linear activation function and an abstract network as shown in Figure A.1. The activation value, o_k, of an output unit, k, is given by:

$$o_k = \sum_j w_{jk} o_j,$$

where the weights, w_{jk}, are between output unit k and the input units designated by j. The o_j are the activation values of the input units. Notice that the output values can be any real value, not just values in the range 0 to 1. As in back-propagation, the goal of the algorithm is to adjust the weights so as to minimize the sum of the squared error on the output units for all the patterns in the training set. We begin the derivation of the delta rule by defining the error for the output units, E_p, for a particular pattern, p, as:

$$E_p = \frac{1}{2}\sum_k (t_{pk} - o_{pk})^2.$$

Here, t_{pk} is the target value for the kth output unit and o_{pk} is the actual value of the unit. For the rest of the derivation of the delta rule we will be dropping the p subscript from the error, target, and unit activation terms. Now, the first task is to find out how the error changes, as a weight, w_{kj}, in the network is changed. To do this, we take the partial derivative of E with respect to w_{kj}:

$$\frac{\partial E}{\partial w_{jk}} = \frac{\partial E}{\partial o_k}\frac{\partial o_k}{\partial w_{jk}} = -(t_k - o_k)o_j.$$

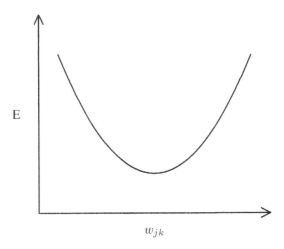

E

w_{jk}

Figure A.2: How the error, E, changes as a function of the weight, w_{jk}, between the input unit j and the output unit, k.

The error, E, is a quadratic function of the weight, w_{jk}, and Figure A.2 shows a generalized plot of how error varies with changes to the weight, w_{jk}. Notice that with a positive derivative, w_{jk} should be decreased by a small amount proportional to the slope in order to decrease the error. With a negative derivative, w_{jk} should be increased by a small amount proportional to the slope. Let this constant of proportionality be η. With this in mind, here is the delta rule:

$$w_{jk} \leftarrow w_{jk} + \eta\frac{\partial E}{\partial w_{jk}} = w_{jk} + \eta(t_k - o_k)o_j.$$

This formula is often shortened to:

$$w_{jk} \leftarrow w_{jk} + \eta\delta_k o_j,$$

where $\delta_k = (t_k - o_k)$ is called the error signal.

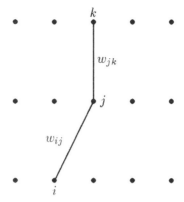

Figure A.3: A general three-layer back-propagation network. When w_{jk} changes, it affects only the error on one output unit, k. When w_{ij} changes, it affects the error on all the output units.

A.1.2 The Generalized Delta Rule, or Back-Propagation

The derivation of the generalized delta rule follows the same pattern, but now with the nonlinear activation function, networks with more than two layers become important. Figure A.3 shows a general outline of a three-layer network. In this network, the following relationships hold:

$$o_k = \frac{1}{1 + e^{-net_k}}$$

$$net_k = \sum_j w_{jk} o_j$$

$$o_j = \frac{1}{1 + e^{-net_j}}$$

$$net_j = \sum_i w_{ij} o_i$$

As before, the object is to minimize the error:

$$E_p = \frac{1}{2} \sum_k (t_{pk} - o_{pk})^2.$$

Also, as before, we will be dropping the p subscript representing the pattern number from all the quantities.

The first part of the problem is to find how E changes as the weight, w_{jk}, leading into an output unit changes. The second part of the problem is to find out how E changes as w_{ij}, a weight leading into a hidden layer unit, j, changes. Finding the formula for the first part is the easiest. We simply write:

$$\frac{\partial E}{\partial w_{jk}} = \frac{\partial E}{\partial o_k} \frac{\partial o_k}{\partial net_k} \frac{\partial net_k}{\partial w_{jk}} = -(t_k - o_k) o_k (1 - o_k) o_j.$$

As in the delta rule, it is common to define the error signal,

$$\delta_k = (t_k - o_k)o_k(1 - o_k).$$

Then the formula for weight changes for weights leading into the output layer becomes:

$$\Delta w_{jk} = \eta \delta_k o_j.$$

Now, for the second part that relates the change in E to the change in w_{ij}. In this case, a change to the weight, w_{ij}, changes o_j and this changes the inputs into each unit, k, in the output layer. The change in E with a change in w_{ij} is therefore the sum of the changes to each of the output units. The chain rule produces:

$$
\begin{aligned}
\frac{\partial E}{\partial w_{ij}} &= \sum_k \frac{\partial E}{\partial o_k} \frac{\partial o_k}{\partial net_k} \frac{\partial net_k}{\partial o_j} \frac{\partial o_j}{\partial net_j} \frac{\partial net_j}{\partial w_{ij}} \\
&= \sum_k -(t_k - o_k)o_k(1 - o_k)w_{jk}o_j(1 - o_j)o_i \\
&= \sum_k -\delta_k w_{jk} o_j (1 - o_j)o_i. \\
&= -o_i o_j (1 - o_j) \sum_k \delta_k w_{jk} \\
&= -o_i \delta_j.
\end{aligned}
$$

If we let:

$$\delta_j = o_j(1 - o_j) \sum_k \delta_k w_{jk},$$

then the weight change can be written:

$$\Delta w_{ij} = \eta \delta_j o_i.$$

These derivations can be generalized to networks with more than three layers. The result is the following set of formulas. The first one specifies the weight changes to unit j, no matter what layer it is in:

$$\Delta w_{ij} = \eta \delta_j o_i.$$

The second formula specifies the error signal for the output layer:

$$\delta_j = (t_j - o_j)o_j(1 - o_j).$$

The third formula specifies the error signal for unit j, in a hidden layer with units k, above:

$$\delta_j = o_j(1 - o_j) \sum_k w_{jk} o_k.$$

Glossary

activation function In the most common type of neural network modeling, a neuron adds up its inputs to produce a single scalar value which is then passed to an activation function to compute a new value. Such a function is often called a squashing function or a transfer function.

artificial intelligence A field of computer science where one goal is to produce intelligent programs or computer architectures that compare favorably with human-level performance and another goal is simply to automate procedures that are normally performed by people.

associative memory A memory system which, when it is given a piece of data as an input, produces another piece of data that has been associated with the input.

auto-associative network A network which produces on its output units the same pattern that was input to it.

back-propagation Also called backprop. The most useful neural networking algorithm, capable of classifying data and fitting data points.

backward chaining Applying rules by guessing that the consequent of a rule is true and then attempting to verify that the antecedents are true.

Boltzman machine An algorithm used on Hopfield-type networks that is designed to find the lowest minimum of an energy landscape.

Chinese room argument An argument by John Searle that supposedly shows that a purely mechanical device or an electronic computer cannot understand the world in the same sense that a human being does.

chunking/chunks In SOAR, a chunk is a rule and chunking is the forming of the rules, however, the term was derived from psychological research by Miller and it is uncertain how the two types of chunks and chunking are related.

computational energy A mathematical quantity defined for Hopfield networks that was designed to help the network fill in the missing states of a partial pattern.

conceptual dependency A notation devised by Schank for representing a sentence.

condensed/compressed knowledge (My term): Symbol processing AI has traditionally believed that programs could be intelligent just by loading them with lots of abstract facts and rules about the world without giving the programs an instance of a complete object. These disembodied facts are condensed knowledge. Compressed knowledge is a special case where rules or networks are used to replace large numbers of instances.

conflict resolution Given a set of rules that may be applied, conflict resolution is applied to determine which rule to apply.

connectionist AI Using neural networks to try to produce human-level performance in a program.

cytoskeleton A skeleton-like structure found in most living cells that may be used for computing. The structure consists of microtubules, hollow tubes made out of a protein called tubulin. There is a theory that the cytoskeleton may be functioning as a computer and/or it may be used to perform quantum mechanical computing.

distributed representation A neural networking representation of data where the characteristics of the object or activity are distributed over a large number of neurons, as opposed to a local representation where a single neuron represents the object or activity.

dualism The idea that people have a physical part and a spiritual part. Quite often it is claimed that the two cannot interact with each other. This claim is a silly idea since then the body could not affect the spirit and the spirit could not affect the mind. Dualism is typically associated with the French philosopher Rene Descartes.

evaluation function A function that computes the goodness of a game board configuration.

expert system A program that supposedly performs as well as a human expert on a very small task. The concept was developed within the framework of rule-based systems so many people tend to think of an expert system as rule-based rather then network-based.

feed-forward network A network arranged in layers where the calculations proceed from an input layer to an output layer, but not in the opposite direction.

field processing A proposed method of neural computing by MacLennan where large numbers of neurons can be treated like fields, that is fields as in electric, magnetic, and gravitational fields.

forward chaining Applying rules from antecedents to produce consequents.

frame A symbolic structure that stores closely coupled patterns so that when a partial pattern is input to the system, the system can complete the pattern in a reasonable way.

generalization The ability to deal correctly with novel patterns that are only similar to patterns the system has dealt with before.

heuristic search A search for a solution to a problem that uses information about the problem to speed the search.

hill climbing Trying to maximize a certain quantity where the "landscape" is not well known with the idea that the highest part of the landscape will give the best solution.

Hopfield network A type of network designed by Hopfield that has a quantity called computational energy associated with it and where the update algorithm for the network will converge to a minimum of the computational energy.

inference rules Algorithms in predicate calculus that allow you to produce new conclusions from existing statements.

intelligence (My standard): A system in some world, either the real world or a small artificial world, that recognizes patterns, solves problems with trial and error behavior, and learns from its experience is intelligent. Other people's standards for intelligence may be tighter or looser.

interactive activation network A network, usually arranged in layers, where the activations are passed from the input layer to the output layer and back from the output layer to the input layer. The updates of neurons take place over and over, normally until the activation values of the neurons stabilize.

learning vector quantization A method for moving a small number of points around within the space of a number of different classes of patterns so that only these very few points, not a large database of points, need to be stored for use in a nearest neighbor algorithm. The points that are moved around are called codebook vectors or prototype points.

linear pattern classifier A simple classification algorithm that will find a line, plane, or hyperplane between two different classes of patterns.

local representation A representation for objects or actions in a neural network where each object or action is represented by a single neuron.

microtubules Very small and thin hairlike structures present in most living cells that form a kind of skeleton for the cell.

minimax An algorithm for game playing where, at one level, one player is trying to minimize the chances for success of the other player, while at the next level, the player is trying to maximize his own chances for success.

monotonic logic A form of reasoning where whatever conclusion is drawn must be correct, this is as opposed to nonmonotonic logic where any conclusion may be withdrawn as new information is revealed.

nearest neighbor algorithms Algorithms that attempt to identify an unknown pattern by finding its nearest pattern that has been correctly identified.

neocognitron Any one of a series of computer models of vision proposed by Fukushima that can recognize distorted and displaced instances of letters of the alphabet and digits.

neural networking The use of large numbers of simple processors that carry out relatively simple calculations in parallel.

parallel distributed processing (PDP) The use of neural networks with distributed representations.

parse tree A tree structure that shows how a statement was derived from the rules of a grammar.

pattern recognition (1) a formal field of academic study where (2) the goal is to classify patterns or otherwise (3) find similarities between patterns.

perceptron A neural network, especially those designed by Frank Rosenblatt, but also a term that is often applied to back-propagation networks.

property list A list of properties that a symbol has. Properties are pretty much the equivalent of microfeatures in distributed neural networking representations.

precomputed features Extra features you can input to a network that are computed from mathematical combinations of other input features with the hope that the precomputed features will help the network learn its task better and/or faster.

predicate calculus A formal functionlike notation for expressing true statements together with a set of inference rules that allow you to prove the truth or falsity of other statements.

prototype points In the LVQ family of algorithms, the small number of representative points are placed in space in such a way as to do a very good job of pattern classification with a nearest neighbor algorithm.

PSSH The Physical Symbol System Hypothesis, the hypothesis that intelligence and thinking can be done using only symbols, structures of symbols, and rules.

quantum mechanics In the early part of the 20th century, physicists had discovered that electrons, protons, and various other very small particles were not obeying the known laws of physics. After a while, it became clear that particles also behave like waves. The main formula found to describe how the particles move around is the Schroedinger wave equation, a second-order partial differential equation that uses complex numbers.

recurrent network A network like the feed-forward network except some data flows back down to the input layer to be used with the next input.

Resolution A specific set of techniques for reaching conclusions in predicate calculus.

rule A formula to compute a consequent given some antecedents.

script Like a frame, however, scripts store a series of events that fall into a pattern, then when the system gets a few events that fit into a script, it can predict that the other events happened as well.

semantics The meaning of things, especially in AI, the meaning of words and sentences.

semantic grammar A grammar for a natural language that is built around semantically meaningful quantities like people, places, and things rather than around the traditional categories of nouns, verbs, and other parts of speech.

semantic network A pictorial representation of facts and relations between the facts.

semantic transparency A term proposed by Andy Clark. A set of rules written in some natural or artificial language can be understood by human beings so the set of rules is said to have semantic transparency. On the other hand, neural networks typically have their "rules" in the form of incomprehensible numbers (weights) between neurons and it is impossible to translate this into neat, understandable natural language.

sigmoid An S-shaped function.

strong AI The belief that running a program on a computer is thinking, as opposed to weak AI, the belief that the program is only simulating thinking.

symbol A unique mark on a piece of paper, although in a computer, a symbol is a unique integer. However, rigorously speaking, the only operations defined for symbols are tests for equality; no ordering relations or arithmetic are defined for symbols.

symbol grounding The idea that symbol-based descriptions of the world are incomplete and must finally be understood, or grounded, in terms of sensory images, principally visual images.

syntax The rules that describe the proper order of words in a sentence.

test/training set Pattern recognition algorithms are typically trained with one set of data, the training set and then evaluated on another set of data, the test set. Generally, the test set performance is worse than the training set results, but it is a better measure of the true performance ability of the program.

transition network grammar A symbolic network whose purpose is to parse a sentence, that is, to turn out a parse tree based on some formal grammar of the language.

Turing test A weak test for thinking proposed by Alan Turing. A program passes the Turing test if it can fool some people for a short period of time into thinking that the program is human, based on a conversation with it.

weak AI The belief that a program running on a computer is only simulating thinking, as opposed to strong AI, the belief that the system is thinking.

Bibliography

[1]   B.P. Allen, "Case-Based Reasoning: Business Applications," *Comm. ACM*, Vol. 37, No. 3, Mar. 1994, pp. 40–42.

[2]   R.B. Allen, "Sequential Connectionist Networks for Answering Simple Questions about a Microworld," *Proc. Cognitive Science Society*, 1988, pp. 489–495.

[3]   R.B. Allen, "Reference in Connectionist Language Users," *Connectionism in Perspective*, Zurich, Oct. 1988.

[4]   T. Ash, "Dynamic Node Creation in Backpropagation Networks," *Connection Science* Vol. 1, pp. 365–375, 1989 and also: Institute for Cognitive Sciences Technical Report 8901, Feb. 1989, University of California, San Diego, La Jolla, CA.

[5]   K. Ashley and E. Rissland, "Compare and Contrast, a Test of Expertise," *Proc. DARPA sponsored Workshop on Case Based Reasoning*, Kolodner, ed., Morgan Kaufmann, 1988, reprinted from *Proceedings of AAAI-87*, 1987, pp. 31–36.

[6]   K. Ashley and E. Rissland, "A Case-Based Approach to Modelling Legal Expertise," *IEEE Expert*, Fall 1988.

[7]   V. Barker and D. O'Connor, "Expert Systems for Configuration at Digital: XCON and Beyond," *Comm. ACM*, Mar. 1989.

[8]   H. Berliner, "Computer Backgammon," *Scientific American*, June 1980, pp. 64–72.

[9]   H. Berliner, "Backgammon Computer Program Beats World Champion," *Artificial Intelligence*, Vol. 14, No. 1, 1980, pp. 205–220.

[10] H. Berliner, "Deep Thought Wins Fredkin Intermediate Prize," *AI Magazine*, Vol. 10, No. 2, Summer 1989, pp. 89–90.

[11] G.H. Bower and D. Winzenz, "Group Structure, Coding and Memory for Digit Series," *J. Experimental Psychology Monograph*, Vol. 80, Part 2, May 1969, pp. 1–17.

[12] G.H. Bower, J.B. Black, and T.J. Turner, "Scripts in Text Comprehension and Memory," *Cognitive Psychology*, Vol. 11, 1979, pp. 177–220.

[13] G. Bradshaw, R. Fozzard, and L. Ceci, "A Connectionist Expert System that Actually Works," *Advances in Neural Information Processing Systems I*, Touretzky, ed., Morgan Kaufmann, 1989, pp. 248–255.

[14] S. Bradtke and W. Lehnert, "Some Experiments with Case Based Search," *Proc. DARPA Sponsored Workshop on Case Based Reasoning*, Kolodner, ed., Morgan Kaufmann, 1988, pp. 80–93.

[15] J.S. Brown and R.R. Burton, "Multiple Representations of Knowledge for Tutorial Reasoning," in *Representation and Understanding*, Bobrow and Collins, eds., Academic Press, 1975, pp. 311-349.

[16] A.E. Bryson and Y. Ho, *Applied Optimal Control*, Blaisdell, New York, 1969.

[17] B. Buchanan, G. Sutherland, and E.A. Feigenbaum, "Heuristic DENDRAL: A Program for Generating Explanatory Hypotheses in Organic Chemistry," *Machine Intelligence 4*, American Elsevier, New York, 1969.

[18] A. Clark, *Microcognition*, MIT Press, 1989.

[19] A. Clark and C. Thornton, "Trading Spaces: Computation, Representation and the Limits of Uninformed Learning," http://www.artsci.wustl.edu/ philos/papers/-clark.trading-spaces.ps, 1994.

[20] W. Clocksin and C. Mellish, *Programming in Prolog*, Springer-Verlag, 1981.

[21] J.E. Collard, "A B-P ANN Commodity Trader," in *Advances in Neural Information Processing Systems 3*, Morgan-Kaufmann, 1991, pp. 551–556.

[22] E. Collins, S. Ghosh, and C. Scofield, "An Application of a Multiple Neural Network Learning System to the Emulation of Mortgage Underwriting Judgments," *Proc. IEEE 2nd Int'l Conf. Neural Networks*, IEEE Press, Piscataway, NJ, 1988.

[23] R.H. Creecy et al., "Trading MIPS and Memory for Knowledge Engineering," *Comm. ACM*, Vol. 35, No. 8, Aug. 1992, pp. 48–63.

[24] R. Davis, B. Buchanan, and E. Shortliffe, "Production Rules as a Representation for a Knowledge-Based Consultation Program," *Artificial Intelligence*, Vol. 8, No. 1, Feb. 1977.

[25] R. Davis, "Teiresias: Applications of Meta-Level Knowledge," *Knowledge-Based Systems in Artificial Intelligence*, McGraw-Hill, 1982, pp. 239–490.

[26] J. Dayhoff et al., "The Neuronal Cytoskeleton: A Complex System that Subserves Neural Learning," *Rethinking Neural Networks: Quantum Fields and Biological Data*, K.H. Pribram, ed., Lawrence Erlbaum, 1993, pp. 389–442.

[27] H. Dreyfus, "Alchemy and Artificial Intelligence," The RAND Corporation Paper P-3244, Dec. 1965.

[28] Dreyfus and Dreyfus, *Mind Over Machine*, Free Press, 1986.

[29] Dreyfus and Dreyfus, "Making a Mind Versus Modeling the Brain," *Daedalus, Proc. American Academy of Arts and Sciences*, Vol. 117, No. 1, Winter 1988, pp. 15–43.

[30] R. Duda et al., "Semantic Network Representations in Rule-Based Inference Systems," *Pattern Directed Inference Systems*, Waterman and Hayes-Roth, eds., Academic Press, 1978, pp. 203–221.

[31] R. Duda and J. Gaschnig, "Knowledge-Based Expert Systems Come of Age," *BYTE*, Sep. 1981, pp. 238–274.

[32] M. Dyer, "Symbolic NeuroEngineering for Natural Language Processing: A Multilevel Research Approach," Technical Report UCLA-AI-88-14, Computer Science Department, University of California, Los Angeles, Aug. 1988.

[33] M. Dyer, M. Flowers, and Y.A. Wang, "Weight Matrix = Patterns of Activation: Encoding Semantic Networks as Distributed Representations in DUAL, a PDP Architecture," Technical Report UCLA-AI-88-5, Computer Science Department, University of California, Los Angeles, Apr. 1988.

[34] J.C. Eccles and K.R. Popper, *The Self and its Brain*, Springer, Berlin, 1977.

[35] J.L. Elman, "Finding Structure in Time," CRL Technical Report 8801, Center for Research in Language, University of California, San Diego, Apr. 1988.

[36] T. Evans, "A Heuristic Program to Solve Geometric Analogy Problems," *Proc. AFIPS Spring Joint Computer Conference*, 1964, pp. 327–338.

[37] S.E. Fahlman, "Faster Learning Variations on Back-Propagation: An Empirical Study," *Proc. 1988 Connectionist Models Summer School*, Touretzky, ed., Morgan Kaufmann, 1989, ftp://archive.cis.ohio-state.edu/pub/neuroprose/-fahlman.quickprop-tr.ps.Z, ftp://ftp.cs.cmu.edu/afs/cs/project/connect/tr/qp-tr.ps.Z[1]

[38] M.A. Fanty, "Context-Free Parsing in Connectionist Networks," TR 174, Computer Science Department, University of Rochester, 1985.

[39] M.A. Fanty, "Context-Free Parsing with Connectionist Networks," *Neural Networks for Computing*, American Institute of Physics, New York, 1986, pp. 140–145.

[40] J.A. Fodor and Z.W. Pylyshyn, "Connectionism and Cognitive Architecture: A Critical Analysis," *Connections and Symbols*, Pinker and Mehler, eds., MIT Press, 1988, pp. 3–71. (Reprinted from *Cognition: International Journal of Cognitive Science*, Vol. 28, 1988.)

[41] K. Fukushima, "Cognitron: A Self-organizing Multilayered Neural Network," *Biological Cybernetics*, Springer-Verlag, Vol. 20, No. 3/4, Nov. 1975, pp. 121–136.

[42] K. Fukushima, "Neocognitron: A Self-organizing Neural Network Model for a Mechanism of Pattern Recognition Unaffected by Shift in Position," *Biological Cybernetics*, Springer-Verlag, Vol. 36, No. 4, Apr. 1980, pp. 193–201.

---

[1]Note: URLs for articles on the Internet change from time to time. See my *Basis of AI* page at http://www.mcs.com/ drt/basisofai.html.

[43]   K. Fukushima, Sei Miykake, and Takayuki Ito, "Neocognitron: A Neural Network Model for a Mechanism of Visual Pattern Recognition," *IEEE Transactions on Systems, Man and Cybernetics*, Vol. SMC-13, No. 5, Sep./Oct. 1983, pp. 826–834.

[44]   K. Fukushima, "A Neural Network Model for Selective Attention," *Proc. IEEE 1st Int'l Conf. on Neural Networks*, Caudill and Butler, eds., Vol. II, IEEE Press, Piscataway, NJ, 1987, pp. 11–18.

[45]   K. Fukushima, "A Neural Network for Visual Pattern Recognition," *Computer*, Vol. 21, No. 3, Mar. 1988, pp. 65–75.

[46]   K. Fukushima, "Neocognitron: A Hierarchical Neural Network Capable of Visual Pattern Recognition," *Neural Networks*, Vol. 1, No. 2, 1988, pp. 119–130.

[47]   K. Funushima and N. Wake, "Handwritten Alphanumberic Character Recognition by the Neocognitron," *IEEE Trans. Neural Networks*, Vol. 2, No. 3, May 1991, pp. 355–365.

[48]   K. Funahashi, "On the Approximate Realization of Continuous Mappings by Neural Networks," *Neural Networks*, Vol. 2, No. 3, 1989, pp. 183–192.

[49]   H. Gelernter, "Realization of a Geometry-Theorem Proving Machine," *Computers and Thought*, Feigenbaum and Feldman, eds., McGraw-Hill, 1963. (Reprinted from *Proc. Intern. Conf. Information Processing*, 1959.)

[50]   H. Gelernter, R. Hansen, and D. Loveland, "Empirical Explorations of the Geometry-Theorem Proving Machine," in *Computers and Thought*, Feigenbaum and Feldman, eds., McGraw-Hill, 1963, pp. 135–151. (Reprinted from *Proc. Western Joint Computer Conf.*)

[51]   S. Geva and J. Sitte, "Adaptive Nearest Neighbor Pattern Classification," *IEEE Trans. Neural Networks*, Vol. 2, No. 2, Mar. 1991, pp. 318–322.

[52]   P. Gorman and T. Sejnowski, "Analysis of Hidden Units in a Layered Network Trained to Classify Sonar Targets," *Neural Networks*, Vol. 1, No. 1, 1988, pp. 75–89.

[53]   P. Gorman and T. Sejnowski, "Learned Classification of Sonar Targets Using a Massively Parallel Network," *IEEE Trans. Acoustics, Speech and Signal Processing*, Vol. 36, No. 7, July 1988, pp. 1135–1140.

[54]   S. Grossberg, *Neural Networks and Natural Intelligence*, Grossberg, ed., MIT Press, 1988.

[55]   J. Hadamard, *The Psychology of Invention in the Mathematical Field*, Princeton University Press, 1945.

[56]   U. Halici and M. Sungur, "Solving SOMA Puzzles by Boltzman Machine," 1993 World Congress on Neural Networks, Lawrence Erlbaum and INNS Press, Vol. 1, 1993, pp. 316–319.

[57]   S. Hameroff et al., "Nanoneurology and the Cytoskeleton: Quantum Signaling and Protein Conformational Dynamics as Cognitive Substrate," *Rethinking Neural Networks: Quantum Fields and Biological Data*, K.H. Pribram, ed., Lawrence Erlbaum, 1993, pp. 317–376.

[58]   K.J. Hammond, "Opportunistic Memory: Storing and Recalling Suspended Goals," *Proc. DARPA Sponsored Workshop on Case Based Reasoning*, Kolodner, ed., Morgan Kaufmann, 1988.

[59]   K.J. Hammond, *Case Based Planning: Viewing Planning as a Memory Task*, Academic Press, 1989.

[60]   S.J. Hanson and J. Kegl, "PARSNIP: A Connectionist Network that Learns natural language grammar frmo exposure to Nature Language Sentences," *Proceedings of the Ninth Annual Cognitive Science Meeting*, Lawrence Erlbaum, Hillsdale, NJ, 1987.

[61]   S. Harnad, "Categorical Perception: A Critical Overview," *Categorical Perception: The Groundwork of Cognition*, Harnad, ed., Cambridge University Press, New York, 1987.

[62]   S. Harnad, "Category Induction and Representation," *Categorical Perception: The Groundwork of Cognition*, Harnad, ed., Cambridge University Press, New York, 1987.

[63]   S. Harnad, "The Symbol Grounding Problem," *Physica D*, Vol. 42, 1990, pp. 335–346. ftp://archive.cis.ohio-state.edu/pub/neuroprose/harnad.symbolgrounding.asc.Z.

[64]   D. Hebb, *Organization of Behavior*, Wiley, 1949.

[65]   R. Hecht-Nielsen, "Theory of the Back-Propagation Network," *Proc. Int'l Joint Conf. Neural Networks*, IEEE Press, Piscataway, NJ, 1989, pp. 593–605.

[66]   G.E. Hinton and T.J. Sejnowski, "Learning and Relearning in Boltzmann Machines," *Parallel Distributed Processing*, MIT Press, Vol. 1, 1986, pp. 282–317.

[67]   G.E. Hinton, J.L. McClelland, and D.E. Rumelhard, "Distributed Representations" *Parallel Distributed Processing*, MIT Press, Vol. 1, 1986, pp. 318–362.

[68]   J. Hopfield, "Neural Networks and Physical Systems With Emergent Collective Computational Abilities," *Proc. Nat'l Acad. Sci. USA* 79, Apr. 1982, pp. 2554–2558.

[69]   J. Hopfield, "Neurons With Graded Response Have Collective Computational Properties Like Those of Two-State Neurons," *Proc. Nat'l Acad. Sci. USA* 81, May 1984, pp. 3088–3092.

[70]   K. Hornik, M. Stinchcombe, and H. White, "Universal Approximation of an Unknown Mapping and its Derivatives Using Multi-Layer Feedforward Networks," *Neural Networks* , Pergamon Press, Vol. 3, No. 5, 1990, pp. 551–560.

[71] E. Hunt, "The Memory We Must Have," *Computer Models of Thought and Language*, Schank and Colby, eds., W.H. Freeman, 1973, pp. 343–371.

[72] W. James, *The Principles of Psychology*, Dover Publications, New York, 1950.

[73] W. James, *Psychology*, Fawcett Publications, 1963.

[74] T. Jochem, D.A. Pomerleau, and C. Thorpe, "MANIAC: A Next Generation Neurally Based Autonomous Road Follower" *Proc. Int'l Conf. Intelligent Autonomous Systems*, C.E. Thorpe, ed., IOS Publishers, Amsterdam, 1993.

[75] M. Johnson, *The Body in the Mind*, University of Chicago Press, 1987.

[76] P.N. Johnson-Laird and R.M.J. Byrne, *Deduction*, Lawrence Erlbaum Associates, 1991.

[77] M.I. Jordan, "Attractor Dynamics and Parallelism in a Connectionist Sequential Machine," *Proc. Cognitive Science Society*, 1986.

[78] M.I. Jordan, "Serial Order: A Parallel Distributed Processing Approach," Institute for Cognitive Science Report 8604, University of California, San Diego, 1986.

[79] B.D. Josephson and F. Pallikari-Viras, "Biological Utilisation of Quantum NonLocality," *Foundations of Physics*, Plenum Press, Vol. 21, 1991, pp. 197–207. http://www.phy.cam.ac.uk/www/research/mm/articles/Bell.psi.

[80] H. Kaindl, "Minimaxing Theory and Practice," in *AI Magazine*, Vol. 9, No. 3, Fall 1988, pp. 69–76.

[81] K. Kamijo and T. Tanigawa, "Stock Price Pattern Recognition," *Proc. IJCNN Conf.*, IEEE Press, Piscataway, NJ, 1990.

[82] E.W. Kent, *The Brains of Men and Machines*, BYTE/McGraw-Hill, 1981.

[83] T. Kimoto et al., "Stock Market Prediction System with Modular Neural Networks," *Proc. IJCNN Conf.*, IEEE Press, Piscataway, NJ, Vol. 1, 1990, pp. 1–7.

[84] T. Kohonen, *Self-Organization and Associative Memory*, 2nd Edition, Springer-Verlag, 1988.

[85] T. Kohonen, "Learning Vector Quantization," *Abstracts of the First Annual INNS Society Meeting, Neural Networks*, Supplement 1, Boston, 1988, p. 303.

[86] T. Kohonen, "Improved Versions of Learning Vector Quantization," *Proc. IEEE Joint Conf. Neural Networks*, IEEE Press, Piscataway, NJ, June 1990, pp. 545–550.

[87] J.L. Kolodner, "Extending Problem Solving Capabilities Through Case Based Inference" *Proc. DARPA Sponsored Workshop on Case Based Reasoning*, Kolodner, ed., Morgan Kaufmann, 1988, pp. 21–30. (Reprinted from *Proc. 4th Ann. Int'l Machine Learning Workshop, 1987.*)

[88] J.L. Kolodner, "Retrieving Events from a Case Based Memory: A Parallel Implementation," *Proc. DARPA Sponsored Workshop on Case Based Reasoning*, Kolodner, ed., Morgan Kaufmann, 1988, pp. 233–249.

[89] B. Kort, "Networks and Learning," *AI Magazine*, Vol. 11, No. 3, Fall 1990, pp. 16–19.

[90] P. Koton, "Reasoning about Evidence in Causal Explanations," *Proc. DARPA Sponsored Workshop on Case Based Reasoning*, Kolodner, ed., Morgan Kaufmann, 1988, pp. 260–270.

[91] J. Laird, P. Rosenbloom, and A. Newell, *Universal Subgoaling and Chunking*, Kluwer Academic Publishers, Boston, 1986.

[92] G. Lakoff and M. Johnson, *Metaphors We Live By*, University of Chicago Press, 1980.

[93] G. Lakoff, *Women, Fire and Dangerous Things*, University of Chicago Press, 1987.

[94] G. Lakoff, "A Suggestion for a Linguistics with Connectionist Foundations," *Proc. 1988 Connectionist Models Summer School*, Touretzky, ed., Morgan Kaufmann, 1989.

[95] J.E. Laird, A. Newell, and P.S. Rosenbloom, "SOAR: An Architecture for General Intelligence," *Artificial Intelligence*, Vol. 33, 1987, pp. 1–64.

[96] Y. Le Cun et al., "Handwritten Digit Recognition with a Back-Propagation Network," *Advances in Neural Information Processing Systems 2*, Morgan-Kaufmann, 1990, pp. 396–404.

[97] W. Lehnert, *Case-Based Reasoning as a Paradigm for Heuristic Search*, COINS 87-107, Department of Computer and Information Science, University of Massachusetts at Amherst, Amherst, MA, 1987.

[98] D. Lenat, "AM: Discovery in Mathematics as Heuristic Search," *Knowledge Based Systems in Artificial Intelligence*, McGraw-Hill, 1982, pp. 1–255.

[99] D. Lenat et al., "CYC: Toward Programs with Common Sense," *Comm. ACM*, Vol. 33, No. 8, Aug. 1990, pp. 30–49.

[100] M. Leshno et al., "Multilayer Feedforward Networks With a Nonpolynomical Activation Function Can Approximate Any Function," *Neural Networks*, Vol. 6, No. 6, 1993, pp. 861–867.

[101] M. Lockwood, *Mind, Brain & the Quantum*, Basil Blackwell, 1989.

[102] B. MacLennan, "Continuous Symbol Systems: The Logic of Connectionism," Technical Report CS-91-145, Computer Science Department, University of Tennessee, Knoxville, 1991, ftp://archive.cis.ohio-state.edu/pub/neuroprose/-maclennan.css.ps.Z.

[103] B. MacLennan, "Characteristics of Connectionist Knowledge Representation," Technical Report CS-91-147, Computer Science Department, University of Tennessee, Knoxville, 1991, ftp://archive.cis.ohio-state.edu/pub/neuroprose/-maclennan.cckr.ps.Z.

[104] B. MacLennan, "Field Computation in the Brain," in *Rethinking Neural Networks: Quantum Fields and Biological Data*, K.H. Pribram, ed., Lawrence Erlbaum, 1993, pp. 199–232, ftp://archive.cis.ohio-state.edu/pub/neuroprose/-maclennan.fieldcompbrain.ps.Z.

[105] D. Marr, *Vision*, W.H. Freeman, New York, 1982.

[106] P.J. McCann and B.L. Kalman, "A Neural Network Model for the Gold Market," Technical Report 94-09, Philosophy and Neuroscience Program, Washington University, St. Louis, 1994, ftp://thalamus.wustl.edu/pub/pnp/papers/mccann.gold-market.ps.

[107] J. McCarthy, "Mathematical Logic in Artificial Intelligence," *Daedalus*, Winter 1988, pp. 297–310.

[108] J. McClelland, "The Programmable Blackboard Model," in *Parallel Distributed Processing*, MIT Press, Vol. 2, 1986.

[109] J. McDermott, A. van de Brug, and J. Bachant, "The Taming of R1," *IEEE Expert*, Fall 1986, pp. 33–39.

[110] C. Mead, "Silicon Models of Neural Computation," in Volume 1 of *Proc. IEEE 1st Int'l Conf. Neural Networks*, Butler and Caudill, eds., IEEE Press, Piscataway, NJ, Vol. 1, 1987, pp. 91–106.

[111] C. Mead, *Analog VLSI and Neural Systems*, Addison-Wesley, 1989.

[112] R. Miikkulainen and M. Dyer, "Building Distributed Representations Without Microfeatures," Technical Report UCLA-AI-87-17, Computer Science Department, University of California, Los Angeles, 1987.

[113] R. Miikkulainen and M. Dyer, "Forming Global Representations with Extended Back-Propagation," *Proc. IEEE 2nd Int'l Conf. Neural Networks*, IEEE Press, Piscataway, NJ, Vol. 1, 1988, pp. 285–292.

[114] R. Miikkulainen and M. Dyer, "Encoding Input/Output Representations in Connectionist Cognitive Systems," *Proc. 1988 Connectionist Models Summer School*, Touretzky, ed., Morgan Kaufmann, 1989, pp. 347–356.

[115] R. Miikkulainen and M. Dyer, "A Modular Neural Network Architecture for Sequential Paraphrasing of Script-Based Stories," *Int'l Joint Conf. Neural Networks*, IEEE Press, Piscataway, NJ, Vol. 2, 1989, pp. 49–56.

[116] R. Miikkulainen and M. Dyer, "A Modular Neural Network Architecture for Sequential Paraphrasing of Script-Based Stories," Technical Report UCLA-AI-89-02, Artificial Intelligence Laboratory, Computer Science Department, University of California, Los Angeles, 1989.

[117] R. Miikkulainen, "Script Recognition with Hierarchical Feature Maps," *Connection Science*, Vol. 2, No. 1–2), 1990, pp. 93–101, ftp://cs.utexas.edu/pub/neural-nets/-papers/miikkulainen.script-recognition.ps.Z.

[118] R. Miikkulainen and M.G. Dyer, "Natural Language Processing with Modular PDP Networks and Distributed Lexicon," *Cognitive Science*, Vol. 15, No. 3, pp. 343–399, 1991, ftp://cs.utexas.edu/pub/neural-nets/papers/miikkulainen.natural-language.ps.Z.

[119] R. Miikkulainen, "Trace Feature Map: A Model of Episodic Associative Memory," *Biological Cybernetics 1991*, 1991, ftp://cs.utexas.edu/pub/neural-nets/papers/-miikkulainen.trace-feature-map.ps.Z.

[120] R. Miikkulainen, "DISCERN: A Distributed Neural Network Model of Script Processing and Memory," *Proc. Third Twente Workshop on Language Technology*, Twente, the Netherlands, 1992, ftp://cs.utexas.edu/pub/neural-nets/papers/-miikkulainen.discern-overview.ps.Z.

[121] R. Miikkulainen, *Subsymbolic Natural Language Processing: An Integrated Model of Scripts, Lexicon, and Memory*, MIT Press, Cambridge, MA, 1993.

[122] G.A. Miller, "The Magical Number Sever, Plus or Minus Two: Some Limits on Our Capacity for Processing Information," *Psychological Review*, Vol. 63, 1956, pp. 81–97.

[123] M. Minsky, "A Framework for Representing Knowledge," *The Psychology of Computer Vision*, Winston, ed., McGraw-Hill, 1975.

[124] M. Minsky, *The Society of Mind*, Simon and Shuster, 1987.

[125] M. Minsky and S. Papert, *Perceptrons*, Expanded Edition, MIT Press, 1988.

[126] T. Mitchell et al., "Learning Problem-Solving Heuristics Through Practice," Technical Report LCSR-TR-15, Laboratory for Computer Science Research, Rutgers University, 1981.

[127] H.P. Moravec, *Mind Children*, University Press, San Francisco, 1987.

[128] J. Moses, "Symbolic Integration," Project MAC technical report MAC-TR-47, Massachusetts Institute of Technology, Dec. 1967.

[129] M.C. Mozer, "RAMBOT: A Connectionist Expert System That Learns by Example," Institute for Cognitive Science Report 8610, University of California, San Diego, Aug. 1986.

[130] M.C. Mozer, "RAMBOT: A Connectionist Expert System That Learns by Example," *Proc IEEE 1st Int'l Conf. Neural Networks*, IEEE Press, Piscataway, NJ, Vol. II, 1987.

[131] M.C. Mozer, "A Focused Back-Propagation Algorithm for Temporal Pattern Recognition," Technical Report CRG-TR-88-3, Departments of Psychology and Computer Science, University of Toronto, June 1988.

[132] M.C. Mozer and P. Smolensky, "Skeletonization: A Technique for Trimming the Fat from a Network via Relevance Assessment," Technical Report CU-CS-421-89, Computer Science Department, University of Colorado at Boulder, Boulder, CO, 1989.

[133] D. Nanopoulos, "Theory of Brain Function, Quantum Mechanics and Superstrings," http://xxx.lanl.gov/hep-ph/9505374, 1995.

[134] D.S. Nau, "Decision Quality as a Function of Search Depth on Game Trees," *J. ACM*, Vol. 30, 1983, p. 687.

[135] A. Newell, J.C. Shaw, and H. Simon, "A Variety of Intelligent Learning in a General Problem Solver," *Self-Organizing Systems*, Yovits and Cameron, eds., Pergamon Press, 1960.

[136] A. Newell and G. Ernst, "Some Issues of Representation in a General Problem Solver," *Proc. AFIPS Spring Joint Computer Conf.*, 1967.

[137] A. Newell and G. Ernst, *A Case Study in Generality and Problem Solving*, ACM Monograph, Academic Press, Inc., New York, 1969.

[138] A. Newell and H.A. Simon, "Computer Science as Empirical Inquiry: Symbols and Search," *Comm. ACM*, Vol. 19, No. 3, Mar. 1976, pp. 113–126.

[139] A. Newell, "Physical Symbol Systems," *Cognitive Science*, Vol. 4, 1980, pp. 135–183.

[140] A. Newell and P. Rosenbloom, "Mechanisms of Skill Acquisition and the Law of Practice," *Cognitive Skills and Their Acquisition*, Anderson, ed., Erlbaum, 1981.

[141] A. Newell, *Unified Theories of Cognition*, Harvard University Press, 1990.

[142] N.J. Nilsson, *Learning Machines*, McGraw-Hill, 1965.

[143] N.J. Nilsson, *Problem Solving Methods in Artificial Intelligence*, McGraw-Hill, 1971.

[144] S. Nolfi et al., "Recall of Sequences by a Neural Network," in *Proc. 1990 Connectionist Models Summer School*, Morgan Kaufmann, 1990, pp. 243–252.

[145] C.V. Page, "Heuristics for Signature Table Analysis as a Pattern Recognition Technique," *IEEE Trans. Systems, Man and Cybernetics*, Vol. SMC-7, No. 2, Feb. 1977, pp. 77–86.

[146] D. Parker, "Learning Logic," Invention Report, S81-64, File 1, Office of Technology Licensing, Stanford University, Oct. 1982.

[147] D. Parker, "Learning-Logic," TR47, Center for Computational Research in Economics and Management Science, MIT, Apr. 1985.

[148] R. Penrose *The Emperor's New Mind*, Oxford University Press, 1989.

[149] R. Penrose, *Shadows of the Mind*, Oxford University Press, 1994.

[150] M.P. Perrone and L.N. Cooper, "When Networks Disagree: Ensemble Methods for Hybrid Neural Networks," ftp://archive.cis.ohio-state.edu/pub/neuroprose/-perrone.MSE-averaging.ps.Z, 1992.

[151] S. Pinker, *Language Learnability and Language Development*, Harvard University Press, Cambridge, MA, 1984.

[152] S. Pinker and A. Prince, "Language and Connectionism," *Cognition: International Journal of Cognitive Science*, Vol. 28, 1988. (also reprinted in *Connections and Symbols*, Pinker and Mehler, eds., MIT Press, 1988, pp. 79–193.)

[153] S. Pinker, "Rules of Language," *Science*, Vol. 253, 2 Aug. 1991, pp. 530–535.

[154] T. Plate, "Holographic Reduced Representations," ftp://ftp.cs.utoronto.ca/pub/tap/-plate/plate.ieee95.ps.Z, 1995.

[155] K. Plunkett and V. Marchman, "Pattern Association in a Back Propagation Network: Implications for Child Language Acquisition," Technical Report 8902, Center for Research in Language, University of California, San Diego, La Jolla, CA, Mar. 1989.

[156] K. Plunkett and V. Marchman, "From Rote Learning to System Building," Center for Research in Language, Technical Report 9020, University of California, San Diego, La Jolla, CA, Sep. 1990.

[157] J. Pollack and D. Waltz, "Interpretation of Natural Language," *Byte*, Vol. 11, No. 2, Feb. 1986, pp. 189–198.

[158] J. Pollack, "Implications of Recursive Distributed Representations" *Advances in Neural Information Processing Systems I*, Touretzky, ed., Morgan Kaufmann, 1989, ftp://archive.cis.ohio-state.edu/pub/neuroprose/pollack.nips88.ps.Z.

[159] J. Pollack, "Recursive Distributed Representations," ftp://archive.cis.ohio-state.edu/-pub/neuroprose/pollack.newraam.ps.Z, 1989.

[160] D.A. Pomerleau, J. Gowdy, and C.E. Thorpe, "Combining Artificial Neural Networks and Symbolic Processing for Autonomous Robot Guidance," *Engineering Applications of Artificial Intelligence*, Vol. 4, No. 4, 1991, pp. 279–285.

[161] D.A. Pomerleau, "Efficient Training of Artificial Neural Networks for Autonomous Navigation," *Neural Computation*, Vol. 3, No. 1, 1991, pp. 88–97.

[162] D.A. Pomerleau, *Neural Network Perception for Mobile Robot Guidance*, Kluwer Academic Publishing, 1993.

[163] D.A. Pomerleau, "Input Reconstruction Reliability Estimation," *Advances in Neural Information Processing Systems 5*, C.L. Giles, S.J. Hanson, and J.D. Cowan, eds., Morgan Kaufmann Publishers, San Mateo, CA, 1993, pp. 279–286.

[164] D.A. Pomerleau and D.S. Touretzky, "Understanding Neural Network Internal Representations through Hidden Unit Sensitivity Analysis," *Proc. Int'l Conf. Intelligent Autonomous Systems*, C.E. Thorpe, ed., IOS Publishers, Amsterdam, 1993.

[165] H. Putnam, "Much Ado About Not Very Much," *Daedalus*, Winter 1988, pp. 269–281.

[166] J.R. Quinlan, "Learning Efficient Classification Procedures and Their Application to Chess End Games," *Machine Learning: An Artificial Intelligence Approach*, Michalski et al., eds., Tioga Publishing, 1983.

[167] A.S. Reber, "Implicit Learning of Artificial Grammars," *J. Verbal Learning and Verbal Behavior*, Vol. 5, 1967, pp. 855–863.

[168] A.N. Refenes, A. Zaparanis, and G. Francis, "Stock Performance Modeling Using Neural Networks: A Comparative Study with Regression Models," *Neural Networks*, Vol. 7, No. 2, 1994, pp. 375–388.

[169] T. Reiger, "Recognizing Image-Schemas Using Programmable Networks," *Proc. 1988 Connectionist Models Summer School*, Touretzky, ed., Morgan Kaufmann, 1989, pp. 315–324.

[170] D.L. Reilly et al., "Learning System Architectures Composed of Multiple Learning Modules," *Proc. IEEE 1st Int'l Conf. Neural Networks*, IEEE Press, Piscataway, NJ, 1987.

[171] B.J. Richmond et al., "Neuronal Encoding of Information Related to Visual Perception, Memory, and Motivation," *Rethinking Neural Networks: Quantum Fields and Biological Data*, K.H. Pribram, ed., Lawrence Erlbaum, 1993, pp. 465–486.

[172] M. Riedmiller and H. Braun, "A Direct Adaptive Method for Faster Backpropagation Learning: The RPROP Algorithm," *Proc. IEEE Int'l Conf. Neural Networks*, IEEE Press, Piscataway, NJ, 1993, ftp://i11s16.ira.uka.de/pub/neuro/papers/riedml.icnn93.ps.Z.

[173] M. Riedmiller, "Rprop—Description and Implementation Details," Technical Report, Institut für Logik, Komplexität und Deduktionssyteme, University of Karlsruhe, W-76128 Karlsruhe, FRG, Jan. 1994, ftp://i11s16.ira.uka.de/pub/neuro/papers/rprop.details.ps.Z.

[174] R.H. Risch, "The Problem of Integration in Finite Terms," in *Trans. American Mathematical Society*, Vol. 139, 1969.

[175] E. Rissland and K. Ashley, "Credit Assignment and the Problem of Competing Factors in Case-Based Reasoning," *Proc. DARPA Sponsored Workshop on Case-Based Reasoning*, Kolodner, ed., Morgan Kaufmann, 1988, pp. 327–344.

[176] H. Robbins and S. Monro, "A Stochastic Approximation Method," *Annals of Mathematical Statistics*, Vol. 22, 1951, pp. 400–407.

[177] F. Rosenblatt, *Principles of Neurodynamics*, Spartan Books, New York, 1962.

[178] D. Ruby and D. Kibler, "Exploration of Case-Based Problem Solving," *Proc. DARPA Sponsored Workshop on Case-Based Reasoning*, Kolodner, ed., Morgan Kaufmann, 1988, pp. 345–356.

[179] D. Rumelhart, "Some Problems With the Notion of Literal Meanings," *Metaphor and Thought*, A. Ortony, ed., Cambridge University Press, Cambridge, England, 1979.

[180] D. Rumelhart, G. Hinton, and R.J. Williams, "Learning Internal Representations," *Parallel Distributed Processing*, MIT Press, Vol. 1, 1986, pp. 318–362.

[181] D. Rumelhart and J. McClelland, "Learning the Past Tense," *Parallel Distributed Processing*, MIT Press, Vol. 2, 1986, pp. 216–271.

[182] D. Rumelhart et al., "Schemata and Sequential Thought Processes in PDP Models," *Parallel Distributed Processing*, MIT Press, Vol. 2, 1986, pp. 8–57.

[183] M.F. St. John and J. McClelland, "Learning and Applying Contextual Constraints in Sentence Comprehension," Technical Report AIP-39, Department of Psychology, Carnegie Mellon University, 1988.

[184] M.F. St. John and J. McClelland, "Applying Contextual Constraints in Sentence Comprehension," *Proc. 1988 Connectionist Models Summer School*, Touretzky, ed., Morgan Kaufmann, 1989, pp. 338–346.

[185] K. Saito and R. Nakano, "Medical Diagnostic Expert System Based on PDP Model" *Proc. IEEE Int'l Conf. Neural Networks*, IEEE Press, Piscataway, NJ, 1988, pp. 255–262.

[186] A. Samuel, "Some Studies in Machine Learning Using the Game of Checkers," *Computers and Thought*, Feigenbaum and Feldman, eds., McGraw-Hill, 1963, pp. 71–105. (Reprinted with minor changes from *IBM J. Research and Development*, Vol. 3, No. 3, July, 1959.)

[187] A. Samuel, "Some Studies in Machine Learning Using the Game of Checkers, II, Recent Progress," *IBM J. Research and Development*, Nov. 1967, pp. 601–617.

[188] E. Santos, "A Massively Parallel Self-Tuning Context-Free Parser" *Advances in Neural Information Processing Systems I*, Touretzky, ed., Morgan Kaufmann, 1989, pp. 537–544.

[189] J. Schaeffer et al., "Checkers Program to Challenge for World Championship," *SIGART Bulletin*, Vol. 2, No. 2, 1991, pp. 3–5.

[190] J. Schaeffer et al., "A World Championship Caliber Checkers Program," *Artificial Intelligence* Vol. 53, Nos. 2–3, 1992, pp. 273–289.

[191] J. Schaeffer, "Man versus Machine: The World Checkers Championship," Technical Report TR-92-19, Dept of Computing Science, University of Alberta, 1992, ftp://ftp.cs.ualberta.ca/pub/TechReports/TR-92-19.

[192] J. Schaeffer et al., "Man Versus Machine for the World Checkers Championship," *AI Magazine*, Vol. 14, No. 2, Summer 1993, pp. 28–35.

[193] R. Schank and C. Riesbeck, *Inside Computer Understanding*, Lawrence Erlbaum, 1981.

[194] R. Schank, *Dynamic Memory: A Theory of Reminding and Learning in Computers and People*, Cambridge University Press, 1982.

[195] Schank and Childers, *The Cognitive Computer*, Addison-Wesley, 1984.

[196] J. Searle, "Minds, Brains and Programs," *Mind Design*, J. Haugeland, ed., MIT Press, 1980.

[197] J. Searle, "Minds, Brains and Programs with Open Peer Commentaries," *Behavioral and Brain Sciences*, Vol. 3, 1989, pp. 417–457.

[198] M. Seidenberg and J. McClelland, "A Distributed Developmental Model of Word Recognition and Naming," *Psychological Review*, Vol. 96, No. 4, 1989, pp. 523–568.

[199] T. Sejnowski and C. Rosenberg, "Parallel Networks That Learn to Pronounce English Text," *Complex Systems 1*, 1987, pp. 148–168.

[200] S. Scown, *The Artificial Intelligence Experience: An Introduction*, Digital Press, 1985.

[201] D. Servan-Schreiber, A. Cleeremans, and J. McClelland, "Encoding Sequential Structure in Simple Recurrent Networks," Technical Report CMU-CS-88-183, Computer Science Department, Carnegie Mellon University, 1988.

[202] L. Shastri, *Semantic Nets: Evidential Formalization and its Connectionist Realization*, Morgan Kaufmann, 1988.

[203] L. Shastri and V. Ajjanagadde, "From Simple Associations to Systematic Reasoning: A Connectionist Representation of Rules, Variables and Dynamic Bindings," Technical Report MS-CIS-90-05 LINC LAB 162, Department of Computer and Information Science, School of Engineering and Applied Science, University of Pennsylvania, Philadelphia, 1990.

[204] J.W. Shavlik, "A Framework for Combining Symbolic and Neural Learning," Technical Report 1123, Computer Sciences Department, University of Wisconsin at Madison, 1992, ftp://ftp.cs.wisc.edu/tech-reports/reports/92/tr1123.ps.Z.

[205] P. Shea and V. Lin, "Detection of Explosives in Checked Airline Baggage Using an Artificial Neural System," *Proc. IJCNN Conf.*, IEEE Press, Piscataway, NJ, Vol. 2, 1989, pp. 31–34.

[206] P. Shea and F. Liu, "Operational Experience with A Neural Network in the Detection of Explosives in Checked Airline Luggage," *Proc. IJCNN Conf.*, IEEE Press, Piscataway, NJ, Vol. 2, 1990, pp. 175–178.

[207] S. Shekhar and S. Dutta, "Using Neural Networks for Generalization Problems," *Abstracts of the First Annual International Neural Network Society Meeting*, Boston, 1988, p. 171.

[208] P. Simard, Y. Le Cun, and J. Denker, "Efficient Pattern Recognition Using a New Transformation Distance," *NIPS 5*, Morgan-Kaufmann, 1993, pp. 50–58.

[209] H. Simon, "Is Thinking Uniquely Human?," University of Chicago Magazine, Fall 1981, pp. 12–16.

[210] J. Slagle, "A Heuristic Program that Solves Symbolic Integration Problems in Freshman Calculus," *J. ACM*, Vol. 10, 1963, pp. 507–520. (Reprinted in *Computers and Thought*, Feigenbaum and Feldman, eds., McGraw-Hill, 1963.)

[211] J. Slagle, "A Multipurpose Theorem Proving Program That Learns," *Proc. IFIP Congress 1965*, Spartan Books, Vol. 2, 1965, pp. 323–328.

[212] J. Slagle and P. Bursky, "Experiments with a Multipurpose Theorem Proving Heuristic Program," *J. ACM*, Vol. 15, No. 1, Jan. 1968, pp. 85–99.

[213] J. Slagle and D. Koniver, "Finding Resolution Proofs and Using Duplicate Goals in AND/OR Trees," *Information Sciences*, Vol. 3, No. 1, Jan. 1971.

[214] J. Slagle, *Artificial Intelligence: The Heuristic Programming Approach*, McGraw-Hill, 1971.

[215] M. Smith, "Loan Underwriting by a Neural Network," *Abstracts of the First Annual International Neural Network Society Meeting*, Boston, 1988, p. 468.

[216] J.W. Smith and T.R. Johnson, "A Stratified Approach to Specifying, Designing, and Building Knowledge Systems," *IEEE Expert*, Vol. 8, No. 3, 1993, pp. 15–25.

[217] P. Smolensky, "Harmony Theory," *Parallel Distributed Processing*, MIT Press, Vol. 1, 1986, pp. 194–281.

[218] P. Smolensky, "On the Proper Treatment of Connectionism," Computer Science Department, University of Colorado, Boulder, Colorado, Technical Report CU-CS-377-87, Oct. 1987. (Also in *Behavioral and Brain Science*, Vol. 11, 1988, pp. 1–74.)

[219] G.S. Snoddy, "Learning and Stability," *J. of Applied Psychology*, 10, 1926, pp. 1–36.

[220] E. Sontag, "On the Recognition Capabilities of Feedforward Nets," Report, SYCON 90-03, Rutgers Center for Systems and Control, Department of Mathematics, Rutgers University, New Brunswick, NJ 08903, 1990, ftp://archive.cis.ohio-state.edu/pub/neuroprose/sontag.capabilities.ps.Z.

[221] R. Sproull and W. Newman, *Principles of Interactive Computer Graphics*, McGraw-Hill, 1973.

[222] C. Stanfill, "Learning to Read: A Memory-Based Model," *Proc. DARPA Sponsored Workshop on Case Based Reasoning*, Kolodner, ed., Morgan Kaufmann, 1988, pp. 402–413.

[223] D.M. Steier et al., "Combining Multiple Knowledge Sources in an Integrated Intelligent System," *IEEE Expert*, Vol. 8, No. 3, 1993, pp. 35–44.

[224] D.G. Stork and J. Hall, "Is Back-Propagation Biologically Plausible?," *Proc. Int'l Joint Conf. Neural Networks*, IEEE Press, Piscataway, NJ, 1989, pp. 241–246.

[225] A. Surkan and J. Singleton, "Neural Networks for Bond Rating Improved by Multiple Hidden Layers," *Proc. IJCNN Conf.*, IEEE Press, Piscataway, NJ, 1990, pp. 157–162.

[226] R. Sun, "Beyond Associative Memories: Logics and Variables in Connectionist Models," ftp://archive.cis.ohio-state.edu/pub/neuroprose/sun.beyond.ps.Z, 1992.

[227] R.S. Sutton, "Learning to Predict by the Methods of Temporal Differences," *Machine Learning*, Vol. 3, 1988, pp. 9–44.

[228] M. Tambe et al., "Intelligent Agents for Interactive Simulation Environments," *AI Magazine*, Spring 1995, pp. 15–39.

[229] G. Tesauro and T.J. Sejnowski, "A 'Neural' Network That Learns to Play Backgammon," *Neural Information Processing Systems*, American Institute of Physics, 1988, pp. 794–803.

[230] G. Tesauro, "Connectionist Learning of Expert Preferences by Comparison Training" *Advances in Neural Information Processing Systems I*, Touretzky, ed., Morgan Kaufmann, 1989, pp. 99–106.

[231] G. Tesauro, "Practical Issues in Temporal Difference Learning," IBM Research Report RC 17223(#76307), Sep. 30, 1991, IBM Research Division, T. J. Watson Research Center, Yorktown Heights, NY. (Reportedly to be reprinted in *Machine Learning, 1992.*)

[232] G. Tesauro, "TD-Gammon, A Self-Teaching Backgammon Program Achieves Master-Level Play," *Neural Computation*, Mar. 1994, ftp://archive.cis.ohio-state.edu/pub/neuroprose/tesauro.tdgammon.ps.Z.

[233] G. Tesauro, "Temporal Difference Learning and TD-Gammon," *Comm. ACM*, Vol. 38, No. 3, 1995, pp. 58–68.

[234] S. Thrun, "Extracting Provably Correct Rules from Artificial Neural Networks," Technical Report, Department of Computer Science, University of Bonn, Bonn, Germany, 1995, available from http://www.informatik.uni-bonn.de/˜thrun/publications.html.

[235] D. Touretzky, "Connectionism and Compositional Semantics," Technical Report CMU-CS-89-147, Computer Science Department, Carnegie-Mellon University, Pittsburgh, 1989.

[236] D. Touretzky, "Connectionism and PP Attachment," *Proc. 1988 Connectionist Models Summer School*, Touretzky, ed., Morgan Kaufmann, 1989, pp. 325–332.

[237] G. Towell and J.W. Shavlik, "Interpretation of Artificial Neural Networks: Mapping Knowledge-Based Neural Networks into Rules," *NIPS 4*, Morgan-Kaufmann, 1992, pp. 977–984.

[238] A.M. Turing, "Computing Machinery and Intelligence," *Mind*, Vol. 59, 1950, pp. 433–460.

[239] D.R. Tveter, "How to Succeed at Back-Propagation," available from the author, 1998.

[240] G. Vitiello, "Dissipation and Memory Capacity in the Quantum Brain Model," *Int'l J. Mod. Physics B*, in print, http://xxx.lanl.gov/abs/quant-ph/9502006, 1995.

[241] A. Waibel and J. Hampshire, "Building Blocks for Speech," *BYTE*, Aug. 1989.

[242] A. Waibel et al., "Phoneme Recognition Using Time-Delay Neural Networks," *IEEE Transactions on Acoustics, Speech and Signal Processing*, Vol. ASSP 37, Mar. 1989.

[243] A. Waibel, "Connectionist Glue: Modular Design of Neural Speech Systems," *Proceedings of the 1988 Connectionist Models Summer School*, Touretzky, ed., Morgan Kaufmann, 1989.

[244] A. Waibel, "Consonant Recognition by Modular Construction of Large Phonemic Time-Delay Neural Networks," *Advances in Neural Information Processing Systems 1*, Touretzky, ed., Morgan Kaufmann, 1989.

[245] D. Walker, "Knowledge Resource Tools for Accessing Large Text Files," Technical Report No. 85-21233-25, Bell Communications Research, Morristown, NJ, 1985.

[246] D. Walker and R.A. Amsler, "The Use of Machine-Readable Dictionaries in Sublanguage Analysis," *Sublanguage: Description and Processing*, R. Grishman and R. Kittredge, eds., Lawrence Erlbaum Associates, Hillsdale, NJ, 1985.

[247] D. Waltz and J. Pollack, "Massively Parallel Parsing: A Strongly Interactive Model of Natural Language Interpretation," *Cognitive Science*, Vol. 9, 1985, pp. 51–74.

[248] Waltz and Stanfill, "Toward Memory Based Reasoning," *Comm. ACM*, Vol. 29, No. 12, Dec. 1986, pp. 1213–1228.

[249] D.L. Waltz, "The Prospects for Building Truly Intelligent Machines," *Daedalus*, Winter 1988, pp. 191–212.

[250] R.M. Warren, "Perceptual restoration of missing speech sounds," *Science*, 167, 1970, pp. 393–395.

[251] R. Washington and P.S. Rosenbloom, *Neomycin-SOAR: Applying Search and Learning to Diagnosis*, Technical Report, Knowledge Systems Laboratory, Stanford University, Stanford, CA, 1988.

[252] P.J. Werbos, "Beyond Regression: New Tools for Prediction and Analysis in the Behavioral Sciences," thesis in applied mathematics, Harvard University, Aug. 1974.

[253] P. Werbos, "Generalization of Backpropagation with Application to a Recurrent Gas Market Model," *Neural Networks*, Vol. 1, No. 4, 1988, pp. 339–356.

[254] P. Werbos, "Backpropagation: Past and Future," *Proc. IEEE 2nd Int'l Conf. Neural Networking*, IEEE Press, Piscataway, NJ, Vol. 1, 1988, pp. 343–353.

[255] H. White, "Economic Prediction Using Neural Networks: The Case of IBM Daily Stock Returns," *Proc. IEEE 2nd Int'l Conf. Neural Networks*, IEEE Press, Piscataway, NJ, 1988, pp. 451–458.

[256] H. White, "Neural Network Learning and Statistics," *AI Expert*, Vol. 4, No. 12, Dec. 1989, pp. 48–52.

[257] G.M. White, "Natural Language Understanding and Speech Recognition," *Comm. ACM*, Vol. 33, No. 8, Aug. 1990, pp. 72–82.

[258] B. Widrow, "Adaline and Madeline—1963," *Proc. IEEE 1st Int'l Conf. Neural Networks*, Caudill and Butler, eds., IEEE Press, Piscataway, NJ, 1987, pp. 143–158.

[259] M.V. Wilks, "Artificial Intelligence as the Year 2000 Approaches," *Communications of the ACM*, Vol. 35, No. 8, Aug. 1992, pp. 17–20.

[260] R.S. Williams, "Learning to Program by Examining and Modifying Cases," *Proc. DARPA Sponsored Workshop on Case Based Reasoning*, Kolodner, ed., Morgan Kaufmann, 1988, pp. 463–474.

[261] T. Winograd, "A Procedural Model of Language Understanding," *Computer Models of Thought and Language*, Schank and Colby, eds., W.H. Freeman and Company, 1973, pp. 152–186.

[262] P. Winston, *Artificial Intelligence*, Addison-Wesley, 1977.

[263] F. Wolf, *Parallel Universes*, Simon and Shuster, New York, 1988.

[264] D.H. Wolpert, "Constructing a Generalizer Superior to NETtalk via a Mathematical Theory of Generalization," *Neural Networks*, Pergamon Press, Vol. 3, No. 4, 1990, pp. 445–452.

[265] W.A. Woods, "Transition Network Grammars for Natural Language Analysis," *Comm. ACM*, Vol. 13, No. 10, Oct. 1970.

[266] W.A. Woods, "Lunar Rocks in English: Explorations in Natural Language Question Answering," *Linguistic Structures Processing*, A. Zampolli, ed., North-Holland, 1977.

[267] L. Wos et al., *Automated Reasoning: Introduction and Examples*, Prentice-Hall, 1984.

[268] L. Zadeh, "Fuzzy Logic," *Computer*, Vol. 21, No. 4, Apr. 1988.

Index

IEEE Computer Society Publications

The world-renowned Computer Society publishes, promotes, and distributes a wide variety of authoritative computer science and engineering texts. These books are available in two formats: 100 percent original material by authors preeminent in their field who focus on relevant topics and cutting-edge research, and reprint collections consisting of carefully selected groups of previously published papers with accompanying original introductory and explanatory text.

Submission of proposals: For guidelines and information on Computer Society books, send e-mail to cs.books@computer.org or write to the Acquisitions Editor, IEEE Computer Society, P.O. Box 3014, 10662 Los Vaqueros Circle, Los Alamitos, CA 90720-1314. Telephone +1 714-821-8380.FAX +1 714-761-1784.

IEEE Computer Society Proceedings

The Computer Society also produces and actively promotes the proceedings of more than 130 acclaimed international conferences each year in multimedia formats that include hard and softcover books, CD-ROMs, videos, and on-line publications.

For information on Computer Society proceedings, send e-mail to cs.books@computer.org or write to Proceedings, IEEE Computer Society, P.O. Box 3014, 10662 Los Vaqueros Circle, Los Alamitos, CA 90720-1314. Telephone +1 714-821-8380. FAX +1 714-761-1784.

Additional information regarding the Computer Society, conferences and proceedings, CD-ROMs, videos, and books can also be accessed from our web site at http://computer.org/cspress